DATE DUE

Mechanisms of Immune Regulation

Chemical Immunology

Vol. 58

KARGER

Basel · Freiburg · Paris · London · New York · New Delhi · Bangkok · Singapore · Tokyo · Sydney

Mechanisms of Immune Regulation

Volume Editor
Richard D. Granstein, Charlestown, Mass.

12 figures and 8 tables, 1994

KARGER

Basel · Freiburg · Paris · London · New York · New Delhi · Bangkok · Singapore · Tokyo · Sydney

Chemical Immunology

Formerly published as 'Progress in Allergy'
Founded 1939 by Paul Kallòs

RC
583
,P7
v.58

Bibliographic Indices
This publication is listed in bibliographic services, including Current Contents[R] and Index Medicus.

Contents

Contents

Suppressor Cells and Immunity

Webb, D.R. (Palo Alto, Calif.); Kraig, E. (San Antonio, Tex.); Devens, B.H.
(Palo Alto, Calif.) 146

Topics of Clinical Interest

T Cell Tolerance: Models for Clinical Application to Allergy and Autoimmunity

Schad, V.C. (Waltham, Mass.) 193

Contents

Oral Tolerance: A Biologically Relevant Pathway to Generate Peripheral Tolerance against External and Self Antigens

Mechanisms of Ultraviolet Radiation Carcinogenesis

Contents

Regulation of Immunity by Ultraviolet Radiation and Photosensitized Reactions

Preface

The revolution in biology of the past two to three decades has led to a dramatic increase in our understanding of immune processes. It is now recognized that the immune response represents the activity of an enormous web of interacting cells and humoral factors. Furthermore, the importance of intracellular signalling in these processes is now appreciated as is the role in immune responses of cells, such as keratinocytes and endothelial cells (amongst others), not usually thought of as immunologically relevant. These insights have led both to major advances in the basic molecular and cellular biology of the immune response and also have important implications for clinical immunology. The purpose of this volume is to present the current understanding of physiologic mechanisms of immune regulation along with the impact of this knowledge on selected clinical questions. Reviewed within this volume are such topics as T- and B-cell tolerance (including neonatal tolerance, clonal anergy and the role of immune complexes in tolerance), clonal deletion, suppressor cells, mechanisms of immune privileged sites, and experimental models of tumor immunity. The possible utility of manipulating the immune response for therapeutic benefits is explored in contributions discussing oral tolerance, ultraviolet radiation and photosensitized effects on immunity, T-cell vaccination and regulation of immunity with T-cell epitopes.

Thus, our understanding of the regulation of the immune response has advanced dramatically in recent years. Of perhaps greater importance, this work has led to new approaches for the understanding and treatment of disease. Hopefully, the reader will find the fascinating articles in this issue of *Chemical Immunology* to be informative, useful and stimulating.

Mechanisms of Regulation

Granstein RD (ed): Mechanisms of Immune Regulation.
Chem Immunol. Basel, Karger, 1994, vol 58, pp 1–33

T Cell Tolerance

Melanie S. Vacchio, Jonathan D. Ashwell

Laboratory of Immune Cell Biology, Biological Response Modifiers Program,
National Cancer Institute, National Institutes of Health, Bethesda, Md., USA

Introduction

The essential characteristic that distinguishes lymphocytes from the other cellular components of the host-protective response (e.g., macrophages/monocytes and neutrophils) is that the former respond in an antigen-specific fashion. T lymphocytes bear cell surface antigen-specific receptors (TCRs) that recognize peptides (antigens) bound to major histocompatibility complex (MHC)-encoded glycoproteins on other cells. The specificity of any given TCR is not encoded in the germline, but is determined when the pre-T cell rearranges its α and β chain gene segments to give rise to the antigen/MHC-specific clonotypic heterodimer [1]. This poses a problem for the organism, since a subset of these newly and stochastically generated TCRs will recognize endogenous (self) antigens. If unchecked, these cells would attack the host's own tissues, resulting in autoimmunity. Mechanisms must therefore exist to prevent the immune system from attacking self (i.e. be tolerant of self), while preserving its ability to respond to foreign antigenic challenge.

The first modern insight into how tolerance of self antigens is established was provided by R.D. Owen in 1945, who was studying graft rejection in dizygotic calf twins (freemartins) that share blood supplies after fusion of the placentas, forming natural chimeras [2]. Freemartin pairs accepted long-term skin grafts from one another, suggesting that exposure to allogeneic cells of the other twin during development caused the immune system to recognize the foreign cells as self. In 1949, Burnet and Fenner [3] expanded the theme of self/nonself discrimination, suggesting that exposure to an antigen during a critical period of development, prior to maturation of the immune system, would render the animal tolerant to the antigen for life. This hypothesis was verified in studies by Billingham et al. [4], in which it was demonstrated that injection of allogeneic cells into neonatal mice resulted in lifelong specific tolerance to cells from the donor strain. These early studies described the phenomenon of self tolerance at the level of the whole animal. This review will describe recent work exploring the mechanism(s) of tolerance.

Tolerance Defined

Lack of an immune response to an antigen is the *sine qua non* of tolerance, but in itself unresponsiveness does not necessarily define the tolerant state. Tolerance exists when lymphocytes do not respond even in the presence of an accessible antigen/MHC ligand. Classical tolerance is a 'learned' phenotype, and thus its induction is an active rather than a passive phenomenon. Tolerance occurs under certain circumstances, detailed later in this chapter, when lymphocytes are exposed to the antigen for which they are specific or to an antigen with which their receptors are cross-reactive. Self-tolerance is the special case where the immune system is rendered unable to respond to tissues endogenous to the host. A clue to the nature of tolerance induction in T cells is that it, like induction of effector function, is MHC restricted [5]. This implies that antigen-specific recognition occurs normally, but that the outcome (tolerance vs. activation) is dependent upon other parameters. Some factors that could influence the outcome are the intensity of the signal due to the affinity of receptor/ligand interaction, the stage of development at which the signal is received, or which cell type presents the antigen.

There are instances in which an animal may not respond to a specific antigen yet not be considered tolerant by this definition. For example, an animal may be a nonresponder because the antigenic peptide does not bind its MHC-encoded molecules (a mechanism of nonresponsiveness known as 'determinant selection' [6]). Likewise, an animal may not respond to an antigen that is sequestered by an anatomical barrier. In these cases, the animal's antigen-specific lymphocytes simply never encounter the antigen in the appropriate context. One could argue that these animals are no more 'tolerant' of these antigens that they are of carbohydrates or lipid molecules that do not generate a T cell response because they are not recognized by TCRs. Another potential mechanism of nonresponsiveness, in which the receptor for a certain antigen is never generated, is even more difficult to categorize. Despite the enormous number of receptor specificities that can be generated by gene rearrangement and its attendant combinatorial diversity, it is possible that in some cases the receptor(s) needed to recognize a certain antigen cannot be produced because of physical limitations or constraints on these processes. These cases of nonresponsiveness may or may not be considered examples of tolerance, depending upon how much one wants to emphasize mechanism over outcome.

How Is Tolerance Established?

Once a self antigen is accessible to the immune system, there are two well-recognized mechanisms by which autoreactivity can be controlled; clonal deletion and anergy. The former occurs when the potentially self-reactive cell is eliminated. The latter, also referred to as clonal inactivation, reflects a state in which the antigen-specific T cell exists but fails to manifest effector function when exposed to the antigen. A third potential mechanism of tolerance is suppression, in which antigen-specific suppressor T cells inhibit responsiveness of autoreactive effector T cells. Since the large majority of recent work supports either deletion or anergy as major mechanisms of self tolerance, only these processes will be considered in this review.

Clonal Deletion

Deletion in the Thymus
Deletion of Antigen-Specific Thymocyte Cohorts by Defined Ligands
The thymus is where T cell precursors initially express an antigen-specific receptor, and it is at this stage of development that mechanisms leading to tolerance of self antigens can first be demonstrated. To determine whether clonal T cell lineages are eliminated in an antigen-specific manner, it is necessary to be able to identify and quantitate T cells bearing TCRs of defined antigen specificity. This approach is usually not feasible due to the low frequency of T cell clones bearing TCRs specific for any given classical antigen. However, the generation of Vβ-specific antibodies, coupled with the discovery of superantigens, has made it possible to follow cohorts of T cells that have a defined ligand in common.

The term superantigen was coined to describe molecules that are recognized by TCRs due to interaction with certain Vβ domains, i.e. independently of Vα and J region usage and recombination site-specific sequences [7]. Since ligand recognition is Vβ-dependent, and 'presentation' of superantigens is not as dependent upon MHC class II polymorphism as that of conventional antigens, it was proposed that superantigens bind to side residues on both the MHC molecule and the TCR β chain, rather than lying in the MHC antigen groove and contacting the TCR $\alpha\beta$ combining site [8]. This hypothesis was supported by analysis of MHC and Vβ residues critical for T cell activation by superantigens [9, 10].

The first clear evidence that antigen-specific T cells could be deleted was provided by taking advantage of an endogenous superantigen. Kappler et al. [11] observed that a high proportion of Vβ17a$^+$ T cell hybridomas responded to the products of the MHC *I-E* locus (I-E). Using Vβ17a-specific monoclonal antibody, they found that I-E$^+$ strains of mice lacked peripheral Vβ17a$^+$ cells [12]. Vβ17a was expressed on CD4$^+$CD8$^+$ (double positive) TCR-intermediate (TCR^{med}) immature thymocytes, but not on the more mature CD4$^+$CD8$^-$ or CD4$^-$CD8$^+$ (single positive) TCR^{hi} cells. Thus, self-reactive thymocytes were being generated, but failed to develop into mature T cells. This could in principal be accounted for by two different mechanisms: clonal deletion or lack of positive selection. To distinguish between these possibilities, F$_1$ animals were produced by mating I-E$^+$ and I-E$^-$ mice. Mature single positive Vβ17a$^+$ thymocytes were missing in these animals as well, indicating that the loss of Vβ17a$^+$ cells was due to active deletion. It was subsequently shown that I-E was necessary but not sufficient for deletion. Vβ17a$^+$ T cell hybridomas responded to I-E expressed on B cells but not I-E$^+$ expressed by macrophages or fibroblasts, implying that Vβ17a recognized a tissue-specific antigen in the context of I-E [13].

This observation was followed by numerous studies that used anti-Vβ-specific antibodies or mRNA expression to show that expression of I-E led to decreased expression of T cells expressing Vβ5, 11, 12 and 16 in different strains of mice; use of F$_1$ animals revealed that these instances of decreased Vβ expression were also examples of clonal deletion [14–19]. Background genes were found to affect deletion, suggesting that like the ligand recognized by Vβ17a$^+$ T cells, the ligand(s) mediating deletion of these additional Vβ's was composed of some non-MHC molecule associated with I-E [16–20]. The superantigenic properties of these unknown gene products are shared by another set of molecules, the minor lymphocyte stimulatory antigens (Mls) [21]. Mls antigens represent a set of non-MHC products capable of stimulating a strong T cell proliferative response in Mls-disparate strains of mice. Mls-responsive T cells have very high precursor frequencies (up to 20%) compared to T cells responsive to classical antigens [8]. This observation was explained by the finding that reactivity with Mls determinants is dictated by Vβ usage. Vβ6$^+$ and Vβ8.1$^+$ T cells, for example, respond to Mls^a [22, 23], while Vβ3$^+$ T cells respond to Mls^c [24–26]. Mls determinants expressed in vivo caused deletion of the same Vβ^+ T cell subsets that they activated in vitro [14, 15, 22, 23, 25–29]. A clue to the origin of the endogenous superantigens was provided when genetic mapping studies showed that one of the gene products involved in the deletion of Vβ5.2 was linked to the endogenous mouse mammary tumor virus, Mtv-9 [17]. Succeeding studies found that both the Mls determinants and the non-MHC encoded products that are required for I-E-mediated Vβ-specific deletion coincide with expression of certain mouse mammary tumor proviruses [17, 30–38]. The region of the provirus coding for the superanti-

Table 1. Recognition of superantigens by Vβ families

Vβ	1	2	3	4	5	6	7	8.1	8.2	8.3	9	10	11	12	13	14	15	16	17a	18	19a	20
Endogenous																						
Mtv-1 (Mls^c)			×																			
Mtv-2																×						
Mtv-3 (Mls^c)			×																×			
Mtv-6 (Mls^c)			×		×																	
Mtv-7 (Mls^a)						×	×	×			×											
Mtv-8													×	×					×			
Mtv-9					×								×	×					×			
Mtv-11													×	×								
Mtv-13 (Mls^c)			×																			
Unknown																		×			×	×
Exogenous																						
SEA	×		×									×	×						×			
SEB			×			×	×	×	×										×			
SEC1			×					×	×					×					×			
SEC2			×						×			×							×			
SEC3			×			×	×	×														
SED			×			×	×	×	×				×						×			
SEE													×					×	×			
TSST-1			×															×	×			
ExFT			×									×	×					×	×			
MAM					×		×	×	×													
C3H-exo																	×	×				
MMTV-SW						×	×	×			×											
Balb/cV-exo		×																				

Information for this table was compiled from the following studies: Vβ recognition of endogenous superantigens [15, 17, 30–33, 35, 37, 38, 40, 157–160]; Vβ recognition of exogenous superantigens [34, 36, 39, 49, 51, 52].

gens has been mapped to the 3′ open reading frame (ORF) [39–42]. Transfection of cDNA encoding the Mtv-7 ORF into a B cell tumor has indicated that its 45-kD protein product, perhaps after proteolysis to a smaller polypeptide, stimulates Vβ6$^+$ and Vβ8.1$^+$ T cell hybridomas [43]. Therefore, both the I-E-associated and the Mls superantigens appear to be retroviral proteins that associate with MHC class II and bind TCR subsets defined by their Vβ usage. A summary of the relationship between endogenously expressed Mtv's and the Vβ deleting ligands is given in table 1.

A number of studies that made use of endogenous superantigen expression support the notion that clonal deletion of T cells occurs primarily in the thymus. Examination of TCR Vβ usage in the peripheral T cells of I-E$^+$ mice found specific deletion of Vβ3$^+$ and Vβ11$^+$ T cells in euthymic animals, but normal proportions of these $V\beta^+$ T cell subsets in the relatively small number of peripheral T cells that develop in athymic animals [44, 45]. Data supporting a primary role for the thymus in clonal deletion came from studies in normal mice, in which intraepithelial lymphocytes that mature outside of the thymus express 'forbidden' $V\beta$s [46–48].

Exogenous superantigens have also been identified, the best character- ized of which are the bacterial exotoxins [49]. Among the commonly studied exogenous superantigens are staphylococcal enterotoxins A, B, C, and D, TSST-1, and the mycoplasma antigen, MAM. Like Mls, these superantigens cause massive stimulation of mature T cells expressing certain TCR $V\beta$s, and $V\beta$-specific thymocyte deletion. For example, neo- natal injection of staphylococcal enterotoxin B (SEB) results in specific disappearance of $V\beta8^+$ mature thymocytes, partial loss of $V\beta8^+$ TCR^{med} thymocytes, and decreased numbers of peripheral $V\beta8^+$ T cells [7]. Addi- tion of SEB to fetal thymic organ culture leads to the specific loss of $V\beta8^+$ T cells with little or no effect on $V\beta11^+$ or $V\beta6^+$ thymocytes [50]. A summary of many studies such as these on the influence of exogenous protein superantigens on clonal deletion is given in table 1. Another type of exogenous superantigen is encoded by the C3H mouse mammary tumor virus (C3H-exo), for which the relevant superantigen is passed from mother to offspring through the milk [34]. This was demonstrated by finding that $V\beta14^+$ T cells are deleted in mice that nursed on virus-produc- ing, but not uninfected, mothers. More recently, additional milk-borne mouse mammary tumor viruses have been described that have distinct $V\beta$-specific superantigenic properties [51, 52].

Superantigens are a valuable tool that has allowed one to study the responses of 'antigen-specific' T cell populations. Superantigens, however, represent a unique category of molecules for which T cell recognition differs significantly from that of classical antigens. TCR transgenic mice are another tool that has made it possible to study antigen-specific T cell populations, and in this case conventional antigen can be used to stimulate an identifiable cohort of T cells. One of the first uses of such animals to study thymocyte selection utilized mice with T cells expressing TCR α and β chains derived from a $CD8^+$ cytotoxic T cell clone that recognizes H-Y (a male antigen) in the context of $H-2D^b$ [53]. Expression of these transgenes in $H-2^b$ male mice caused a dramatic reduction in the fraction of $\alpha\beta$ clonotype$^+$ $CD8^+$ single positive cells in the thymus and periphery, and depletion of double positive clonotype-expressing T cells in the thymus [54]. Furthermore, culture of thymocytes from female TCR anti-H-Y transgenic mice with male antigen-presenting cells (APCs) resulted in the death of clonotype$^+$ $CD4^+CD8^+$ thymocytes [55]. Similar transgenic mice were created in which the TCR was specific for the H-2L^d molecule [56]. Co-expression of L^d in these mice also led to the loss of clonotype$^+$ peripheral T cells, and clonotype$^+$ thymocytes were substantially reduced at, or before, the CD4$^+$CD8$^+$ stage of development.

In addition to mice with TCRs that recognize self antigens, mice with transgenic TCRs specific for foreign antigens have been created; in these animals the quantification and timing of antigen exposure can be carefully controlled. In one example, injection of antigen into mice expressing a chicken ovalbumin (OVA)-specific TCR resulted in the rapid loss of clonotype$^+$ CD4$^+$CD8$^+$ thymocytes [57]. In another, addition of antigen to fetal organ cultures of thymuses from mice transgenic for a pigeon cytochrome c-specific TCR resulted in a dose-dependent reduction in clonotype$^+$ cells at the CD4$^+$CD8$^+$ stage of development [58, 59].

Variability in TCR/Ligand Interactions that Affect Clonal Deletion

The deletion observed in TCR transgenic mice occurred at an earlier developmental stage (in CD4$^+$CD8$^+$ cells) than that observed in nontransgenic mice with superantigen, a finding supported by another experimental approach. Injection of anti-CD4 antibodies into irradiated mice during reconstitution with syngeneic bone marrow prevented the normal deletion of CD8$^+$Vβ17a$^+$ T cells [60], suggesting that thymic clonal deletion occurs during the CD4$^+$CD8$^+$ stage of development. However, other studies using superantigens and nontransgenic mice again gave a different answer. Addition of the superantigen SEB to fetal thymic organ culture resulted in the deletion of only single positive, not CD4$^+$CD8$^+$, thymocytes [50]. Furthermore, a study in which 'transitional' CD4$^+$CD8$^+$ intermediates were examined found that CD4$^+$CD8$^+$ thymocytes of the CD4$^+$CD8^{lo} TCR^{med} phenotype (a stage between CD4$^+$CD8$^+$ and single positive thymocytes) still expressed Vβ6 and Vβ17a in Mls^a and I-E$^+$ mice, respectively, but no longer did so when they had attained the more mature CD4$^+$CD8$^-$TCR^{hi} phenotype [61]. Therefore, in mice exposed to exogenous or endogenous superantigens, deletion occurs just prior to or during the transition to single positive thymocytes, while in cases in which a transgenic TCR recognizes a self antigen it occurs somewhat earlier. It is possible that this difference is related to the fact that cell surface TCR expression in transgenic mice occurs earlier in T cell development than in normal animals [62]. Another possible explanation is that the avidity of the TCR/ligand interaction may be responsible for the exact stage of development at which clonal deletion can first be detected. Data in support of this hypothesis comes from studies of K^b-specific TCR transgenic mice in which the density of antigen had a striking affect on clonal deletion [63]. When the K^b-specific TCR was expressed in H-2^b homozygous mice, deletion of antigen-specific thymocytes occurred early in the CD4$^+$CD8$^+$ stage. Expression of this TCR in H-2$^{k \times b}$F$_1$ mice, on the other hand, resulted in little deletion at the CD4$^+$CD8$^+$ stage, although clonotype$^+$ single positive cells were still missing. These data suggest that deletion of cells bearing TCRs with relatively high avidity (or those exposed to high concentrations of ligand) occurs at an earlier stage than that of T cells bearing TCRs of lower avidity. It may be that the up-regulation of TCR expression that accompanies transition from the CD4$^+$CD8$^+$ to the CD4$^+$CD8^{lo} state allows relatively low avidity TCR/ligand interactions to result in clonal deletion.

One way of modulating the apparent avidity of the TCR/ligand interaction is to alter expression or function of adhesion or accessory molecules that either qualitatively or quantitatively enhance TCR-mediated cellular activation. This was done in experiments in which CD8 was prevented from binding its ligand, the MHC class I α3 domain [64–66]. Transgenic mice expressing mutant H-2K^b molecules with α3 domains incapable of being recognized by murine CD8 were crossed with mice transgenic for TCRs that recognize K^b. The K^b-reactive T cells were not deleted in these mice, as evidenced by the generation of anti-K^b cytolytic T cells upon in vitro culture with splenocytes from H-2^b mice [67]. In another study, transgenic mice that expressed HLA-A2 (which interacts poorly with murine CD8) or a chimeric molecule composed of the α1 and α2 domains of HLA-A2 and the α3 domain of H-2K^b were immunized with influenza virus. Although both mice generated CTLs specific for the influenza matrix peptide, CTLs from the HLA-A2 mice were substantially less effective than CTLs from HLA-A2/K^b transgenic mice at lysing autologous targets. Furthermore, CTLs from HLA-A2/K^b animals lysed targets from HLA-A2 mice poorly, while CTLs from HLA-A2 transgenic mice lysed targets from HLA-A2/K^b transgenic mice even in the absence of the matrix peptide [68]. The presence of antigen-specific CTLs in HLA-A2 transgenic mice that recognized HLA-A2/K^b targets in the absence of antigen ('autoreactive CTLs') was taken to indicate that some thymocytes that are normally deleted escape this fate when the contribution of the CD8 'co-receptor' is negated. The effects of varying the TCR/ligand interaction on clonal deletion was addressed from another angle in a study in the TCR transgenic mice with receptors specific for L^d were bred to mice transgenic for CD8 [69]. Under normal conditions, L^d-specific TCR transgenic mice are positively selected in the presence of H-2^b. Expression of the CD8 transgene increased CD8 levels at least twofold compared to normal animals. When the two transgenes were simultaneously expressed in H-2^b F$_1$ mice, a decrease in the total number of thymocytes, particularly CD4$^+$CD8$^+$ cells, was observed. The clonotype-expressing T cells that escaped clonal deletion were either CD8$^-$, or CD8^{lo}, both in the thymus and the periphery. Therefore, by increasing the contribution of CD8 to the TCR/ligand interaction, positive selection became in effect negative selection. Results such as these suggest that quantitative aspects of the interaction between the TCR and its ligand, such as avidity of binding and ligand concentration, determine whether a T cell will be deleted.

The role of other accessory molecules on induction of clonal deletion is currently being analyzed. An in vitro model system has been used to evaluate the effect of LFA-1, ICAM-1, and B7/BB1 on clonal deletion. Overnight culture of transgenic thymocytes expressing a TCR specific for H-Y/D^b on male APCs resulted in a decrease in the number of CD4$^+$CD8$^+$ thymocytes. Introduction of either anti-LFA-1α or anti-CD8 antibodies prevented this clonal deletion. Anti-ICAM-1 also caused a partial decrease in the deletion, whereas anti-CD4 had no effect [70]. Inclusion of the soluble CTLA4-Ig fusion protein, which binds the co-stimulatory B7 molecule found on APCs [71, 72], had no effect on clonal deletion, suggesting that while B7 can act as a co-stimulatory molecule for proliferation, it may not be essential for induction of thymocyte clonal deletion [73].

Types of Thymic APC Responsible for Clonal Deletion

Not all MHC class I- or class II-bearing cells in the thymus act as APCs that mediate clonal deletion. There is a sizable body of evidence that cells originally of bone marrow origin, most likely macrophages and dendritic cells, perform this function. An early study that suggested thymic epithelial cells were inefficient at causing clonal deletion used either I-E$^-$ mice into which a thymus from an I-E$^+$ strain was engrafted, or mice transgenic for an I-Eα chain that was expressed exclusively on thymic epithelium [74]. Clonal deletion was found to be inefficient in such animals, since the percentage of Vβ17a$^+$ T cells was only slightly lower than in I-E$^-$ control animals. Subsequently, Mls^a-induced Vβ6$^+$ T cell deletion was studied in bone marrow chimeras in which the donor and host differed in expression of Mls (donor Mls^b, host Mls^a), and the donor cells or the host either did or did not express I-E [41, 75–77]. Since deletion of Vβ6$^+$ T cells is dependent upon expression of both Mls^a and I-E, it was possible to determine which cell type, bone marrow-derived or thymic epithelium, presented the deleting ligand for the Vβ6 TCR. Deletion of Vβ6$^+$ cells occurred only when I-E was expressed by the bone marrow-derived cells; it did not occur when only the host cells expressed I-E. The Mls^a component of the antigen could be contributed by either the radioresistant epithelial cells or by bone marrow-derived cells. The conclusion from this study was that Mls^a peptides could be transferred in vivo from host epithelium to

I-E$^+$ bone marrow-derived cells, which in turn induced negative selection. Similar findings have been reported by others [78, 79]. There is at least one report, however, that finds under certain circumstances thymic epithelium may also be able to support clonal deletion [80]. In TCR transgenic mice, LCMV-specific T cells can be positively selected on either D^b or D^{bm13}, but are negatively selected only by LCMV presented by D^b. LCMV infection during reconstitution of b→(bm13 × bm14)F$_1$ bone marrow chimeras (H-2^b expressed on bone marrow-derived cells) resulted in loss of TCR^{hi} and TCR^{med} cells; only TCR^{lo} clonotype-expressing cells survived. LCMV infection of bm13→(b × bm13)F$_1$ chimeras (H-2^b expressed only on radiation-resistant host tissue), resulted in loss of even TCR^{lo} cells, which was interpreted to mean that even very early T cell progenitors were deleted. One possible reason for these very different results is that most studies demonstrating a predominant role for bone marrow-derived cells in deletion have used MHC class II-restricted antigens or superantigens, while in this study an MHC class I-restricted antigen was employed. One other study utilizing the H-Y-specific class I-restricted TCR transgenic mice has been reported, and while relatively ineffective compared to dendritic cells, H-Y$^+$ thymic epithelium did cause partial deletion of TCR transgenic thymocytes in bone marrow chimeras in which only they expressed the correct H-2 haplotype [81]. It seems likely, therefore, that although in certain circumstances thymic epithelium may be capable of presenting antigens so that thymocyte deletion results, it is relatively poor at doing so in vivo compared to bone marrow-derived cellular elements.

Exactly which cell can induce deletion in vitro may be another matter. Clonal deletion (apoptosis) has been studied in thymocytes from OVA-specific TCR transgenic mice that were cultured with antigen peptide plus various MHC class II-bearing APC lines [82]. Fibroblasts, B cells, thymic nurse cells (epithelial cells), and even thymocytes that bear a transgenic MHC class II molecule were capable of supporting antigen-induced deletion. Antigen dose-response curves revealed that B cell tumors were more effective in this process than the other cells tested. Data such as this indicate that sufficient TCR occupancy, regardless of the origin of the APC, allows thymocytes to undergo programmed cell death in vitro. Why this does not seem to be the case in vivo is unclear, but may have to do with anatomical localization, MHC class I and class II density, distribution of the TCR's ligand, and presence of co-stimulatory and/or accessory molecules. It seems likely that these factors (and others) will eventually account for the general observation that bone marrow-derived APCs are predominantly responsible for clonal deletion in the thymus.

Cellular Mechanisms of Clonal Deletion

Clearly, the most efficient way to eliminate autoreactive clones from the immune system is to have them die. There are a variety of means by which this might be accomplished, such as effector-mediated cytolysis, exposure to cytotoxic cytokines, or by programmed cell death (PCD), a process in which the affected cell participates in its own demise, as evidenced by a requirement for de novo mRNA and synthesis [83]. Cells undergoing PCD manifest the morphologic signs of apoptosis: blebbing of the plasma membrane, condensation of the cytoplasm, and most notably, chromatin clumping and formation of crescent-shaped caps near the nuclear envelope [84]. The most frequently noted biochemical event accompanying apoptosis is the fragmentation of nuclear DNA into oligonucleosomal-sized pieces, often demonstrated as a 'DNA ladder' by agarose gel electrophoresis [85].

There is now ample evidence that autoreactive thymocytes die in the thymus by a process resulting in programmed cell death. Cyclosporin A prevents TCR-mediated PCD in T cell hybridomas and thymocytes [86, 87], and injection of cyclosporin A into sublethally irradiated mice that normally delete $V\beta 11^+$ T cells results in the survival of these cells in the thymus [88, 89]. The first direct data implicating PCD in thymocyte clonal deletion was generated using anti-TCR antibodies to cross-link thymocyte TCRs. Such treatment induced apoptosis in $CD4^+CD8^+$ thymocytes in fetal thymic organ culture [90]. Similar results were obtained by addition of SEB to fetal thymic organ culture, resulting in the specific apoptosis of thymocytes bearing $V\beta 8.1$ [50]. Finally, antigen-specific T cell deletion was directly demonstrated to be due to PCD using TCR transgenic mice. Exposure of thymocytes bearing receptors for either OVA [57] or the male antigen H-Y [55] to the relevant antigen resulted in specific loss of $CD4^+CD8^+$ cortical thymocytes, accompanied by DNA fragmentation [57]. Evidence thus far strongly supports activation-induced apoptosis as the predominant mechanism of clonal deletion.

Peripheral T Cell Deletion

Although the thymus is required for normal antigen-specific negative selection, there is mounting evidence that clonal deletion can occur in the peripheral immune system. Treatment of mice with anti-I-E monoclonal antibodies from birth resulted in the accumulation of 'forbidden' $V\beta 6^+$ T cells in the periphery [91]. When the antibody treatment was halted and I-E was re-expressed in the periphery, $V\beta 6^+$ T cells slowly disappeared. A similar observation was made in athymic mice, demonstrating that the specific loss of $V\beta 6^+$ cells was indeed due to their deletion and not their 'dilution' by new thymic emigrants lacking a population of $V\beta 6^+$ T cells.

In another study, female H-Y-specific TCR transgenic mice were injected with male (H-Y$^+$) lymph node cells [92]. A decrease in the number of CD8$^+$ clonotype-expressing T cells was observed within 7 days of injection, and was maintained for at least 42 days. It was argued that these T cells were not simply being anatomically sequestered, because the decrease was observed in lymph nodes, spleen, and peripheral blood lymphocytes. Likewise, the decrease in clonotype$^+$ T cells was not due to down-regulation of the TCR, since rather than an increase in the fraction of clonotype$^-$ T cells there was a decrease in the absolute number of T cells. Furthermore, in similar types of studies using the H-Y-specific TCR transgenic mice, CD8$^+$ T cells recovered from female TCR transgenic mice 4 days after injection with male lymphoid cells were found to be actively degrading their DNA (i.e. undergoing apoptosis), indicating that active cell death (clonal deletion) was occurring in the periphery [93]. Double transgenic mice that simultaneously expressed L^d on pancreatic acinar cells and a TCR specific for L^d were also studied [94]. While thymic development appeared to be normal, there was a dramatic (80%) decrease in clonotype-expressing T cells in the periphery.

A characteristic proliferation/deletion response by peripheral T cells has been observed after in vivo administration of superantigens. Injection of Mls^{a+} lymphocytes into Mls^{a-} mice resulted in expansion of Vβ6$^+$ T cells within 4 days, followed by specific depletion of Vβ6$^+$ T cells within 2 weeks [95]. Similarly, expansion of female H-Y-specific TCR transgenic T cells was observed within 5 days after injection of female lymphoid cells into male nude recipients, followed by the rapid disappearance of these cells by day 9 [96]. In another study, within 4 days of injection of SEB into mice there was an increase in the fraction of Vβ8$^+$ T cells, but by day 7 they had largely disappeared [97]. In this case, DNA fragmentation, indicating ongoing apoptosis, was detected in splenocytes on day 4 after injection and was no longer observed by day 7. This response to SEB occurred in both euthymic and athymic animals. The finding of DNA fragmentation directly demonstrated that exposure of peripheral T cells to SEB resulted in their death. The consistent finding of early T cell proliferation raises the possibility that cellular activation may be a factor in the decision to undergo deletion. In fact, T cell utilization of IL-2 has been implicated in this process; SEB-induced Vβ8 T cell deletion in vivo was largely prevented by injection of anti-IL-2 but not anti-IL-4 antibodies [98]. Further support for this possibility comes from a study in which in vitro exposure to anti-TCR antibodies resulted in the death of splenic T cells preactivated with antigen and IL-2 but not naive T cells [99].

In the studies described above, the mechanism of peripheral deletion appears to be consistent with activation-induced T cell apoptosis. It is also possible that another type of peripheral clonal deletion might be mediated by veto cells, a term used to describe lymphoid cells that kill T cells whose TCRs recognize antigen on the veto cell itself [100]. Veto cells were originally defined functionally in mixed lymphocyte cultures of MHC 'B' bone marrow cells and MHC 'A' anti-'B' lymphocytes. This resulted in the suppression of anti-'B' precursor CTL (pCTL) generation [101, 102]. Veto activity was found not only in bone marrow cells, but also in fetal thymus and liver from normal mice, spleen cells from nude mice, and CTL clones [100]. The ability to suppress pCTL by CTL clones was not mediated through the TCR of the CTL, but rather through the TCR of the pCTL that recognized MHC class I on the surface of the CTL. Therefore, veto activity is MHC-restricted in the sense that an 'A' CTL clone can suppress anti-'A' pCTL but not anti-'B' pCTL. Recent experiments using transgenic mice that express an L^d-specific TCR have shown that veto cells actually cause the death of their 'targets' [103, 104]. Culture of transgenic spleen cells expressing the receptor for L^d with activated bone marrow cells from BALB/c ($H-2^d$), but not B10.BR ($H-2^k$), mice resulted in suppression of cytotoxic responses. Analysis of TCR clonotype expression showed that the pCTL had been clonally deleted, and not simply functionally silenced. There is now experimental data to suggest how veto cells might work. Apoptosis was induced in TCR-activated pCTLs by ligation of the MHC class I $\alpha 3$ domain with either CD8 or anti-$\alpha 3$ antibodies [105]. Thus, CD8 on the veto cell could bind the pCTL MHC class I while the pCTL TCR is occupied, delivering a lethal signal to the pCTL. However, in another study, antibodies to CTL granules also inhibited the veto effect [103]. It has yet to be resolved whether veto cell activity is solely due to activation-induced apoptosis of pCTLs, granule-mediated cytotoxicity by veto cells, or a combination of both. Whatever the mechanism, however, it is an interesting possibility that veto cells could play a role in maintaining tolerance to MHC class I-restricted antigens in peripheral T cells.

Anergy

Thymocyte Anergy
Although clonal deletion appears to be the primary means of inducing tolerance in the thymus, clonal anergy occurs as well, presumably as a 'backup' mechanism to prevent autoreactive cells that escape deletion from causing harm. Anergy in thymocytes was independently demonstrated by two groups using radiation bone marrow chimeras. To produce mice bearing an antigen recognized by a particular Vβ family, (Mls^b) → F$_1$ (Mls$^{a \times b}$) bone marrow chimeras that express I-E (recognized by Vβ17)

only on radioresistant host epithelial cells and not bone marrow-derived APC were created [78]. These mice delete neither $V\beta 6^+$, which recognizes Mls^a in association with I-E, nor $V\beta 17^+$ T cells. A telling result was that T cells and thymocytes from the $V\beta 6^+$ bone marrow chimeras proliferated poorly in response to stimulation with F_1 APCs or anti-$V\beta 6$ antibodies, although they did proliferate when stimulated with anti-$\alpha\beta$ antibodies. Another group employed a similar strategy and obtained virtually identical results [79]. Thus, thymic epithelium has the ability to induce anergy in developing thymocytes.

Induction of Anergy in Peripheral T Cells

Tolerance to extrathymic antigens has been extensively studied in transgenic mice that express an immunogenic antigen and/or MHC class I or class II at an extrathymic site. For example, double transgenic animals were generated by crossing mice expressing the K^b transgene in pancreatic β cells with mice expressing a TCR specific for K^b [106]. Such F_1 animals were tolerant of K^b skin grafts, and minimal CTL activity was generated to K^b by their T cells. This was not because of antigen-specific clonal deletion, since clonotype-expressing T cells were present in both the thymus and the periphery at levels comparable to the K^b-TCR-only transgenic mice. In another study, expression of I-E on pancreatic β cells in an I-E$^-$ strain resulted in tolerance to I-E, as judged by lack of lymphoctyic infiltrates in the pancreas and failure of T cells from these animals to proliferate when stimulated by I-E$^+$ splenocytes in vitro [107]. Surprisingly, thymocytes were also tolerant, even though I-E could not be detected in the thymus. The same group constructed transgenic mice in which the elastase promoter was used to express I-E on pancreatic acinar cells in otherwise I-E$^-$ mice [108]. Again, both peripheral and thymic T cells were shown to be unresponsive to I-E in in vitro stimulation assays. The lack of T cell responsiveness to I-E was not thought to be due to clonal deletion, since both $V\beta 5^+$ and $V\beta 17a^+$ T cells (known to be deleted in I-E$^+$ strains of mice) were found in the periphery [109]. Furthermore, the $V\beta 5^+$ and $V\beta 17a^+$ T cells could not be activated even by anti-$V\beta$-specific monoclonal antibodies, an empirically useful test of clonal anergy. Culture of an antigen-specific T cell clone with the relevant peptide and pancreatic islet β cells expressing the I-E transgene induced long-term in vitro nonresponsiveness to restimulation, one of the hallmarks of clonal anergy [110]. How can thymocytes be tolerant of an antigen that is expressed only in the periphery? One possibility is that MHC-encoded molecules are shed from the peripheral pancreatic cells and carried to the thymus, where actual induction of tolerance induction occurs. Such a possibility was tested for MHC class I in a study in which transgenic H-2$^{b \times d}$ mice were made to produce and secrete soluble H-2K^k [111]. It was found that these mice were not tolerant of cell surface K^k in that they proliferated when cultured with K^k-bearing APC. They did not respond, however, to soluble K^k plus APCs expressing H-2D^b. This indicates that there exist two distinct populations of cells, one that recognizes processed K^k in association with cell surface MHC class I (tolerized in these mice), and one that recognizes cell surface K^k (not tolerized). Furthermore, the results suggest that, at least for MHC class I, shed MHC molecules do not tolerize thymocytes to their endogenous transmembrane counterparts. A similar experiment has not yet been performed with MHC class II. Another possible explanation of thymocyte unresponsiveness to a solely peripheral antigen is that tolerance occurs in the thymus due to incomplete tissue restriction of the transgene, resulting in lows levels of thymic expression that are undetectable by standard assay methods. Arguing against this theory is the finding that in transgenic mice that express low but detectable levels of I-E on the thymic medullary epithelium, Vβ17a$^+$ T cells responded to stimulation with anti-Vβ17a antibodies nearly as well as mice that do not express I-E at all [109]. However, PCR has recently been used to show that mice transgenic for K^b driven by the rat insulin promoter (to restrict K^b expression to pancreatic β cells) do indeed express K^b mRNA in the thymus, and furthermore that the level of K^b could induce skin graft tolerance when thymuses from transgenic mice were transferred to K^b mice [112]. This underscores the lack of certainty with which one can be sure that a gene is expressed only in the targeted tissue, and raises the possibility that expression in the thymus may have accounted for the results of others who found tolerance to antigens thought to be expressed only in the periphery.

It should be noted that other experiments using transgenic animals, although of similar design, have yielded somewhat different answers. For example, transgenic mice that expressed H-2K^b on pancreatic islet β cells [113, 114] were tolerant of K^b in vivo. Although the islets were aberrant and the mice developed diabetes, there were no lymphocytic infiltrates in the islets, the mice accepted skin grafts from K^b mice, and their lymphocytes failed to lyse of K^b target cells. This tolerance, however, did not extend to the thymus, since thymocytes from these animals had the capacity to lyse K^b-bearing targets. When T cells from transgenic mice were transferred to syngeneic, thymectomized, and irradiated recipients, the tolerant status did not change over time; recipients were still tolerant of K^b-bearing skin grafts [115]. Interestingly, the tolerance could be overcome by the addition of IL-2 to cultures during generation of effector CTL, similar to the observation made in an in vitro model for anergy (see below), and arguing against clonal deletion of CTL precursors as the mechanism of tolerance. The lack of tolerance in the thymus is in contrast to the study in which I-E expression was directed to pancreatic β cells [107]. One possible resolution for this difference is that while MHC class II presents exogenous antigens, MHC class I normally present endogenously synthesized antigens; thus, antigens shed by the pancreas would not be expected to be presented by class I molecules in the thymus [114].

It is possible to create a situation in which partial clonal deletion occurs, resulting in a tolerant state that resembles, but is not, anergy. Such a situation can arise if antigen-specific high avidity T cells only are deleted, leaving the low avidity clones that respond relatively poorly to the antigen. In one study, for example, transgenic C3H mice expressing a chimeric MHC class I molecule, $Q10/L^d$, solely on hepatic tissue were analyzed [116, 117]. These transgenic animals were tolerant of this hybrid molecule, exhibiting no lymphocytic hepatic infiltrates and relatively low, although detectable, $Q10/L^d$-specific CTL activity. Notably, the decreased CTL activity was not due to a decrease in the number of CTL precursors, and the relatively decreased cytotoxic response to $Q10/L^d$ persisted even when the transgenic spleen cells were transferred into non-$Q10/L^d$-expressing recipients. Although in this example one cannot distinguish between clonal deletion and long-lived clonal anergy, the normal numbers of $Q10/L^d$-specific CTL precursors led to the proposal that tolerance may have been established by the 'inactivation/deletion' of a high avidity T cell subset, leaving relatively low avidity T cells incapable of mounting an autoimmune response. In another example, T cells from human insulin transgenic mice proliferated in response to human insulin, albeit more weakly than non-transgenic controls, and did not provide help for B cell antibody production [118, 119]. Human insulin-specific hybridomas made from these mice required higher insulin concentrations to achieve their maximal response, and had lowered maximal production of lymphokines. Interestingly, while $V\beta 1$ expression was prevalent in nontransgenic insulin-specific hybridomas, other $V\beta$'s predominated in those of transgenic origin. These results were interpreted to mean that the high avidity, $V\beta 1$-utilizing, T cell clones were lost, either by deletion or anergy, leaving relatively lower avidity T cells that were not representative of the normal repertoire. Similar findings were reported in T cells from TCR β chain transgenic mice; expression of $V\beta 8.1$ in transgenic Mls^{a+} animals resulted in partial deletion of $V\beta 8.1^+$ T cells [120, 121]. T cell clones made from the few remaining $V\beta 8.1^+$ T cells in these mice were responsive to Mls^a, but their activation required more Mls^a-bearing APC than T cell clones from Mls^a-transgenic animals, and had lower proliferative responses and less lymphokine secretion upon stimulation than clones isolated from Mls^{a-} mice. These studies demonstrate that clonal deletion of a subset of antigen-specific T cells can also contribute to a tolerant state.

Many of the studies analyzing the relationship between extrathymic antigen expression and tolerance are summarized in table 2.

Mechanism of Anergy

Antigen-specific T cell clones have provided an excellent model system with which to study T cell anergy. Exposure of normal helper T cell clones to the appropriate antigen/MHC typically results in lymphokine production and proliferation. In certain situations, however, TCR occupancy fails to induce IL-2 secretion, and upon restimulation the T cell clone does not respond [122]. This unresponsive state is induced when the T cell encounters antigen in the absence of cellular co-stimulatory, or 'second', signals. Prevention of co-stimulation can be achieved by treating APCs with chemical fixatives, such as 1-ethyl-3-(3-dimethylaminopropyl)carbodiimide (ECDI) [123, 124], or by stimulation with antigen presented by purified MHC class II molecules incorporated into planar membranes or with anti-TCR antibodies bound to plastic [125–127]. Co-stimulation can be added in 'trans'; that is, the addition of third-party antigen-nonbearing cells with co-stimulatory activity can restore the T cells' proliferative response in these cultures, although with lesser efficiency than when the co-stimulatory molecules are on the cell presenting the antigen [123, 128]. TCR occupancy without co-stimulatory signals induces normal Ca^{2+} increases, phosphatidylinositol hydrolysis, and protein kinase C activation, yet lymphokine secretion, most notably of IL-2, is prevented [129]. The anergized state of the T cell can be overcome by the addition of exogenous IL-2 [130, 131]. The simplest interpretation of these findings is that anergy results when sufficient TCR occupancy occurs in the absence of IL-2. Two reports have addressed the issue of how transcription factors required for IL-2 production are regulated upon the receipt of a tolerizing stimulus, each reaching a different conclusion. In one study, gel mobility shift analysis of nuclear extracts from T cells stimulated with chemically fixed APCs found a failure to up-regulate NF-AT and one form of NF-κB; AP-1 induction was normal [132]. Surprisingly, restimulation of the anergized T cells with irradiated APC and antigen resulted in normal induction of the transcription factors known to be involved in IL-2 gene regulation, although no IL-2 protein was made. Another study using both gel mobility shift and promoter-regulated reported gene transcription analyses found that stimulation of previously anergized T cells did not induce AP-1 activity, while the induction of NF-AT was normal [133]. IL-2 treatment of anergized clones, which reverses the anergic state, also resulted in the normal antigen-driven up-regulation of AP-1. Further studies will be necessary to reconcile the different findings concerning anergy and the activation of nuclear transcription factors.

Table 2. Examples of tolerance in transgenic mice with peripheral expression of antigen

Transgene	Tissue expression	Tolerance		Reference
		in vivo	in vitro	
Anergy				
I-E	islet β cells	+	+	107
I-E	pancreatic acinar cells	+	+	108
K^b	islet β cells	+	+	113,114
K^b × IL-2	islet β cells	+	ND	161
K^b × TCR[1]	islet β cells	+	+	106
L^d × TCR	acinar cells	+	+	94
Influenza HA[2]	islet β cells	+	+	162
Deletion				
Q10/L^d [3]	liver	+	±	116,117
K^b × TCR	islet β cells	+[4]	ND	112
L^d × TCR	acinar cells	+	+	94
Receptor down-regulation				
K^b × TCR	liver	+	+	146
K^b × TCR	neuroectoderm	+	−	146
Discordance between in vivo and in vitro responses				
I-A^d	islet β cells	+	−	148
I-A^k	islet β cells	+	−	149
I-A^d	pancreatic acinar cells	+	−	150
K^b	RBC	+	±	155
K^b × TCR	epithelial	+	−	147
Nonrecognition				
K^b × TCR	islet β cells	+[4]	ND	112
LCMV-GP[5]	islet β cells no virus	+	−	153
	+ virus	−	−	153
LCMV-GP × TCR	islet β cells no virus	+	−	154
	+ virus	−	−	154
No tolerance				
Influenza HA	islet β cells	−	ND	156

[1] (Peripheral MHC transgenic × TCR transgenic)F$_1$ animals.
[2] Influenza hemaglutinin.
[3] Deletion suggested but not proven.
[4] Tolerant of pancreas, but rejected K^b-bearing skin grafts.
[5] LCMV glycoprotein.

It has recently become clear that there are a number of different co-stimulatory molecule(s), counter-receptors expressed on both T cells and the APCs. Two APC co-stimulatory molecules have been identified: B7 [71, 72] and heat-stable antigen (HSA) [134]. The T cell counter-receptors for B7 are CD28 and CTLA-4; the counter-receptor for HSA is unknown at this time. B7 is present at high levels on dendritic cells (the most potent of APC), and although expressed only at low levels on resting B cells, it is induced upon B cell activation. Induction of B7 in B cells has been reported to depend upon the intracellular portion of the MHC class II molecule [135]. The implication is that an MHC class II molecule bound by a TCR signals the B cell to up-regulate B7, which in turn binds its counter-receptors on the T cell. B cell tumors such as Nalm-6 lack B7 and fail to stimulate freshly purified T cells in an allogeneic mixed lymphocyte culture; Nalm-6 does so, however, after transfection of B7 [136]. Unlike stimulation via the TCR, the 'first' signal, direct cross-linking of CD28 with antibodies, does not yield phosphatidylinositol hydrolysis or Ca^{2+} increases. It does, however, induce tyrosine kinase activity [137, 138]. The result of simultaneous signaling via the TCR and CD28 is synergistic induction of IL-2 (apparently by both increased transcription and stabilization of IL-2 mRNA [139, 140]). Furthermore, antibodies against CD28 substitute for an APC-derived co-stimulatory signal, so that in their presence antigen presented by chemically fixed APCs results in IL-2 secretion and T cell proliferation [141]. As presented above, T cells with a phenotype very similar to that of anergic T cell clones can be generated in vivo. It is likely that in vivo as well this is due to antigen presentation by tissues that lack adequate co-stimulation. Keratinocytes, for example, can be made to express MHC class II by in vivo treatment with interferon-γ. Using these cells as APCs results in T cell anergy rather than activation [142]. Lack of B7 (and/or other co-stimulatory ligands) may be responsible for the ability of keratinocytes as well as thymic epithelium and other tissues to induce anergy.

The importance of B7 as a counter-receptor in peripheral T cell activation and tolerance has recently been demonstrated with the use of a CTLA-4-Ig soluble fusion protein. The CTLA-4 molecule binds B7 with higher affinity than CD28, and in this soluble form prevents the CD28 contribution to T cell activation [72]. Injection of CTLA-4-Ig into mice bearing xenogeneic pancreatic islet grafts under the kidney capsule resulted not only in acceptance of the graft during the course of treatment, but also long-term graft survival after the CTLA-4-Ig therapy was stopped [143]. Based upon the in vitro anergy model, the speculation was that the T cells recognizing the xenogeneic graft did not become effector cells, but rather became anergic, because delivery of the co-stimulatory signal was blocked by the fusion protein. CTLA-4-Ig was also used to suppress rejection of a vascular cardiac allograft [144]. In this case the fusion protein was effective only during the course of its administration; after discontinuing the treatment the graft was rejected. The failure to induce long-term tolerance in this experimental system may have been due to the much greater antigenic potency and the highly vascular nature of the graft. Another study utilizing CTLA-4-Ig evaluated its effect on recognition of a soluble antigen, keyhole-limpet hemocyanin (KLH) [145]. Similar to the cardiac allograft study, administration of CTLA-4-Ig led to suppression of anti-KLH responses during treatment, but not long-term tolerance once treatment was stopped. The finding that cell surface molecules other than the TCR *qualitatively* alter the T cell response has provided an intriguing mechanism for regulating the absolute nature of the T cell response, inducing or silencing effector function.

Why Peripheral Anergy?

Although probably secondary to elimination in the thymus, induction of anergy could play an important role in preventing autoaggressive immune behavior. Under stress the thymus involutes, just as it does with old age. It is possible that under these circumstances the thymus may not function as efficiently when screening for potentially autoreactive clones. There is also a need to prevent T cells from responding to self antigens present in the periphery but not the thymus. Although some of the studies with transgenic mice indicate that thymocytes can be tolerized even when an antigen is expressed only extrathymically, this is not a universal finding. It seems likely that there are self antigens that thymocytes never encounter, and there must exist a means of silencing them. There are also circumstances (sites of inflammation, for example) in which MHC class I and class II expression may be elevated, especially in 'aberrant' locations such as epithelium, providing sufficient ligand for recognition by low to moderate avidity anti-self T cells that were positively selected in the thymus. A design in which T cell recognition of MHC molecules (perhaps in the absence of appropriate secondary co-stimulatory ligands) leads to anergy would provide an efficient means of preventing deleterious T cell activation.

Other Mechanisms of Unresponsiveness

Receptor Down-Regulation

Down-regulation of TCR expression in peripheral T cells has been demonstrated, providing a mechanism for preventing unresponsiveness distinct from clonal deletion and clonal anergy. Transgenic mice that express H-2K^b only on cells of neuroectodermal origin were crossed with TCR K^b-specific transgenic mice [146]. The double transgenic F$_1$ mice were shown to be tolerant of K^b, since they failed to reject a K^b-expressing tumor, EL-4. While TCR clonotype expression was normal in the thymus, it was significantly decreased in the periphery. There did exist, however, a peripheral TCR$^{neg/lo}$ Thy-1$^+$ population, suggesting that the decrease in clonotype expression was not due to clonal deletion. The peripheral T cells re-expressed the clonotype when cultured in vitro with K^b-expressing spleen cells, which correlated with the development of anti-clonotype inhibitable K^b-specific CTL activity. In another case, mice expressing K^b only on hepatocytes were crossed with transgenic mice with TCRs specific for K^b [147]. Here, also, there was a large decrease in clonotype expression in the periphery compared to the thymus, accompanied by an increase in peripheral T cells with little or no TCR. However, this example of TCR down-regulation was irreversible, since culture of these cells with K^b spleen cells did not result in up-regulation of the TCR. It is unclear why these antigens caused TCR down-regulation rather than clonal anergy. It is possibly because of some as yet unknown property of the cells expressing the antigens, perhaps because of tissue-specific accessory or co-stimulator molecules, or density of antigen expression. In any case, receptor down-regulation is yet another means of maintaining a tolerant state.

Discordance Between in vivo and in vitro T Cell Responsiveness

Some studies of extrathymic antigen expression in transgenic mice have found tolerance in vivo yet potent T cell antigen-specific responses in vitro [148–150]. Studies in which expression of an MHC class II (I-A) molecule was restricted to pancreatic islet β cells or pancreatic exocrine tissue found that while no lymphocytic infiltrates were present, the T cells isolated from these animals were normally responsive to I-A-expressing APC in vitro (see table 2). The autoreactive T cells were therefore neither deleted nor anergic. One difficulty in interpretation is that in vivo tolerance was defined by lack of lymphocytic infiltrates in the tissues expressing the antigen. None of these studies used an assay such as skin graft acceptance, so it is not known whether the animals would be tolerant of the antigen when expressed on typical APCs. Only one study demonstrated that the transgenic I-A molecule could be recognized in vivo by I-A^d-reactive T cells adoptively transferred into the transgenic mice, leading to subsequent destruction of the I-A-expressing pancreatic islets [150]. The autoreactive T cells in these mice may have down-regulated their antigen-specific TCRs in vivo and recovered TCR expression when cultured in vitro. While this cannot be ruled out, another study with a similar functional outcome showed this not to be the case [147]. Transgenic mice that express K^b only on certain types of epithelial cells were bred with mice that express a transgenic TCR specific for K^b. It was found that the F_1 mice expressed equivalent numbers of clonotype$^+$ cells in the thymus and periphery, and their T cells responded to K^b in vitro. Nonetheless, these animals failed to reject K^b-expressing skin grafts. An alternative explanation to account for the discrepancy between in vitro and in vivo tolerance is that these T cells were 'silenced' due to constant TCR occupancy and overstimulation. Exposure of T cells to high concentrations of antigen in vitro results in 'high dose suppression' of proliferation [151, 152]. In contrast to anergy, this nonresponsive phenotype is short-term, occurs despite the presence of IL-2, and is reversible, with recovery occurring after the removal of antigen.

Another reason an animal might fail to respond to peripheral self antigens is that they may be anatomically inaccessible to the immune system (sequestered antigen), although it is not clear why this should be the case in some but not all mice in which antigen expression was targeted to pancreatic islet cells. Several studies have shown that certain antigens appear to be 'ignored' by the immune system. Transgenic mice that express LCMV-encoded proteins on pancreatic islet cells had no lymphocytic infiltration of the pancreas, but nevertheless manifested in vitro CTL responses when cultured with LCMV-infected target cells [153]. Interestingly, infection of the transgenic mice with LCMV caused progressive immune-mediated damage to the pancreatic islets, demonstrating that LCMV-specific T cells existed in vivo, and that they were capable of effector function when adequately stimulated. Another group used a similar approach, and in addition bred the LCMV transgenic mice with transgenic mice whose T cells bore LCMV-specific TCRs [154]. No TCR down-regulation occurred in vivo, anti-LCMV CTL responses were inducible in vitro, and LCMV infection led to destruction of the pancreatic islets. Why were the viral antigens apparently 'ignored' in the absence of active viral infection? It is possible that the T cells recognizing the LCMV antigens are of relatively low avidity, and moving them from a resting to an active state requires interactions with virally infected APCs that express high levels of adhesion molecules and co-stimulatory activity. Furthermore, viral infection may lead to inflammatory cytokine production, which might promote endothelial permeability and lymphocyte migration to sites where the antigen is expressed (in this case, the pancreas). In another example, transgenic mice expressing K^b under control of the rat insulin promoter were bred to K^b-specific TCR transgenic mice [112]. Low expression of K^b in the thymus resulted in decreased expression of T cells with high levels of the transgenic TCR in thymus and periphery. However, the presence of T cells expressing lower levels of the clonotype did not result in a response against K^b-bearing pancreatic β cells, even after these animals had rejected K^b-bearing skin grafts. This suggests that T cells with low levels of (or low avidity) anti-self TCRs may persist without being inactivated. Crossing these animals to transgenic mice that express the IL-2 gene in pancreatic β cells resulted in infiltration of the pancreas and severe diabetes occurring as early as 1 week after birth. These results support the idea that low avidity, but functional, autoreactive T cells can persist in the peripheral immune system. Ignoring the presence of antigen in the periphery would seem to be a risky means of tolerance, since introduction of an antigen that cross-reacts with the 'ignorant' TCR, or local generation of cytokines, can induce an immune response.

The split between in vivo and in vitro T cell antigen-specific responses is a problem whose eventual solution should further refine our notions about requirements for T cell activation in vivo and mechanisms of T cell tolerance.

Conclusion

It is clear that there exist highly specific mechanisms that prevent immunocomponent animals from responding to certain antigens. The major mechanisms are clonal deletion and clonal anergy, the relative contribution of each no doubt differing in different circumstances. Other mechanisms, such as receptor down-regulation, are less well characterized, and their occurrence in the course of the physiologic generation of tolerance is less certain. As detailed in this review, the wealth of studies on this subject has provided a generally consistent portrait of different types of tolerance, although there are sufficient contradictory and unexplained observations to preclude a simple synthesis. There is, of course, a great deal of difficulty in comparing so many experimental approaches, since many parameters that one would like to control for were not, or could not be, consistently provided. Another problem is that different criteria are often used to characterize tolerance. In some studies failure to detect in vitro T cell proliferation was taken as proof of tolerance, while in others lack of T cell cytotoxicity was used. At least one study that looked at both proliferation and cytotoxicity in transgenic mice expressing K^b on erythroid cells found the former to be intact while antigen-specific cytotoxicity was lacking [155]. The same is true for assessing in vivo tolerance. Some studies used lack of lymphocytic infiltration as evidence of tolerance, while others used the additional measure of skin graft rejection. An example of discordance here is that transgenic mice expressing K^b on pancreatic islet cells and TCRs specific for K^b have no lymphocytic infiltrates in the pancreas, but nevertheless reject skin grafts from K^b mice [112]. Finally, it should be mentioned that not all studies found that expression of an antigen in the pancreas led to immune tolerance; expression of influenza hemaglutinin on pancreatic islet β cells resulted in spontaneous autoimmunity, manifested by lymphocyte infiltration of the pancreas and decreased pancreatic function [156]. Why these animals were not tolerant when so many other transgenic animals expressing foreign antigens in the same tissues were is not known, but may be due to critical, and untested, variables such as TCR avidity and antigen density. Regardless of the finer points, given the present understanding of tolerance (deletion vs. anergy, TCR signaling pathways and co-stimulatory signaling pathways, accessory receptors and counter-receptors, etc.) and the rapid advances that are being made in this area, it is inevitable that before long one will be able to manipulate ligand induction of effector function or tolerance in therapeutically beneficial ways.

References

1 Bjorkman, P.J.; Davis, M.M.: T-cell receptor genes and T-cell recognition. Nature *334:* 395–398 (1988).

2 Owen, R.D.: Immunogenetic consequences of vascular anastomoses between bovine twins. Science *102:* 400–401 (1945).

3 Burnet, F.M.; Fenner, R.: The production of antibodies. 2nd. ed. MacMillan, London (1949).

4 Billingham, R.E.; Brent, L.; Medawar, P.B.: Actively acquired tolerance of foreign cells. Nature *172:* 603–606 (1953).

5 Matzinger, P.; Zamoyska, R.; Waldmann, H.: Self tolerance is H-2-restricted. Nature *308:* 738–741 (1984).

6 Rosenthal, A.S.; Barcinski, M.A.; Blake, J.T.: Determinant selection is a macrophage-dependent immune response gene function. Nature *267:* 156–158 (1977).

7 White, J.; Herman, A.; Pullen, A.M.; Kubo, R.; Kappler, J.W.; Marrack, P.: The Vβ-specific superantigen staphylococcal enterotoxin B: Stimulation of mature T cells and clonal deletion in neonatal mice. Cell *56:* 27–35 (1989).

8 Janeway, C.A.; Yagi, J.J.; Conrad, P.J.; Katz, M.E.; Jones, B.; Vroegrop, S.; Buxer, S.: T cell responses to Mls and to bacterial proteins that mimic its behavior. Immunol. Rev. *107:* 61–68 (1989).

9 Pullen, A.M.; Wade, T.; Marrack, P.; Kappler, J.W.: Identification of the region of T cell receptor β chain that interacts with the self superantigen Mls-1^a. Cell *61:* 1365–1374 (1990).

10 Dellabona, P.; Peccoud, J.; Kappler, J.W.; Marrack, P.; Benoist, C.; Mathis, D.: Superantigens interact with MHC class II molecules outside of the antigen groove. Cell *62:* 1115–1121 (1990).

11 Kappler, J.W.; Wade, T.; White, J.; Kushnir, E.; Blackman, M.; Bill, J.; Roehm, N.; Marrack, P.: A T cell receptor Vβ segment that imparts reactivity to a class II major histocompatibility complex product. Cell *49:* 263–271 (1987).

12 Kappler, J.W.; Roehm, N.; Marrack, P.: T cell tolerance by clonal elimination in the thymus. Cell *49:* 273–282 (1987).

13 Marrack, P.; Kappler, J.: T cells can distinguish between allogeneic major histocompat-
    ibility complex products on different cell types. Nature 332: 840–842 (1988).
14 Okada, C.Y.; Weissman, I.L.: Relative Vβ transcript levels in thymus and peripheral
    lymphoid tissues from various mouse strains. Inverse correlation of I-E and Mls
    expression with relative abundance of several Vβ transcripts in peripheral lymphoid
    tissues. J. Exp. Med. 169: 1703–1719 (1989).
15 Vacchio, M.V.; Hodes, R.J.: Selective decreases in T cell receptor Vβ expression.
    Decreased expression of specific Vβ families is associated with expression of multiple
    MHC and non-MHC genes. J. Exp. Med. 170: 1335–1346 (1989).
16 Tomonari, K.; Lovering, E.: T-cell receptor-specific monoclonal antibodies against a
    Vβ11-positive mouse T-cell clone. Immunogenetics 28: 445–451 (1988).
17 Woodland, D.; Happ, M.P.; Bill, J.; Palmer, E.: Requirement for cotolerogenic gene
    products in the clonal deletion of I-E reactive T cells. Science 247: 964–967 (1990).
18 Bill, J.; Kanagawa, O.; Woodland, D.L.; Palmer, E.: The MHC molecule I-E is
    necessary but not sufficient for the clonal deletion of Vβ11-bearing T cells. J. Exp. Med.
    169: 1405–1419 (1989).
19 Vacchio, M.S.; Ryan, J.J.; Hodes, R.J.: Characterization of the ligand(s) responsible for
    negative selection of Vβ11- and Vβ12-expressing T cells: Effects of a new Mls determi-
    nant. J. Exp. Med. 172: 807–813 (1990).
20 Abe, R.; Kanagawa, O.; Sheard, M.A.; Malissen, B.; Foo-Philips, M.: Characterization
    of a new minor lymphocyte stimulatory system. I. Cluster of self antigens recognized by
    'I-E-reactive' Vβs, Vβ5, Vβ11 and Vβ12 T cell receptors for antigen. J. Immunol. 147:
    739–749 (1991).
21 Abe, R.; Hodes, R.J.: T cell recognition of minor lymphoctye stimulating (Mls) gene
    products. Annu. Rev. Immunol. 7: 683–708 (1989).
22 MacDonald, H.R.; Schneider, R.; Lees, R.K.; Howe, R.C.; Acha-Orbea, H.; Festen-
    stein, R.; Zinkernagel, R.M.; Hengartner, H.: T cell receptor Vβ use predicts reactivity
    and tolerance to Mlsᵃ-encoded antigens. Nature 332: 40–45 (1988).
23 Kappler, J.W.; Staerz, U.; White, J.; Marrack, P.: Self-tolerance eliminates T cells
    specific for Mls-modified products of the major histocompatibility complex. Nature 332:
    35–40 (1988).
24 Abe, R.; Vacchio, M.S.; Fox, B.; Hodes, R.J.: Preferential expression of the T-cell
    receptor Vβ3 gene by tolerance to Mlsᶜ reactive T cells. Nature 335: 827–830 (1988).
25 Fry, A.M.; Matis, L.A.: Self-tolerance alters T-cell receptor expression in an antigen-
    specific MHC-restricted immune response. Nature 335: 831–833 (1988).
26 Pullen, A.M.; Marrack, P.; Kappler, J.W.: The T-cell repertoire is heavily influenced by
    tolerance to polymorphic self-antigens. Nature 335: 796–801 (1988).
27 MacDonald, H.R.; Pedrazzini, T.; Schneider, R.; Louis, J.A.; Zinkernagel, R.M.;
    Hengartner, H.: Intrathymic elimination of Mlsᵃ-reactive (Vb6⁺) cells during neonatal
    tolerance induction of Mlsᵃ-encoded antigens. J. Exp. Med. 167: 2005-2013 (1988).
28 Okada, C.Y.; Holzmann, B.; Guidos, C.; Palmer, E.; Weissman I.L.: Characterization of
    a rat monoclonal antibody specific for a determinant encoded by the Vβ7 gene segment.
    J. Immunol. 144: 3473–3477 (1990).
29 Happ, M.P.; Woodland, D.L.; Palmer, E.: A third T-cell receptor β-chain variable
    region gene encodes reactivity to Mls-1ᵃ gene products. Proc. Natl Acad. Sci. USA 86:
    6293–6296 (1989).
30 Six, A.; Jouvin-Marche, E.; Loh, D.Y.; Cazenave, P.A.; Marche, P.N.: Identification of
    a T cell receptor β chain variable region, Vβ20, that is differentially expressed in various
    strains of mice. J. Exp. Med. 174: 1263–1266 (1991).

31 Woodland, D.L.; Happ, M.P.; Gollob, K.J.; Palmer, E.: An endogenous retrovirus mediating deletion of αβ T cells? Nature 349: 529–530 (1991).

32 Dyson, P.J.; Knight, A.M.; Fairchild, S.; Simpson, E.; Tomonari, K.: Genes encoding ligands for deletion of Vβ11 T cells cosegregate with mammary tumor virus genomes. Nature 349: 529–530 (1991).

33 Frankell, W.N.; Rudy, C.; Coggin, J.M.; Huber, B.T.: Linkage of Mls genes to endogenous mouse mammary tumor viruses of inbred mice. Nature 349: 526–528 (1991).

34 Marrack, P.; Kushnir, E.; Kappler, J.: A maternally inherited superantigen encoded by a mammary tumour virus. Nature 349: 524–526 (1991).

35 Fairchild, S.; Knight, A.M.; Dyson, P.J.; Tomonari, K.: Cosegregation of a gene encoding a deletion ligand for Tcrb-V3+ T cells with Mtv-3. Immunogenetics 34: 227–230 (1991).

36 Wei, W.Z.; Fiscor-Jacobs, R.; Tsai, S.J.; Pauley, R.: Elimination of Vβ2 bearing T-cells in BALB/c mice implanted with syngeneic preneoplastic and neoplastic mammary lesions. Cancer Res. 51: 3331–3333 (1991).

37 McDuffie, M.; Schweiger, D.; Reitz, B.; Ostrowska, A.; Knight, A.M.; Dyson, P.J.: I-E-independent deletion of Vβ17a+ T cells by Mtv-3 from the nonobese diabetic mouse. J. Immunol. 148: 2097–2102 (1992).

38 Foo-Phillips, M.; Kozak, C.A.; Principato, M.C.; Abe, R.: Characterization of the Mlsf system. II. Identification of mouse mammary tumor virus proviruses involved in the clonal deletion of self-Mlsf-reactive T cells. J. Immunol. 149: 3440–3447 (1992).

39 Choi, Y.; Kappler, J.W.; Marrack, P.: A superantigen encoded in the open reading frame of the 3′ long terminal repeat of mouse mammary tumor virus. Nature 350: 203–207 (1991).

40 Acha-Orbea, H.; Shakhov, A.N.; Scarpellino, L.; Kolbe, E.; Mueller, V.; Vessaz-Shaw, A.; Fuchs, R.; Bloechlinger, K.; Rollini, P.; Billotte, J.; Sarafidou, M.; MacDonald, H.R.; Diggelmann, H.: Clonal deletion of Vβ14-bearing T cells in mice transgenic for mammary tumor virus. Nature 350: 207–211 (1991).

41 Pullen, A.M.; Choi, Y.; Kushnir, E.; Kappler, J.; Marrack, P.: The open reading frames in the 3′ long terminal repeats of several mouse mammary tumor virus integrants encode Vβ3-specific superantigens. J. Exp. Med. 175: 41–48 (1992).

42 Beutner, U.; Frankel, W.N.; Cote, M.S.; Coffin, J.M.; Huber, B.T.: Mls-1 is encoded by the LTR open reading frame of the mouse mammary tumor provirus Mtv-7. Proc. Natl Acad. Sci. USA 89: 5432–5436 (1992).

43 Winslow, G.M.; Scherer, M.T.; Kappler, J.W.; Marrack, P.: Detection and biochemical characterization of the mouse mammary tumor virus 7 superantigen (Mls-1a). Cell 71: 719–730 (1992).

44 Hodes, R.J.; Sharrow, S.O.; Solomon, A.: Failure of T cell receptor Vβ negative selection in an athymic environment. Science 246: 1041–1044 (1989).

45 Fry, A.M.; Jones, L.A.; Kruisbeek, A.M.; Matis, L.A.: Thymic requirement for clonal deletion during T cell development. Science 246: 1044–1046 (1989).

46 Huang, L.; Crispe, I.N.: Distinctive selection mechanisms govern the T cell receptor repertoire of peripheral CD4−CD8− α/β T cells. J. Exp. Med. 176: 699–706 (1992).

47 Rocha, B.: Characterization of Vβ-bearing cells in athymic (nu/nu) mice suggests an extrathymic pathway for T cell differentiation. Eur. J. Immunol. 20: 919–925 (1990).

48 Murosaki, S.; Yoshikai, Y.; Ishida, A.; Nakamura, T.; Matsuzaki, G.; Takimoto, H.; Yuuki, H.; Nomoto, K.: Failure of T cell receptor Vβ negative selection in murine intestinal intra-epithelial lymphocytes. Int. Immunol. 3: 1005–1013 (1991).

49 Marrack, P.; Kappler, J.: The staphylococcal enterotoxins and their relatives. Science
    248: 705–711 (1990).
50 Jenkinson, E.J.; Kingston, R.; Smith, C.A.; Williams, G.T.; Owens, J.J.T.: Antigen-in-
    duced apoptosis in developing T cells: A mechanism for negative selection of the T cell
    receptor repertoire. Eur. J. Immunol. 19: 2175–2177 (1989).
51 Held, W.; Shakhov, A.N.; Waanders, G.; Scarpellino, L.; Leuthy, R.; Kraehenbuhl, P.;
    MacDonald, H.R.; Acha-Orbea, H.: An exogenous mouse mammary tumor virus with
    properties of Mls-1ᵃ (Mtv-7). J. Exp. Med. 175: 1623–1634 (1992).
52 Hodes, R.J.; Novick, M.B.; Palmer, L.D.; Kneppert, J.E.: Association of a Vβ2-specific
    superantigen with a tumorigenic milk-borne mouse mammary tunor virus. J. Immunol.
    150: 1422–1428 (1993).
53 Teh, H.S.; Kisielow, P.; Scott, B.; Kishi, H.; Uematsu, Y.; Blüthmann, H.; von
    Boehmer, H.: Thymic MHC antigens and the specificity of the αβ T cell receptor
    determine the CD4/CD8 phenotype of T cells. Nature 335: 229–234 (1988).
54 Kisielow, P.; Blüthmann, H.; Staerz, U.D.; Steinmetz, M.; von Boehmer, H.: Tolerance
    in T-cell receptor transgenic mice involves deletion of nonmature CD4⁺CD8⁺ thymo-
    cytes. Nature 333: 742–746 (1988).
55 Swat, W.; Ignatowicz, L.; von Boehmer, H.; Kisielow, P.: Clonal deletion of immature
    CD4⁺8⁺ thymocytes in suspension culture by extrathymic antigen-presenting cells.
    Nature 351: 150–153 (1991).
56 Sha, W.C.; Nelson, C.A.; Newberry, R.D.; Kranz, D.M.; Russell, J.H.; Loh, D.Y.:
    Positive and negative selection of an antigen receptor on T cells in transgenic mice.
    Nature 336: 73–76 (1988).
57 Murphy, K.M.; Heimberger, A.B.; Loh, D.Y.: Induction by antigen of intrathymic
    apoptosis of CD4⁺CD8⁺TCRˡᵒ thymocytes in vivo. Science 250: 1720–1723 (1990).
58 Spain, L.M.; Berg, L.J.: Developmental regulation of thymocyte susceptibility to dele-
    tion by 'self'-peptide. J. Exp. Med. 176: 213–223 (1992).
59 Vasquez, N.J.; Kaye, J.; Hedrick, S.M.: In vivo and in vitro clonal deletion of
    double-positive thymocytes. J. Exp. Med. 175: 1307–1316 (1992).
60 Fowlkes, B.J.; Schwartz, R.H.; Pardoll, D.M.: Deletion of self-reactive thymocytes
    occurs at a CD4⁺CD8⁺ precursor stage. Nature 334: 620–623 (1988).
61 Guidos, C.J.; Danska, J.S.; Fathman, C.G.; Weissman, I.L.: T cell receptor-mediated
    negative selection of autoreactive T lymphocyte precursors occurs after commitment to
    the CD4 or CD8 lineages. J. Exp. Med. 172: 835–845 (1990).
62 Kisielow, P.; Swat, W.; Rocha, B.; von Boehmer, H.: Induction of immunological
    unresponsiveness in vivo and in vitro by conventional and super-antigens in developing
    and mature T cells. Immunol. Rev. 122: 69–85 (1991).
63 Auphan, N.; Schönrich, G.; Malissen, M.; Barad, M.; Hämmerling, G.; Arnold, B.;
    Malissen, B.; Schmitt-Verhulst, A.M.: Influence of antigen density on degree of clonal
    deletion in T cell receptor transgenic mice. Int. Immunol. 4: 541–547 (1992).
64 Potter, T.A.; Bluestone, J.A.; Rajan, T.V.: A single amino acid substitution in the α3
    domain of an H-2 class I molecule abrogates reactivity with CTL. J. Exp. Med. 166:
    956–962 (1987).
65 Connolly, J.M.; Wormstall, E.M.; Hansen, T.H.: The Lyt2 molecule recognizes residues
    in the class I α3 domain in allogeneic cytotoxic T cell responses. J. Exp. Med. 168:
    352–331 (1988).
66 Salter, D.R.; Norment, A.M.; Chen, B.P.; Krensky, A.M.; Littman, D.R.; Parham, P.:
    Polymorphism in the α3 domain of HLA-A molecules affects binding to CD8. Nature
    338: 345–348 (1988).

67 Ingold, A.L.; Landel, C.; Knall, C.; Evans, G.A.; Potter, T.A.: Co-engagement of CD8 with the T cell receptor is required for negative selection. Nature *352:* 721–723 (1991).

68 Sherman, L.A.; Hesse, S.V.; Irwin, M.J.; La Face, D.; Peterson, P.: Selecting T cell receptors with high affinity for self-MHC by decreasing the contribution of CD8. Science *258:* 815–818 (1992).

69 Robey, E.A.; Ramsdell, F.; Kioussis, D.; Sha, W.; Loh, D.; Axel, R.; Fowlkes, B.J.: The level of CD8 expression can determine the outcome of thymic selection. Cell *69:* 1087–1096 (1992).

70 Carlow, D.A.; van Oers, N.S.C.; Teh, S.J.; Teh, H.S.: Deletion of antigen-specific immature thymocytes by dendritic cells requires LFA-1/ICAM interactions. J. Immunol. *148:* 1595–1603 (1992).

71 Linsley, P.S.; Clark, E.A.; Ledbetter, J.A.: T-cell antigen CD28 mediates adhesion with B cells by interacting with activation antigen B7/BB1. Proc. Natl Acad. Sci. USA *87:* 5031–5035 (1990).

72 Linsley, P.S.; Brady, W.; Urnes, M.; Grosmaire, L.S.; Damle, N.K.; Ledbetter, J.A.: CTLA-4 is a second receptor for the B cell activation antigen B7. J. Exp. Med. *174:* 561–569 (1991).

73 Tan, R.; Teh, S.J.; Ledbetter, J.A.; Linsley, P.S.; Teh, H.S.: B7 costimulates proliferation of CD4⁻CD8⁺ T lymphocytes but is not required for the deletion of immature CD4⁺CD8⁺ thymocytes. J. Immunol. *149:* 3217–3224 (1992).

74 Marrack, P.; Lo, D.; Brinster, R.; Palmiter, R.; Burkly, L.; Flavell, R.H.; Kappler, J.: The effect of thymus environment on T cell development and tolerance. Cell *53:* 627–633 (1988).

75 Speiser, D.E.; Kolb, E.; Schneider, R.; Pircher, H.; Hengartner, H.; MacDonald, H.R.; Zinkernagel, R.M.: Tolerance to Mlsᵃ by clonal deletion of Vβ6⁺ T cells in bone marrow and thymus chimeras. Thymus *13:* 27–33 (1989).

76 Rammensee, H.G.; Hügin, D.: Elimination of self-reactive CD8⁺, but not CD4⁺, T cells by a peripheral immune mechanism. Transplantation *49:* 565–571 (1990).

77 Speiser, D.E.; Lees, R.K.; Hengartner, H.; Zinkernagel, R.M.; MacDonald, H.R.: Positive and negative selection of T cell receptor Vβ domains controlled by distinct cell populations in the thymus. J. Exp. Med. *170:* 2165–2170 (1989).

78 Ramsdell, F.; Lantz, T.; Fowlkes, B.J.: A nondeletional mechanism of thymic self-tolerance. Science *246:* 1039–1041 (1989).

79 Roberts, J.L.; Sharrow, S.O.; Singer, A.: Clonal deletion and clonal anergy in the thymus induced by cellular elements with different radiation sensitivities. J. Exp. Med. *171:* 935–940 (1990).

80 Speiser, D.E.; Pircher, H.; Ohashi, P.S.; Kyburz, D.; Hengartner, H.; Zinkernagel, R.M.: Clonal deletion induced by either radioresistant thymic host cells or lymphohemopoietic donor cells at different stages of class I-restricted T cell ontogeny. J. Exp. Med. *175:* 1277–1283 (1992).

81 Carlow, D.A.; Teh, S.J.; Teh, H.S.: Altered thymocyte development resulting from expressing a deleting ligand on selecting thymic epithelium. J. Immunol. *148:* 2988–2995 (1992).

82 Iwabuchi, K.; Nakayama, K.; McCoy, R.L.; Wang, F.; Nishimura, T.; Habu, S.; Murphy, K.M.; Loh, D.Y.: Cellular and peptide requirements for in vitro clonal deletion of immature thymocytes. Proc. Natl Acad. Sci. USA *89:* 9000–9004 (1992).

83 Ashwell, J.D.: Lymphocyte programmed cell death; in Snow, E.C. (ed.): T and B Cell Handbook. San Diego, Academic Press, 1993, in press.

84 Wyllie, A.H.; Kerr, J.F.R.; Currie, A.R.: Cell death: The significance of apoptosis. Int. Rev. Cytol. *68:* 251–306 (1980).

85 Wyllie, A.H.: Glucocorticoid-induced thymocyte apoptosis is associated with endoge-
    nous nuclease activation. Nature *284:* 555–556 (1980).
86 Mercep, M.; Noguchi, P.D.; Ashwell, J.D.: The cell cycle block and lysis of an activated
    T cell hybridoma are distinct processes with different Ca^{2+} requirements and sensitivity.
    J. Immunol. *142:* 4085–4092 (1989).
87 Shi, Y.; Sahai, B.M.; Green, D.R.: Cyclosporin A inhibits activation-induced cell death
    in T-cell hybridomas. Nature *339:* 625–626 (1989).
88 Jenkins, M.K.; Schwartz, R.H.; Pardoll, D.M.: Effects of cyclosporine A on T cell
    development and clonal deletion. Science *241:* 1655–1657 (1988).
89 Gao, E.K.; Lo, D.; Cheney, R.; Kanagawa, O.; Sprent, J.: Abnormal differentiation of
    thymocytes in mice treated with cyclosporin A. Nature *336:* 176–179 (1988).
90 Smith, C.A.; Williams, G.T.; Kingston, R.; Jenkinson, E.J.; Owen, J.J.T.: Antibodies to
    CD3/T cell receptor complex induce death by apoptosis in immature T cells in thymic
    cultures. Nature *337:* 181–184 (1989).
91 Jones, L.A.; Chin, L.T.; Longo, D.L.; Kruisbeek, A.M.: Peripheral clonal elimination of
    functional T cells. Science *250:* 1726–1729 (1990).
92 Zhang, L.; Martin, D.R.; Fung-Leung, W.P.; Teh, H.S.; Miller, R.G.: Peripheral
    deletion of mature CD8$^+$ antigen-specific T cells after in vivo exposure to male antigen.
    J. Immunol. *148:* 3740–3745 (1992).
93 Carlow, D.A.; Teh, S.J.; van Oers, N.S.C.; Miller, R.G.; Teh, H.S.: Peripheral tolerance
    through clonal deletion of mature CD4$^-$CD8$^+$ T cells. Int. Immunol. *4:* 599 (1992).
94 Fields, L.E.; Loh, D.Y.: Organ injury associated with extrathymic induction of immune
    tolerance in doubly transgenic mice. Proc. Natl Acad. Sci. USA *89:* 5730–5734 (1992).
95 Webb, S.; Morris, C.; Sprent, J.: Extrathymic tolerance of mature T cells: Clonal
    elimination as a consequence of immunity. Cell *63:* 1249–1256 (1990).
96 Rocha, B.; von Boehmer, H.: Peripheral selection of the T cell repertoire. Science *251:*
    1225–1228 (1991).
97 Kawabe, Y.; Ochi, A.: Programmed cell death and extrathymic reduction of Vβ8$^+$
    CD4$^+$ T cells in mice tolerant to *Staphylococcus aureus* enterotoxin B. Nature *349:*
    245–248 (1991).
98 Lenardo, M.J.: Interleukin-2 programs mouse $\alpha\beta$ T lymphocytes for apoptosis. Nature
    353: 858–861 (1991).
99 Russell, J.H.; White, C-L.; Loh, D.Y.; Meleedy-Rey, P.: Receptor-stimulated death
    pathway is opened by antigen in mature T cells. Proc. Natl Acad. Sci. USA *88:*
    2151–2155 (1991).
100 Rammensee, H.G.: Veto function in vitro and in vivo. Int. Rev. Immunol. *4:* 175–191
    (1989).
101 Muraoka, S.; Miller, R.G.: Cells in bone marrow and in T cell colonies grown from
    bone marrow can suppress generation of cytotoxic T lymphocytes directed against their
    self antigens. J. Exp. Med. *152:* 54–59 (1980).
102 Miller, R.G.: An immunological suppressor cell inactivating cytotoxic T-lymphocyte
    precursor cells recognizing it. Nature *287:* 544–547 (1980).
103 Hiruma, K.; Nakamura, H.; Henkart, P.A.; Gress, R.E.: Clonal deletion of postthymic
    T cells: Veto cells kill precursor cytotoxic T lymphocytes. J. Exp. Med. *175:* 863–868
    (1992).
104 Hiruma, K.; Gress, R.E.: Cyclosporine A and peripheral tolerance. Inhibition of veto
    cell-mediated clonal deletion of postthymic precursor cytotoxic T lymphocytes. J Im-
    munol. *149:* 1539–1547 (1992).
105 Sambhara, S.R.; Miller, R.G.: Programmed cell death of T cells signaled by the T cell
    receptor α3 domain of class I MHC. Science *252:* 1424–1427 (1991).

106   Morahan, G.; Hoffmann, M.W.; Miller, J.F.: A nondeletional mechanism of peripheral tolerance in T-cell receptor transgenic mice. Proc. Natl Acad. Sci. USA 88: 11421–11425 (1991).

107   Lo, D.; Burkly, L.; Widera, G.; Cowing, C.; Flavell, R.A.; Palmiter, R.D.; Brinster, R.L.: Diabetes and tolerance in transgenic mice expressing class II MHC molecules in pancreatic beta cells. Cell 53: 159–168 (1988).

108   Lo, D.; Burkly, L.C.; Flavell, R.A.; Palmiter, R.D.: Brinster, R.L.: Tolerance in transgenic mice expressing class II major histocompatibility complex on pancreatic acinar cells. J. Exp. Med. 179: 87–104 (1989).

109   Burkly, L.C.; Lo, D.; Kanagawa, O.; Brinster, R.L.; Flavell, R.A.: T-cell tolerance by clonal anergy in transgenic mice with nonlymphoid expression of MHC class II I-E. Nature 342: 564–566 (1989).

110   Markmann, J.; Lo, D.; Naji, A.; Palmiter, R.D.; Brinster, R.L.; Heber-Katz, E.: Antigen presenting function of class II MHC expressing pancreatic beta cells. Nature 336: 476–479 (1988).

111   Arnold, B.; Messerle, M.; Jatsch, L.; Küblbeck, G.; Koszinowski, U.: Transgenic mice expressing a soluble foreign H-2 class I antigen are tolerant to allogeneic fragments presented by self class I but not the the whole membrane-bound alloantigen. Proc. Natl Acad. Sci. USA 87: 1762–1766 (1990).

112   Heath, W.R.; Allison, J.; Hoffman, M.W.; Schönrich, G.; Hämmerling, G.; Arnold, B.; Miller, J.F.A.P.: Autoimmune diabetes as a consequence of locally produced inter-leukin-2. Nature 359: 547–549 (1992).

113   Allison, J.; Campbell, I.L.; Morahan, G.; Mandel, T.E.; Harrison, L.C.; Miller, J.F.A.P.: Diabetes in transgenic mice resulting from over-expression of class I histocom-patibility molecules in pancreatic β cells. Nature 333: 529–533 (1988).

114   Morahan, G.; Allison, J.; Miller, J.F.: Tolerance of class I histocompatibility antigens expressed extrathymically. Nature 339: 622–624 (1989).

115   Miller, J.F.A.P.; Morahan, G.; Allison, J.; Hoffman, M.: A transgenic approach to the study of peripheral T-cell tolerance. Immunol. Rev. 122: 103–116 (1991).

116   Wieties, K.; Hammer, R.E.: Jones-Youngblood, S.; Forman, J.: Peripheral tolerance in mice expressing a liver-specific class I molecule: Inactivation/deletion of a T-cell subpop-ulation. Proc. Natl Acad. Sci. USA 87: 6604–6608 (1990).

117   Jones-Youngblood, S.L.; Wieties, K.; Forman, J.; Hammer, R.E.: Effect of the expres-sion of a hepatocyte-specific MHC molecule in transgenic mice on T cell tolerance. J. Immunol. 144: 1187–1195 (1990).

118   Poindexter, N.J.; Landon, C.; Whiteley, P.J.; Kapp, J.A.: Comparison of the T cell receptors on insulin-specific hybridomas from insulin transgenic and nontransgenic mice. J. Immunol. 149: 38–44 (1992).

119   Whiteley, P.J.; Poindexter, N.J.; Landon, C.; Kapp, J.A.: A peripheral mechanism preserves self-tolerance to a secreted protein in transgenic mice. J. Immunol. 145: 1376–1381 (1990).

120   Yui, K.; Komori, S.; Katsumata, M.; Siegel, R.M.; Greene, M.I.: Self-reactive T cells can escape clonal deletion in T-cell receptor Vβ8.1 transgenic mice. Proc. Natl Acad. Sci. USA 87: 7135–7139 (1990).

121   Yui, K.; Katsumata, M.; Komori, S.; Gill-Morse, L.; Greene, M.I.: Response of Vβ8.1[+] T cell clones to self Mls-1[a]: Implications for the origin of autoreactive T cells. Int. Immunol. 4: 125–133 (1992).

122   Lamb, J.R.; Skidmore, B.J.; Green, N.; Chiller, J.M.; Feldman, M.: Induction of tolerance in influenza virus-immune T lymphocyte clones with synthetic peptides of influenza hemagglutinin. J. Exp. Med. 157: 1434–1437 (1983).

123 Jenkins, M.K.; Ashwell, J.D.; Schwartz, R.H.: Allogeneic non-T spleen cells restore the responsiveness of normal T cell clones stimulated with antigen and chemically-modified antigen-presenting cells. J. Immunol. *140:* 3324–3330 (1988).

124 Jenkins, M.K.; Schwartz, R.H.: Antigen presentation by chemically modified splenocytes induces antigen-specific T cell unresponsiveness in vitro and in vivo. J. Exp. Med. *165:* 302–319 (1987).

125 Quill, H.; Schwartz, R.H.: Stimulation of normal inducer T cell clones with antigen presented by purified Ia molecules in planar lipid membranes: Specific induction of a long-lived state of proliferative nonresponsiveness. J. Immunol. *138:* 3704–3712 (1987).

126 Watts, T.H.; McConnell, H.M.: Biophysical aspects of antigen recognition by T cells. Annu. Rev. Immunol. *5:* 461–475 (1987).

127 Jenkins, M.K.; Chen, C.; Jung, G.; Mueller, D.L.; Schwartz, R.H.: Inhibition of antigen-specific proliferation of type 1 murine T cell clones after stimulation with immobilized anti-CD3 monoclonal antibody. J. Immunol. *144:* 16–21 (1990).

128 Liu, Y.; Janeway, C.: Cells that present both specific ligand and costimulatory activity are the most efficient inducers of clonal expansion of normal CD4 T cells. Proc. Natl Acad. Sci. USA *89:* 3845–3849 (1992).

129 Schwartz, R.H.: A cell culture model for T lymphocyte clonal anergy. Science *248:* 1349–1356 (1990).

130 Essery, G.; Feldmann, M.; Lamb, J.R.: Interleukin-2 can prevent and reverse antigen-induced unresponsiveness in cloned human T lymphocytes. Immunology *64:* 413–417 (1988).

131 Desilva, D.R.; Urdahl, K.B.; Jenkins, M.K.: Clonal anergy is induced in vitro by T cell receptor occupancy in the absence of proliferation. J. Immunol. *147:* 3261–3267 (1991).

132 Go, C.; Miller, J.: Differential induction of transcription factors that regulate the interleukin-2 gene during anergy induction and restimulation. J. Exp. Med. *175:* 1327–1336 (1992).

133 Kang, S.M.; Beverly, B.; Tran, A.C.; Brorson, K.; Schwartz, R.H.; Lenardo, M.J.: Transactivation of AP-1 is a molecular target of T cell clonal anergy. Science *257:* 1134–1138 (1992).

134 Liu, Y.; Jones, B.; Aruffo, A.; Sullivan, K.M.; Linsley, P.S.; Janeway, C.A.: Heat-stable antigen is a costimulatory molecule for CD4 T cell growth. J. Exp. Med. *175:* 437–441 (1992).

135 Nabavi, N.; Freeman, G.J.; Gault, A.; Godfrey, D.; Nadler, L.M.; Glimcher, L.H.: Signalling through the MHC class II cytoplasmic domain is required for antigen presentation and induces B7 expression. Nature *360:* 266–269 (1992).

136 Norton, S.D.; Zuckerman, L.; Urdahl, K.B.; Shefner, R.; Miller, J.; Jenkins, M.K.: The CD28 ligand, B7, enhances IL-2 production by providing a costimulatory signal to T cells. J. Immunol. *149:* 1556–1561 (1992).

137 Lu, Y.; Granelli-Piperno, A.; Bjorndahl, J.M.; Phillips, C.A.; Trevillyan, J.M.: CD28-induced T cell activation. Evidence for a protein-tyrosine kinase signal transduction pathway. J. Immunol. *149:* 24–29 (1992).

138 Vanderberghe, P.; Freeman, G.J.; Nadler, L.M.; Fletcher, M.C.; Kamoun, M.; Turka, L.A.; Ledbetter, J.A.; Thompson, C.B.; June, C.H.: Antibody and B7/BB1-mediated ligation of the CD28 receptor induces tyrosine phosphorylation in human T cells. J. Exp. Med. *175:* 951–960 (1992).

139 June, C.H.; Ledbetter, J.A.; Gillespie, M.M.; Lindsten, T.; Thompson, C.B.: T-cell proliferation involving the CD28 pathway is associated with cyclosporine-resistant interleukin-2 gene expression. Mol. Cell. Biol. *7:* 4472–4481 (1987).

140   Fraser, J.D.; Newton, M.E.; Weiss, A.: CD28 and T cell antigen receptor signal transduction coordinately regulate interleukin-2 gene expression in response to super-antigen stimulation. J. Exp. Med. *175*: 1131–1134 (1992).

141   Harding, F.A.; McArthur, J.G.; Gross, J.A.; Raulet, D.H.; Allison, J.P.: CD28-mediated signalling costimulates murine T cells and prevents induction of anergy in T-cell clones. Nature *356*: 607–609 (1992).

142   Gaspari, A.A.; Jenkins, M.K.; Katz, S.I.: Class II MHC-bearing keratinocytes induce antigen-specific unresponsiveness in hapten-specific TH1 clones. J. Immunol. *141*: 2216–2220 (1988).

143   Lenschow, D.J.; Zeng, Y.; Thistlewaite, J.R.; Montag, A.; Brady, W.; Gibson, M.G.; Linsley, P.S.; Bluestone, J.A.: Long-term survival of xenogeneic pancreatic islet grafts induced by CTLA4Ig. Science *257*: 789–791 (1992).

144   Turka, L.A.; Linsley, P.S.; Lin, H.; Brady, W.; Leiden, J.M.; Wei, R.Q.; Gibson, M.G.; Zheng, X.G.; Myrdal, S.; Gordon, D.; Bailey, T.; Bolling, S.F.; Thompson, C.B.: T-cell activation by the CD28 ligand B7 is required for cardiac allograft rejection in vivo. Proc. Natl Acad. Sci. USA *89*: 11102–11105 (1992).

145   Linsley, P.S.; Wallace, P.M.; Johnson, J.; Gibson, M.G.; Greene, J.L.; Ledbetter, J.A.; Singh, C.; Tepper, M.A.: Immunosuppression in vivo by a soluble form of the CTLA-4 T cell activation molecule. Science *257*: 792–795 (1992).

146   Schönrich, G.; Kalinke, U.; Momburg, F.; Malissen, M.; Schmitt-Verhulst, A.M.; Malissen, B.; Hämmerling, G.J.; Arnold, B.: Down-regulation of T cell receptors on self-reactive T cells as a novel mechanism for extrathymic tolerance induction. Cell *65*: 293–304 (1991).

147   Schönrich, G.; Momburg, F.; Malissen, M.; Schmitt-Verhulst, A.M.; Malissen, B.; Hämmerling, G.J.; Arnold, B.: Distinct mechanisms of extrathymic T cell tolerance due to differential expression of self antigen. Int. Immunol. *4*: 581–590 (1992).

148   Miller, J.; Daitch, L.; Rath, S.; Selsing, E.: Tissue-specific expression of allogeneic class II MHC molecules induces neither tissue rejection nor clonal inactivation of alloreactive T cells. J. Immunol. *144*: 334–341 (1990).

149   Böhme, J.; Haskins, K.; Stecha, P.; van Ewijk, W.; LeMeur, M.; Gerlinger, P.; Benoist, C.; Mathis, D.: Transgenic mice with I-A on islet cells are normoglycemic but immunologically intolerant. Science *244*: 1179–1183 (1989).

150   Murphy, K.M.; Weaver, C.T.; Elish, M.; Allen, P.M.; Loh, D.Y.: Peripheral tolerance to allogeneic class II histocompatibility antigens expressed in transgenic mice: Evidence against a clonal-deletion mechanism. Proc. Natl Acad. Sci. USA *86*: 10034–10038 (1989).

151   Matis, L.A.; Glimcher, L.H.; Paul, W.E.; Schwartz, R.H.: Magnitude of response of histocompatibility-restricted T-cell clones is a function of the product of the concentrations of antigen and Ia molecules. Proc. Natl Acad. Sci. USA *80*: 6019–6023 (1983).

152   Suzuki, G.; Kawase, Y.; Koyasu, S.; Yahara, I.; Kobayashi, Y.; Schwartz, R.H.: Antigen-induced suppression of the proliferative response of T cell clones. J. Immunol. *140*: 1359–1365 (1988).

153   Oldstone, M.B.; Nerenberg, M.; Southern, P.; Price, J.; Lewicki, H.: Virus infection triggers insulin-dependent diabetes mellitus in a transgenic model: Role of anti-self (virus) immune response. Cell *65*: 319–331 (1991).

154   Ohashi, P.S.; Oehen, S.; Buerki, K.; Pircher, H.; Ohashi, C.T.; Odermatt, B.; Malissen, B.; Zinkernagel, R.M.; Hengartner, H.: Ablation of 'tolerance' and induction of diabetes by virus infection in viral antigen transgenic mice. Cell *65*: 305–317 (1991).

155   Yeoman, H.; Mellor, A.L.: Tolerance and MHC restriction in transgenic mice expressing a MHC class I gene in erythroid cells. Int. Immunol. *4*: 59–65 (1992).

156 Roman, L.M.; Simons, L.F.; Hammer, R.E.; Sambrook, J.F.; Gething, M.H.: The expression of influenza virus hemagglutinin in the pancreatic beta cells of transgenic mice results in autoimmune diabetes. Cell *61:* 383–387 (1991).

157 Woodland, D.L.; Lund, F.E.; Happ, M.P.; Blackman, M.A.; Palmer, E.; Corley, R.B.: Endogenous superantigen expression is controlled by mouse mammary tumor proviral loci. J. Exp. Med. *174:* 1255–1258 (1991).

158 Blackman, M.A.; Lund, F.E.; Surman, S.; Corley, R.B.; Woodland, D.L.; Major histocompatability complex-restricted recognition of retroviral superantigens by Vβ17a⁺ T cells. J. Exp. Med. *176:* 275–280 (1992).

159 Gollob, K.J.; Palmer, E.: Divergent viral superantigens delete Vβ5⁺ T lymphocytes. Proc. Natl Acad. Sci. USA *89:* 5138–5141 (1992).

160 Singer, P.A.; Balderas, R.S.; Theofilopoulos, A.N.: Thymic selection defines multiple T cell receptor Vβ 'repertoire phenotypes' at the CD4/CD8 subset levels. EMBO J. *9:* 3641–3648 (1990).

161 Allison, J.; Malcolm, L.; Chosich, N.; Miller, J.F.: Inflammation but not autoimmunity occurs in transgenic mice expressing constitutive levels of interleukin-2 in islet beta cells. Eur. J. Immunol. *22:* 1115–1121 (1992).

162 Lo, D.; Freedman, J.; Hesse, S.; Palmiter, R.D.; Brinster, R.L.; Sherman, L.A.: Peripheral tolerance to an islet cell-specific hemagglutinin transgene affects both CD4⁺ and CD8⁺ T cells. Eur. J. Immunol. *22:* 1013–1022 (1992).

Dr. Jonathan D. Ashwell, Room 1B-40, Building 10, National Institutes of Health, Bethesda, MD 20892 (USA)

Granstein RD (ed): Mechanisms of Immune Regulation.
Chem Immunol. Basel, Karger, 1994, vol 58, pp 34–66

Pathways and Regulation of B-Cell Responsiveness and Tolerance

David W. Scott[a], *Garvin L. Warner*[b], *Xiao-Rui Yao*[a], *Sally C. Kent*[a]

[a]Division of Immunology and Immunotherapy, University of Rochester Cancer Center and Department of Microbiology and Immunology, Rochester, N.Y.;
[b]Pharmaceutical Research Institute, Bristol-Myers Squibb, Syracuse, N.Y., USA

Historical Overview

The Need for B-Cell Tolerance

The prevention of autoreactivity depends upon mechanisms to eliminate anti-self reactive lymphocyte clones. The failure to eliminate such clones was recognized as early as 1901 by Paul Ehrlich [1]. Owen [2] provided a basis for self-tolerance in his analysis of dizygotic cattle twins that exchanged blood in utero and developed tolerance to their non-identical twins red cell antigens. Deliberate experimental induction of tolerance was first achieved in the 1950s by Billingham et al. [3].

Despite extensive further studies in the 1950s and early 1960s, it was not until 1967–1971 that Weigle and co-workers [4, 5] formally demonstrated the kinetics of tolerance induction and that both T cells and B cells could be rendered unresponsive in vivo by deaggregated human γ-globulin. These results provided both experimental and theoretical explanations for the establishment of tolerance in the whole animal. That is, if either B cells or T cells are rendered tolerant, then the lack of one responsive 'link' in the chain of humoral immunity would lead to a failure to produce a potentially autoreactive antibody.

Prior to these studies, the need for B-cell tolerance has been doubted. In fact, it was well established that low affinity autoantibodies, reactive with self-markers, could readily be observed in ostensibly normal, healthy individuals. This suggested that tolerance in the B-cell lineage did not exist. The general lack of autoreactivity could then be readily explained, provided that no T cells recognizing self-antigens were available and activated in the host organism. Therefore, these experiments of Weigle's group were important because they established that B cells, or at least their precursors in the bone marrow [5], could be rendered unresponsive.

During the course of these studies, it became apparent that animals rendered partially unresponsive with haptenated antigens often produced low affinity antibodies against the tolerated epitope [6]. This result could be explained based on the fact that higher affinity B cells might more readily interact with a tolerogen and would be rendered unresponsive with lower doses of antigen. On the other hand, it is possible that T-cell tolerance could also explain the lower affinity in partially unresponsive animals [7, 8]. That is, with a lack of T-cell help, there would be less driving the system towards the switching to higher affinity clones under normal conditions of affinity maturation. Using purified hapten-specific B cells (see below), Venkataraman and Scott [9], demonstrated that higher affinity B cells were rendered unresponsive *prior to exposure to antigenic challenge* even with a thymus-independent challenge.

The fact that viral infections could create neoantigens and the cytokines produced in response to such infections could up-regulate class I and/or class II histocompatibility antigen expression provided a mechanism by which potentially 'autoreactive' T cells might readily be activated. That is, T cells may recognize such a neoantigen and provide help for self-reactive B cells. Therefore, the existence of B-cell tolerance provided a safety net to further ensure the lack of anti-self reactivity.

In the 1970s and 1980s, the recognition of somatic mutation during immune responsiveness, especially after class switching, raised another potential danger signal for autoreactivity. If B-cell clones mutate during the generation of immune responsiveness, another window of tolerance susceptibility would have to be proposed to maintain the integrity of the organism. Indeed, this has now been observed [10] (see below).

Finally, a number of model systems had developed for the in vitro induction of tolerance in B cells in thymus-independent model systems that validated the concept of B-cell unresponsiveness. Formal proof came from the elegant transgenic models that have been utilized recently (see below, 'Transgenic Mice in the Analysis of Tolerance') [reviewed in 11, 12]. Thus, while T-cell unresponsiveness accounts for many of the properties of self-tolerance, tolerance in the B-cell pool is not only a fail-safe, but also is a requirement for the avoidance of 'horror autotoxicus'. A number of these model systems are discussed below.

Models of B-Cell Tolerance

Anti-IgM as a Surrogate for Antigen

B cells with specificity for a given antigen occur at a frequency of less than 1 in 1,000 [13]. Hence, detailed analysis of the signaling pathways towards tolerance induction has been impeded by the rare frequency of these cells. While it has been possible to isolate these cells by hapten affinity purification techniques, it has been extremely difficult to grow and clone these cells to provide the large numbers necessary for such studies. Despite these hurdles, significant work has been done in this area and is reviewed below in the section 'Fate of Hapten-Specific B Cells in Tolerance'.

As an alternative to these analyses, we have used anti-immunoglobulin as a surrogate for antigen in order to analyze the biochemical pathways of tolerance induction. This approach is based on the fact that all B cells express immunoglobulin (IgM $\pm$ IgD) and that signal transduction via cross-linking of surface Ig is thought to mimic the physiological interactions that occur in vivo when antigen is encountered. While there is no formal proof that this assumption is correct, many experimental systems have established the correlation between antigen and anti-Ig cross-linking. Thus, recent work with transgenic mice, as well as studies with B-cell lines transfected with specific receptors for known epitopes, have provided verification that this model provides accurate physiological insights. Moreover, the fact that treatment of fetal or neonatal mice with anti-IgM can cause a dose-dependent state of unresponsiveness and even deletion, and that appropriate manipulation of adult B cells in vitro mimics what is seen in transgenic mice exposed to tolerogen as adults, also provides evidence that these systems are self-validating.

The original observation that B cells could be rendered nonresponsive with anti-μ antibodies was made by Lawton and Cooper [14]. These workers found that immunoglobulin synthesis of all classes was ablated by pretreatment with anti-μ antibodies and indeed that B cells expressing surface IgM were depleted in these mice [14]. B-cell depletion induced by anti-IgM at birth was dose-dependent and of finite duration, just like experimental immunologic tolerance. Therefore, this system will first be discussed as a model for neonatal B-cell tolerance.

Neonatal Depletion

Following the observations of Lawton and Cooper [14] that treatment of neonatal mice and anti-Ig led to elimination of B cells, Raff et al. [15] and Sidman and Unanue [16] reported that the exposure of neonatal B cells to anti-Ig led to the modulation of surface IgM expression. Interestingly, the re-synthesis of new Ig did not occur in neonatal B cell treated in this manner in contrast to the re-emergence of surface IgM in anti-Ig-capped and modulated adult B cells. Although one could infer that apparent B-cell depletion is, in fact, due to the lack of the re-expression of surface IgM in developing B cells in such treated animals, additional evidence indicates that these cells are actually missing from these mice. For example, examination of lymphoid tissues of anti-μ-treated animals indicated that typical

T-cell-independent regions in the superficial cortex and the germinal centers were devoid of B cells, and plasma cells were virtually absent in the medullary cords of the lymph nodes; parallel observations were made in the periarteriolar lymphocyte sheath of the splenic white pulp. More recently, we observed [Borrello and Scott, unpubl. data, and 17] that the numbers of B220[+] cells in neonatal mice treated with anti-IgM actually decreased after approximately 24–28 h of treatment in vivo. This paralleled the decrease in surface IgM expression, but eliminated the potential artifact that IgM had been modulated off the newly emerging immature IgM[+] B cells. Initial attempts to demonstrate apoptosis in this population by genomic DNA analysis failed. However, in collaboration with Brown et al. [18], it was demonstrated that a subpopulation of apoptotic cells appeared in neonatal B cells treated with anti-IgM as monitored by forward and 90° light scattering techniques [19].

Nossal et al. [20] utilized this system to examine the dose dependence of this process. Interestingly, they found that neonatal B cells treated with anti-IgM at low concentrations did not lead to a depletion of B cells, but rather led to a state of anergy in that the potential to produce immunoglobulin was depressed. In contrast, higher concentrations caused complete modulation of IgM and depletion of B cells was apparent. This process, which they initially termed 'clonal abortion', paved the way to later observations of anergy in more mature B-cell populations. These studies also clearly validated the observations of Metcalf and Klinman [21] that neonatal B cells treated with antigen in a splenic focus analysis were far more susceptible to tolerance induction than adult B cells, treated similarly. This exquisite sensitivity to tolerance induction disappeared by 7 days of age. Thus, there is a critical window of susceptibility to tolerance induction during B-cell development, as with T cells [cf. 22].

While the signaling pathways leading to apoptosis in neonatal B cells are not clear yet, a number of insights exist based on studies in normal developing, immature B cells, as well as B-cell lymphomas. It should be noted that the neonatal spleen population contains approximately 50% CD5[+] (B-1) B cells, although it is unclear whether this early expression of the CD5 marker is relevant to the function of these cells. Analysis of signal transduction pathways in B lymphomas used as models for immature cells will be discussed later. Suffice it to say that Monroe and co-workers [23] have recently analyzed the signaling potential of neonatal B cells and found a defect in phospholipase C-mediated activation of the phosphatidylinositol (PI) pathway and a lack of protein kinase C (PKC) activation leading to a silencing of *egr-1* gene expression (see below). Indeed, recent studies from the Monroe laboratory have demonstrated that neonatal B cells may lack certain IgM-associated proteins and fail to demonstrate a critical kinase activity, which they have tentatively identified only as a p56 molecule [24]. In addition, as mentioned above, neonatal cells may be defective in the expression of certain competence genes such as *egr-1*; the role of the bcl-2 antiapoptotic product in this model has not been reported, but may be inferred from the long lifespan of B cells in E*μ-bcl-2* transgenic mice [25]. Further analysis in this area should prove informative and exciting.

Adult B-Cell Tolerance with Anti-Ig

The need for tolerance induction in adult B cells can be implied from the fact that B cells are capable of undergoing somatic mutation of their variable region genes thereby broadening their specificity while at the same time allowing for affinity maturation [10]. Problems potentially arise since this process is random and therefore may allow for the appearance of new self-reactive binding specificities in the mature B-cell population (i.e., IgM^+, IgD^+). The mature B-cell population, compared with neonatal B cells, appears to be inherently less susceptible to tolerance induction [21]. However, mature B cells are clearly able to be rendered tolerant [26].

Positive versus Negative Signaling

A number of groups have been using the fact that cross-linking of surface Ig on mature B cells results in a marked reduction in the amount of antibody produced in response to stimulation with LPS [28–31]. We have employed a two-stage model [27] involving an induction phase during which nonresponsiveness is induced in B cells by preincubation (4–24 h) with anti-Ig and a challenge phase during which differentiation is induced using either LPS or fluoresceinated-*Brucella abortus* (FL-*BA*). The advantage of this system is that it allows differential manipulation of the induction and the challenge phase so that we can address questions concerning each phase independently. The induction of non-responsiveness toward LPS challenge can be accomplished using either intact or $F(ab')_2$ anti-Ig; however, induction of nonresponsiveness to challenge with FL-*BA* (a specific T-independent antigen) was much more efficient when intact anti-Ig is used. These data imply that responsiveness to antigen-specific T-independent antigens such as FL-*BA*, whose activity is mediated at least partially via sIg, can only be inhibited when the FcR is ligated with sIg. The FcR can modulate signaling via sIg since $F(ab')_2$ anti-Ig causes a sustained rise in intracellular Ca^{2+} whereas intact anti-Ig does not [32]. In addition, PI hydrolysis was abortively shut down with intact anti-Ig [32].

There is evidence that there is an intrinsic difference between the negative and positive signaling pathways mediated by sIg. For instance, pretreatment with cholera toxin (CT) for 2 h prior to stimulation with anti-Ig inhibits positive signaling as defined by the ability of anti-Ig to stimulate an increase in cell size, and cause B cells to enter the S phase of the cell cycle [33]. However, CT only partially inhibits anti-Ig-induced increases in Ia expression and has very little effect on the ability of anti-Ig to inhibit LPS driven B-cell differentiation [33]. Cyclosporin A (CSA) is well known to inhibit the activation of T cells, but has also been shown to inhibit the activation of B cells via sIg [34, 35]. In contrast to CT, CSA is able to effectively inhibit anti-Ig induced B-cell nonresponsiveness [34]. As such, it appears that anti-IgM is able to stimulate at least two divergent pathways in the B cell. One is a positive signaling pathway leading to cell enlargement and entry into the cell cycle which can be inhibited by both CT and CSA; the other is a negative signaling pathway which results in the down-regulation of T-cell-independent B-cell differentiation, but not proliferation. The latter is inhibited by CSA, but is not affected by CT. The fact that anti-Ig-induced increases in Ia expression correlate with negative signaling suggests that once a B cell encounters specific antigen, it absolutely requires T-cell help in order to progress to antibody production.

The negative signal mediated by anti-Ig, as read-out by the decrease in LPS-induced antibody formation, has been studied by several groups at the molecular level. These workers have established that treatment of normal B cells with anti-IgM results in decreased steady-state levels of the secretory form of μ-chain mRNA [28–31], a result suggesting that anti-Ig initiates events that ultimately affect 5′ regulatory regions (and respective transcription factors) for the Ig locus.

Role of Cytokines in Preventing Tolerance

The role of cytokines in affecting the induction of and/or reversing nonresponsiveness has also been examined [27]. In our system IL-4 was unable to overcome the nonresponsive state once it had been established. However, pretreatment with IL-4 (prior to encounter with anti-Ig/antigen) inhibits the induction of B-cell nonresponsiveness. In contrast, recent work suggests that when B cells are cultured in the presence of anti-IgM for 72 h they become sensitive to the antibody production promoting activity of IL-4 and IL-5 [36]. It is interesting to note that exposure to anti-IgD antibodies has neither a stimulatory nor an inhibitory effect on the ability of the B cells to produce antibody when subsequently exposed to IL-4 and IL-5. Taken together, these data clearly demonstrate that not only is the presence or absence of lymphokines important in determining the outcome of such an interaction, but also the temporal sequence of exposure to both the cytokine and the potential antigen are critical.

sIgM is capable of mediating two opposing functional consequences. One outcome is the delivery of a positive signal (i.e., stimulation of growth and/or maturation) while the other is the delivery of a negative signal (inhibition of the above). Whether a given B cell is capable of receiving either type of signal or whether certain B-cell subsets are capable of receiving only one type of signal at any given developmental time remains open to speculation. Recent evidence suggesting that the association of a 56-kD member of the sIgM receptor complex is developmentally regulated supports the latter notion [24, 37]. However, if sIgM is capable of mediating two separate and distinct biochemical signalling events then it would be interesting to determine whether sIgM is linked exclusively to the biochemical machinery responsible for mediating either a positive or negative signalling event at a given time. Alternatively, a single sIgM molecule may associate with either the positive or negative signalling apparatus depending on the relative abundance of each type of apparatus in development. Finally, the ultimate result of a given B cell's encounter with antigen also may be determined by some distal event, such as the presence or absence of appropriate T-cell help at the proper time pre- or postcontact with antigen.

Differential Roles of IgM and IgD

Surface IgM and IgD are receptors for antigen on naive, mature B cells. Early work, based on the observation that immature B cells initially express only IgM, suggested that this receptor delivered a negative signal whereas IgD was responsible solely for the delivery of positive growth-promoting signals to mature B cells. In our model, the polyclonal anti-mouse Ig that is used to induce B-cell nonresponsiveness theoretically cross-links IgM to IgM, IgD to IgD, and IgM to IgD. The use of monoclonal and/or isotype-specific antibodies allows for the cross-linking of selected isotypes on the surface of the B cell. Recent work in this lab suggests that of the monospecific reagents, only anti-IgM is able to mediate a negative signal [38]. However, simultaneously independent cross-linking of sIgD and sIgM (using monospecific reagents) results in synergy in terms of the ability to induce B-cell nonresponsiveness. Interestingly, in the 72-hour priming model, anti-IgD also has not been able to prime for subsequent lymphokine-mediated antibody production [36]. Taken together, these data suggest that IgD by itself is unable to affect B-cell differentiation in any way, either positive or negative, but can cooperate with IgM-mediated signals.

Cross-linking either IgM or IgD (or both) leads to very similar early signaling events [39–42]. The events have been well characterized (see below, 'Analysis of Signal Transduction Pathways in B-Cell Tolerance') and include PI hydrolysis to inositol trisphosphate and diacylglycerol, Ca^{2+} mobilization from intracellular stores, as well as mobilization across the plasma membrane, PKC activation and tyrosine phosphorylation [39–42].

The involvement of G proteins in coupling antigen receptor occupancy with transmembrane signaling has also been implicated in this process [39]. However, work by Mond and co-workers [41] has called into question whether any of these events are *required* for the delivery of a signal since dextran-coupled anti-Ig does not induce detectable changes in any of these parameters at concentrations clearly able to cause B-cell entry into cell cycle.

The fact that both IgM and IgD appear to mediate similar early biochemical events [41, 42] seems to suggest that there may be no biochemical basis to explain why cross-linking of IgM is able to deliver either a positive or negative signal; clearly, however, IgD appears to be capable of delivering positive signals for growth. In addition, membrane IgM and IgD are associated with different accessory proteins and it may be these proteins, acting as substrates for PKC and/or tyrosine kinases, that differentiate between signals delivered via IgM and IgD. In this regard, it is interesting to point out that inhibitors of tyrosine kinases seems to be able to inhibit the delivery of a negative signal by anti-IgM to splenic B cells [38] and that anti-IgD and anti-IgM lead to subtly differential patterns of tyrosine phosphorylation [42]. Therefore, the downstream events coupled to these receptors will be an important area to understand.

Are CD5+ B Cells Sensitive to Tolerance Induction?

CD5+ B cells make up greater than 50% of the resident peritoneal B cell population. These B cells appear to be self-renewing, have restricted V gene region usage, and are thought to play a role in immunoregulation [43]. Peritoneal B cells appear to be resistant to tolerance induction using anti-Ig as a surrogate for antigen in vitro [44]. However, CD5+ B cells are as sensitive as conventional splenic B cells to 'tolerance', using agents such as activators of PKC and Ca^{2+} ionophores which bypass sIg [44]. These data suggest that there is something different between CD5+ B cells and conventional B cells in regard to the ability of their antigen receptors to transmit signals across the plasma membrane. It is interesting to note that CD5+ peritoneal B cells can be induced to enter the cell cycle with activators of PKC alone, i.e., there is no requirement for elevated (Ca^{2+}), suggesting that these B cells normally exist in a different state of differentiation such that they are less Ca^{2+} dependent and are less sensitive to tolerance induction. Cohen and Rothstein [45] have suggested recently that this may be due to the activation of different PKC isozymes.

CD5+ B cells have been implicated in immunoregulation and the induction of autoimmune disease [43]. Recent work suggests that CD5 expression on normal B cells can be induced by prolonged exposure to anti-Ig [46]. Indeed, we recently found that primed B cells are resistant to tolerance induction [47]. Assuming that antigen priming induced CD5 expression, it also alters the functional fate of these cells. Moreover, further data indicate that CD5+ B cells are capable of producing IL-10 and are depleted from animals that are administered anti-IL-10 antibodies probably due to increased levels of IFN-γ [48]. Taken together, these data suggest that CD5+ B cells differ from conventional B cells in that they require IL-10, are sensitive to elevated levels of IFN-γ, and are less sensitive to tolerance induction.

Fate of Hapten-Specific B Cells in Tolerance

Role of B-Cell Maturational Stage

The first efforts to determine the fate of antigen-specific B cells in immunologic tolerance were performed by Diener and Paetkau [13] in the early 1970s. Using polymerized flagellin as an antigen, they took advantage of an in vitro model for tolerance in which there is a dose-dependent antibody response followed, in antigen excess, by an unresponsive state [13]. Using tritium-labeled flagellin to track antigen-binding cells (ABC), these workers observed that specific ABC were detectable under conditions in which unresponsiveness was observed. However, these ABC failed to cap and shed antigen as was observed under immunogenic culture conditions. In this entirely in vitro system, it was difficult to measure the persistence of these ABC and it was assumed at that time that such cells ultimately were deleted because they were unable to function.

Subsequently, our laboratory [8, 26, 49] utilized fluorescein coupled to heterologous γ-globulins as antigen in order to trace the ultimate fate of antigen-binding, hapten-specific cells under conditions of tolerance. This followed studies that initially failed to trace any significant activation steps (DNA, RNA, protein synthesis, ability to replicate virus) in thoracic duct lymphocytes from animals rendered tolerant to a carrier-like sheep γ-globulin [50]. Venkataraman and Scott [49], went on to find that ABC detectable by virtue of the fluorescein tag on FITC-coupled sheep γ-globulin (FL-SGG) were detectable in the spleens of tolerogen-injected rats for several days after the administration of this conjugate. The numbers of FL-SGG-binding cells declined by 7 days, at which time their frequency was undetectable. Interestingly, when antigen was re-administered to these tolerant rats, normal levels of ABC were observed [49]. Since FL-SGG induced tolerance for both FL-specific B cells and SGG-specific B and helper T cells [26], it was not known whether these ABC were T or B cells. However, hindsight tells us that the nature of antigen binding by T cells would tend to obviate the expression of antigen-binding capacity in that population.

These data suggested that antigen-reactive B cells were present in ostensibly tolerant rodents, and these cells were capable of re-binding antigen; nonetheless, these cells were anergic to further stimulation by the same antigen. Indeed, so-called 'tolerant' ABC had a reduced cloning efficiency at 1 week but were normal at approximately 21 days after tolerance induction. Newly derived B cells did not appear in tolerant mice that had a reduced burst size in response to antigen [51], but a normal cloning efficiency. Thus, it appeared that the original ABC had disappeared by some deletional mechanism and newly derived B cells had replaced them. Evidence supporting such an occurrence was indeed obtained [51].

When Venkataraman and Scott [49] repeated these experiments in a neonatal tolerance model, completely different results were obtained. In this case, no cells capable of re-binding antigen were present approximately 1 week after tolerogen administration. Coupled with the failure of hapten-specific B cells to respond to LPS even via proliferation, these results suggested that in the neonatal period, tolerogen interaction with newly arising B cells led to deletion of these cells [49]. These data were completely consistent with the Nossal model using high doses of anti-IgM in vitro [20] and with results [Borrello and Scott, unpubl. data, and 17, 27] suggesting that deletion of B cells is a common event during development, whereas anergy with persistence of ABC occurs during adult life.

Importance of the Carrier
In 1974, Aldo-Benson and Borel [52] reported that mice injected with DNP coupled to isologous γ-globulins were rendered specifically unresponsive and that antigen-*bearing* B cells could be detected in the peripheral lymphoid organs of these animals. Such antigen-bearing cells persisted for several weeks after the induction of tolerance: their presence reflected the duration of the unresponsive state in these animals. Since these results contrasted with those of Venkataraman and Scott [49], we collaborated with Borel's group to examine the basis of the difference. Mice were injected with either FL-mouse IgG or FL-SGG (sheep) and examined for antigen-binding/bearing cells in the spleens at various times. Interestingly, it was found that ABC rapidly disappeared in the adult mice injected with FL coupled to heterologous γ-globulins, but persisted with FL-isologous γ-globulins. These results emphasize that the nature of the carrier plays an important role, not only in the persistence in ABC [53], but also was critically important in the ability to induce unresponsiveness per se.

The potential role of Fc receptor binding in the induction of unresponsiveness with immunoglobulin carriers (as well as with anti-immunoglobulin) is not understood. While both isologous and heterologous γ-globulins are able to bind mouse-Fcγ-RII receptors, subtle differences in the affinity or ability to down-regulate antigen-specific signals are not clear. The effects of Fc receptor cross-linking of anti-Ig signaling are discussed above and elsewhere in this volume [cf. E. Fedyk et al., this volume].

Interestingly, Chace and Scott [54] demonstrated that hapten-specific B cells could be isolated from tolerogen-injected and normal, adult mice in equal numbers, confirming the earlier work of Venkataraman and Scott [49]. Although these cells proliferated normally with LPS, they failed to incorporate thymidine when exposed to specific antigen and were defective in their response to anti-immunoglobulin signaling. Further studies demonstrated that these cells did demonstrate early biochemical reactions to surface IgM cross-linking in that they increased their surface Ia induction and entered the G_1 phase of the cell cycle. However, they were completely unable to progress from G_1 into S, thus demonstrating a critical blockage in the barrier between G_1 and DNA synthesis [54, 55].

Transgenic Mice in the Analysis of Tolerance

Lysozyme versus H-2: Anergy and Deletion

Transgenic mice have literally revolutionized the study of tolerance in both B and T cells! Initially, these mice were used to engineer foreign antigens into genes under the control of different promoters so that antigen expression could be controlled in a tissue-specific manner. These studies allowed the induction of tolerance to be studied to the same foreign epitope presented in different tissues and at different times. However, the fate of ABC in these mice could not be evaluated. An important step was made when Goodnow et al. [56] and Nemazee and Bürki [57] combined the transgenic approach to introduce a foreign antigen with transgenic mice possessing rearranged genes for the immunoglobulin receptors to that antigen. This allowed analysis of the ultimate fate of antigen-specific B cells.

Lysozyme Transgenics

In the first model, the gene for an extrinsic antigen, hen egg white lysozyme (HEL), was driven by the metallothionein promoter such that different amounts of HEL were produced in transgenic lines in the presence or absence of heavy metals, such as zinc in the drinking water. That is, when transgenic mice were fed zinc sulfate, significant increases in serum levels of HEL of greater than 100-fold could be observed [11, 56, 58]. Thus, one could examine the extent of tolerance when a self-antigen was expressed at relatively low to high levels. These mice displayed, as expected, a dose-dependent expression of unresponsiveness at the B-cell level, which was confirmed by challenge with HEL coupled to T-dependent antigens for which sufficient help was present in these animals. Such a model system allowed one to look at affinity questions and timing of exposure to an antigen for the induction of tolerance. The fate of epitope-specific B cells was then examined when Goodnow et al. [56] bred HEL-transgenic mice with lines possessing IgM and IgD receptors displaying high affinity interactions with HEL. When anti-HEL mice were bred with HEL-transgenic animals, tolerance was again observed as with single transgenic mice, although the dose requirements for tolerance was slightly different due to the fact that virtually all B cells expressed anti-HEL of high affinity. Importantly, the fate of the HEL-specific B cells could now be traced by flow cytometry using fluorescein coupled to HEL. To their surprise, the levels of antigen-specific B cells were similar in anti-HEL transgenic mice bred to HEL-expressing animals and in those anti-HEL-transgenic mice bred to normal controls. Therefore, tolerance at the B-cell level was manifested by the persistence of antigen-binding B cells which were anergic to specific challenge [56], as in the results of Chace and Scott [54]. In addition, more recent data suggested these animals fail to respond even to LPS challenge [11].

Interestingly, the surface phenotype of these B cells was IgD^{high}, IgM^{low}, $B220^+$, $J11D^+$, Ia^+. That is, these cells appear to have modulated selectively their membrane IgM, a phenotype which persisted even upon transfer and 'parking' in an HEL-free host. Subsequent studies demonstrated that the induction of this phenotypic change could be observed in adult animals expressing low levels of HEL after they were exposed to drinking water containing zinc. This response was relatively fast, occurring over 2–4 days [11, 58]. Subsequently, they also observed that B-cell tolerance could be induced within 2 days with adult mature B cells from single-HEL-transgenic mice transferred into irradiated mice possessing high levels of HEL. Anergy appeared rapidly and was accompanied by the similar change in phenotypic expression, that is, down-regulation of IgM. Recent data [59] suggests that this modulation is manifested at the molecular level in terms of transcription of membrane IgM message.

The distribution of B cells in the follicles appears to be different in double transgenic mice that were tolerant to HEL. These animals lack HEL-binding B cells in the marginal zone of the spleen, as well as plasma cells, presumably due to an inability to stimulate B-cell differentiation in these anergic populations. This also implies, however, that stimulation through cross-reacting environmental antigens or mitogens plays a role in the distribution of ABCs observed in nontolerant (HEL-) mice [60].

The original anti-HEL-transgenic mice were constructed to express both surface IgM and IgD specific for HEL. Subsequently, IgM-only and

IgD-only anti-HEL transgenics were produced which were then bred with HEL transgenics. The IgM-only transgenics were similarly anergic and showed the same 10- to 20-fold reduction in membrane IgM levels, as reported in the IgM + IgD transgenics, thus creating a situation which looked like specific B cells had disappeared [11, 58]. Clearly, however, there were normal levels of B220$^+$ cells and low levels of IgM specific for HEL on the surface of these anergic populations. This is an important point when we consider other transgenic models below. Interestingly, IgD-only transgenic mice were also anergic, although not to the same extent. This suggests that IgD, when expressed as the sole receptor for a B cell, can transmit a negative signal, although clearly it is not as effective as IgM alone. Whether IgD transmits a negative signal in normal B cells for tolerance cannot be extrapolated from these data, especially since our studies that IgD cross-linking alone would contradict this result (see above and Gaur et al. [38]). Indeed, one must consider that in a normal B cell, IgD exceeds IgM quantitatively. However, antigen has the potential to cross-link both IgM and IgD, although for a given B cell, the amount of IgM-IgM, IgM-IgD, on IgD-IgD cross-linking cannot be distinguished since the relative local concentration of these receptors on the surface exposed to antigen is unknown. Nevertheless, it is clear that IgD per se does not render a B cell resistant to tolerance induction, as was implied from earlier work by us [61], but it may decrease the sensitivity of such cells to negative signals [62].

H-2 Transgenics

In a similarly conceived model, Nemazee and co-workers [12, 57] generated transgenic mice which expressed IgM receptors specific for H-2^b and H-2^k. When these mice were bred with H-2^b or H-2^k animals, tolerance was observed which was manifested by a disappearance of anti-H-2-specific B cells (measured by anti-idiotype and total B220$^+$ cells). This experimental result could be challenged because B cells that were expressing anti-H-2 also expressed the 'tolerated' antigen on the surface of the very same cells and could possibly be eliminated by a process of 'intracellular indigestion' [63]. To resolve this question, Nemazee and Bürki [64] transferred anti-H-2-transgenic bone marrow cells into irradiated mice expressing the relevant H-2 antigens. Similar to their results in double transgenic mice, they found that these mice were not only unresponsive, but also deleted antigen-specific B cells. Hence, the interaction during development of a B cell with an H-2 antigen for which it is specific can lead to a deletional event. Interestingly, immature B cells in the bone marrow could be found in tolerant mice, but these cells failed to be detected in the spleen. This result suggests that there is window before which antigen-specific B cells can be deleted.

Notably, when transgenic mice were constructed that expressed a truncated, soluble H-2 molecule, no tolerance was observed. Whether this reflects a lack of ability of the soluble antigen to sufficiently cross-link epitope-specific B cells or an unusual feature of soluble H-2 in vivo is unknown. Nonetheless, one must question why in one experimental model (HEL) there is clear-cut anergy and no deletion, whereas, in another (H-2), there is deletion. An obvious difference is that in the later case IgM-only transgenics were bred such that the B cells that could interact with the tolerogen expressed only IgM. As noted above, this was tested by Goodnow's group and has now been examined by Nemazee and co-workers [12]. In their experience, IgM + IgD transgenics also are rendered unresponsive and undergo deletional events. Hence, while IgD may not transmit as strong a negative signal for B-cell tolerance, it does not prevent a deletional event when encountering certain kinds of antigens.

This left the nature of the antigen as a major difference between the two systems. Recently, Goodnow and co-workers [65] began to create HEL transgenics in which HEL is expressed as a membrane antigen driven by the H-2 promoter. When these mice were bred with anti-HEL mice, there was a complete absence of HEL-reactive B cells in the spleen, but the presence of some immature HEL-specific bone marrow B cells that had not been deleted. These results suggest that deletion occurs at a level between the immature B220^{low} cells and the peripheralization of these cells as mature IgM$^+$ circulating B cells.

What is the mechanism of deletion in the H-2-transgenic mice? Recent data from Nemazee's group [12] suggests that in these mice a significant increase in RAG-1 and RAG-2 transcription occurs and that the opportunity to somatically mutate receptors is engaged. Thus, when encountering a potentially tolerogenic signal, B cells either are deleted (by a process that presumably involves apoptosis, but has not yet been observed in their model) or undergo somatic rearrangements to allow the expression of a receptor that does not bind the antigen with sufficient affinity to cause deletion. This fail-safe procedure permits the survival of the B cell and could be expected to be occurring under normal conditions of B-cell tolerance such that tolerance to self-antigens may drive the appearance of anti-non-self specificities with very much lower affinity for self. This is curiously similar to one of the original manifestations of Jerne's network theory [66].

Haptens, DNA and Other Epitopes

While these model systems offer the opportunity to examine the induction of tolerance to genetically controlled, newly introduced self-antigens, significant information has accumulated from the use of transgenic mice expressing antihapten specificities and the introduction of hapten at various times during their life. This system gains advantage from the extensive data already in existence involving hapten-specific tolerance induction in vivo, although it may lack the physiological events of the double transgenic models. Thus, Zöller [67] had demonstrated that anti-TNP-transgenic mice exposed to TNP conjugated to different protein carriers were rendered unresponsive and appeared to become anergic. Köhler and co-workers [62] have extended this and demonstrated that the time of introduction of antigen is critical for the induction of tolerance. Mice exposed to TNP conjugates early in life underwent an apoptotic form of deletional tolerance, whereas those exposed in later life were far more resistant to tolerance induction and became anergic. Similarly, IgM-only transgenics were the most susceptible to tolerance induction and IgD-only the least [62]. These data fully support the previous models, but are distinguished by the fact the early exposure to an antigen which is introduced systemically in a controlled fashion can either lead to deletion or anergy, and does not appear to depend on whether a membrane or soluble form is seen. In fact, a simple interpretation of all these systems is that both deletion and anergy can occur and that the form of the antigen plays a dominant, but not a completely dictatorial role in this process. Indeed, we previously stated that 'all self antigens are not tolerated equally' [68] and that we must consider, not only the chemical form of the antigen, but also its serum concentration and membrane association. Added to this list of parameters would include the time at which it is exposed to the immune system, where in the immune system it may be exposed, and whether it has the capacity to cross-link IgM to itself or to IgD.

Several other interesting transgenic models for tolerance need mention here. In one, the β-galactosidase molecule was engineered to be an integral membrane protein of B cells in transgenic lines by Theopold and Köhler [69]. These mice were profoundly tolerant to β-galactosidase in terms of antibody production, but possessed a normal number of anti-β-galactosidase-specific B cells. This suggests that, in this model system, tolerance is reflected as a form of anergy (since anti-β-galactosidase B cells are present, but unresponsive). While this is difficult to prove since these were not double transgenics, they raise the important point that expression of epitopes in a membrane form, even on the B cells themselves, does not lead to a deletional form of tolerance. In contrast, Zinkernagel et al. [70] used the vesicular stomatitis virus as a source of antigen that was expressed in transgenic mice in a number of peripheral tissues, although not in lymphoid organs. A lack of antibody formation occurred that appeared to be dependent on B-cell tolerance. In contrast, infection with living virus led to antibody production, suggesting that potentially autoreactive B cells were present and could turn on the immune response by T cells that may recognize altered antigen. This experiment establishes that a variety of consequences of the expression of antigen in transgenic mice may occur. Therefore, a lack of B-cell tolerance could lead to a potentially extremely dangerous scenario.

An elegant model for autoimmunity and tolerance was recently developed by Erickson et al. [71]. They created transgenic mice with the V_H gene of an anti-DNA heavy chain alone or in combination with various chain specificities. Depending on the combinatorial make-up of the transgenic receptor, the cells would have either high or low affinity for single-stranded or double-stranded DNA. In summary, deletion was present, especially for those B cells expressing V_H and an endogenous light chain reactive with anti-double-stranded DNA. Interestingly, when these mice were crossed with MRL/lpr animals (which manifests a lupus-erythematosus-like syndrome), a lack of tolerance was observed. Thus, additional genetic factors may modulate the potential for self-tolerance.

A Cautionary Note on B-Cell Subsets in Transgenic Mice

The enormous impact of transgenic mice for studies of tolerance cannot be overstated. Nonetheless, it is important to provide certain caveats with respect to the use of these models and to emphasize that additional studies in normal animals are required to validate each system. The first point to be considered is that normal rearrangements are pre-empted in transgenic mice by definition, although the amount of leakiness for endogenous rearrangements varies considerably from line to line. How this would impact on the evolution of the B-cell repertoire and the potential purging that occurs when environmentally expressed 'self'-antigens are encountered is not understood. In addition, the predominance of a given idiotype-expressing cell in much greater numbers than would normally be expressed led to the possibility that significant amounts of transgenic antibody could be released into the circulation or into the local milieu either by normal shedding or due to environmental stimulation with microorganisms or with cytokine stimulation. Therefore, when antigen is first encountered, a significant amount of immune complexes (albeit mostly IgM) may form that could have a significant impact on the subsequent degree of unresponsiveness engendered.

Finally, and most importantly, the number of transgene-expressing cells and the frequency of CD5 (B-1) subsets in transgenic mice is extremely variable. However, one pattern has begun to emerge (although this is a biased view by the authors) is that transgenic immunoglobulins specific for 'conventional' antigens leads to mice with relatively normal numbers of B cells in their spleen and very few CD5$^+$ B cells expressed outside of the peritoneal cavity. In contrast, transgenic mice expressing V_H genes often observed in the primary B-1 repertoire, with potentially auto-reactive possibilities, lead to an exaggerated expression of $CD5^+$ B cells in the spleen. Concomitantly, there is a decrease in the total number of B cells in these peripheral organs. This leads to the possibility that exposure to the transgene-encoded favored antigen during development may have different consequences depending on whether it is expressed in the putative B-1 (CD5) lineage or in the conventional B-2 lineage. This is important, because we have found that the peritoneal $CD5^+$ B cells are relatively resistant to tolerance induction [44] and, moreover, that transgenic mice expressing high numbers of $CD5^+$ B cells in the spleen were relatively resistant to in vitro tolerance induction, compared to non-transgenic litter-mate controls [Liau and Scott, in preparation]. In contrast, Murakami et al. [72] find that surviving anti-self red blood cell transgenic B cells may be saved from apoptosis in the peritoneum. Therefore, one must consider a variety of parameters in the evaluation of transgenic models for B-cell tolerance that include the nature of the antigen, its membrane expression, the ontogenic state of development, and whether the antigen is recognized by the repertoire of B-1 or B-2 V_H genes. Not withstanding these caveats, enormous progress has been made utilizing transgenic mice as models for tolerance.

B-Lymphoma Models for Tolerance

Cell Cycle Control by Oncogenes, Cyclins and the Retinoblastoma Gene Product, pRB

During the last decade, we have used a series of B-cell lymphomas as model systems for understanding tolerance and growth control in lymphomagenesis [17, 73]. Using anti-Ig as well as anti-idiotype as surro-gates for antigen, we found that a subset of these lymphomas received a negative signal for growth in early G_1 and arrested at/near the $G_1 : S$ border [74]. These cells subsequently undergo apoptosis and die [75]. It is our hypothesis that these events mimic a deletional form of tolerance and provide the groundwork for understanding growth control of B-cell lymphomas. Importantly, these systems focus on the mediators of cell cycle controls, which will be reviewed briefly at this point.

Regulation of cell cycle progression in actively dividing cells is a crucial mechanism by which growth is controlled by exogenous factors or in the differentiation process. In recent years, much has been learned about the proteins involved in cell cycle control in budding yeast, *Saccharomyces cerevisiae*, and in higher eukaryotes, such as clams and humans. Phases of the cell cycle are controlled by classes of cyclins and serine/threonine kinases (*cdk* family members) [for reviews, see 76, 77]. The major decision points at which regulation can occur are in G_1 (when a commitment to DNA synthesis is made) and at the G_2/M border (for mitosis). In humans, cyclin A is active in G_1, S and G_2 and is important for the progression from G_1 to S and G_2 to M. Cyclin B (B1 and B2 are known) is important for the progression from G_2 to M. Cyclins C, D, and E are active in G_1. A family of at least ten serine-threonine kinases has been identified. Of these, *cdc2* kinase is active at both of these decision points and can act on cyclin A and cyclin B.

The cyclin proteins are associated with other growth-controlling proteins. Cyclin A, which is actively involved in the progression from G_1 to S, is associated with tumor suppressor gene products, p53, p107 and pRB, in the G_1 phase [78, 79]. The importance of the inactivation of pRB was first identified in patients with retinoblastoma, a cancer of the retina primarily found in young children [80]. In tumor cells of this and other types, mutations in both of the *RB-1* alleles leads to inactive pRB and apparently uncontrolled proliferation by the tumor. Re-introduction of the wild-type *RB-1* gene into tumor-derived cell lines (with mutated *RB-1* alleles) results in reduction of tumor-forming ability of these cells in nude mice and decreased cellular growth rate in vitro [81, 82].

The tumor suppressor gene product, p53, is also intimately involved in growth control. Mutations in p53 or the loss of heterozygosity or over-expression has been found to occur in tumor progression in many human cancers and in model murine systems [for reviews, see 83, 84]. Transfection of and expression of wild-type p53 allows cells to progress from G_1 to S and shows a decrease in gene amplification events [85, 86].

Recently, insight into the mechanism of cell cycle progression has been gained by analyzing pRB. The extent of phosphorylation of pRB, a 110-kD nuclear phosphoprotein, controls its activity as an antioncogene and, thus, varies throughout the cell cycle. Underphosphorylated pRB is found in early G_1 and becomes phosphorylated in mid- to late G_1. Heavily phosphorylated pRB persists through S phase [87–91] to the onset of M, when it is dephosphorylated [92]. Additional evidence that pRB functions as an element in cell cycle control comes from studies using viral oncogene products, such as adenovirus E1A, SV40 large T antigen and human papillomavirus E7. These viral oncogene products have been shown to bind to a 'pocket' in underphosphorylated pRB in G_1 and promote its phosphorylation. This modification inactivates pRB as a growth suppressor and promotes uncontrolled cell cycle progression [87, 92–94]. Moreover, the activity of TGF-β to prevent pRB phosphorylation is prevented by transfection of the indicated viral oncogenes (see below).

Phosphorylation of pRB is also important for the localization of pRB during stages of the cell cycle and its association with other proteins. There is tight association of underphosphorylated pRB with the nucleus during G_1, whereas loosely nuclear-associated, low-salt extractable hyperphosphorylated pRB is found during the rest of the cell cycle [95]. This tight anchoring is hypothesized to be important in regulating the localization of underphosphorylated pRB. Weinberg and co-workers [96] found that expression of *RB-1* in pRB-deficient osteosarcoma cells leads to growth arrest of these cells. However, when cyclins A and E were overexpressed, cell cycle progression commenced and pRB was hyperphosphorylated. This points to multiple interactions between pRB, cyclins, and associated kinases.

In addition, numerous cellular proteins have been found to be involved with RB-mediated cell cycle progression. The transcription factor, E2F, has been found to interact with pRB in G_1 and later dissociates from pRB at approximately the time of pRB phosphorylation at the G_1/S border; this presumably allows E2F to actively participate in transcriptional events [97, 98]. E2F also complexes with cyclin A/*cdc2* kinase (p33) and another RB-like protein, p107, in S phase [99]. While these interactions are complex at many stages of the cell cycle, it is clear that regulation of pRB phosphorylation is critical for cell cycle progression.

Role of c-myc and TGF-β in Cell Cycle Control: Overview

Which factors influence the phosphorylation state of pRB and thus control cell cycle progression? One of the most potent inhibitors of cell cycle progression is TGF-β, a paracrine polypeptide that is prototypical of functionally and structurally related peptides important in growth, differentiation, and morphogenesis. Within the TGF-β family, three highly homologous genes, TGF-β1, TGF-β2, TGF-β3, are expressed as proteins ubiquitously in humans and all other mammals. All TGF-β forms are growth inhibitory for most cells of nonmesenchymal origin in vitro [for reviews, 100, 101].

The growth inhibitory properties of TGF-β1 have been implicated in directing cell cycle progression. For example, addition of TGF-β1 to mink lung epithelial cells in mid-G_1, but not at the G_1/S border, leads to prevention of pRB phosphorylation and cell cycle arrest [102]. Modulation of early response genes by TGF-β may also lead to the pRB-mediated growth suppressive state. In these mink lung cells, the abundance of *c-myc* mRNA is reduced 4 h after TGF-β addition; steady-state mRNA levels of *junB* were also modulated by TGF-β. In cells released from a G_1 block (blocked by contact inhibition), TGF-β addition decreased *c-myc* mRNA expression significantly at 6 h as compared to untreated, G_1 block-released controls. Thus, the downstream effects of TGF-β include modulation of

early response genes such as *c-myc* and *junB* and control of phosphorylation of tumor suppressor gene products. However, in DU145 cells (human
prostate carcinoma cells bearing a mutant, nongrowth-suppressive pRB),
TGF-β1 also decreased *c-myc* mRNA expression at the transcription
initiation level [103]. This indicates that TGF-β can down-regulate *c-myc*
expression independently of the growth-suppressive function of pRB.

A further link between TGF-β-induced growth arrest, down-regulation of *c-myc* mRNA expression, and pRB binding has been established in
human keratinocytes [104–106]. As noted above, TGF-β1-induced growth
arrest and down-regulation of *c-myc* expression was blocked by expression
of SV40 large T, HPV16 E7 and adenovirus E1A proteins, all of which
bind to pRB in the underphosphorylated form. These authors proposed
that TGF-β acts through a tumor suppressor gene product, in this case
pRB, to negatively regulated *c-myc* transcription and subsequent cell cycle
progression [104–106].

Regulation of Cell Growth by c-myc and Anti-IgM in B-Cell Lymphomas

Anti-IgM treatment of murine B-cell lymphomas has been known for
many years to lead to growth arrest. Such inhibited cells accumulate in G_1
and presumably are blocked at or near the G_1/S border [73, 74]. Subsequently, these cells display apoptotic features, such as DNA laddering and
apoptotic bodies by 24–48 h [75, 107–109]. TGF-β has also been demonstrated to growth inhibit these murine B-cell lymphomas [75] and lead to
cell cycle blockage in G_1, as well as apoptosis [17, 75].

Another feature of anti-IgM treatment of murine B-cell lymphomas is
the down-regulation of *c-myc* transcription and the apparent loss of *myc*
protein [110, 111]. Although the function of myc is largely unknown, the
structure and regulation of *c-myc* provide some clues. The *c-myc* locus is
complex, containing four transcriptional start sites with five detected
mRNA species arising from these sites. Of the four myc protein products
detected in a variety of cell types, the 64- and 67-kD forms are the most
common and found in all normal cells [112]. DNA-binding sites for
transcription factors such as Oct-1, Fos/Jun, E2F, NFκB, a TGF-β response element, and AP-2 [112–114] are located 5′ of the most frequently
utilized promoters, P1 and P2. The myc protein regulates its own synthesis
transcriptionally via sequences in exon 3 and post transcriptionally by an
unknown mechanism. In fact, myc belongs to a family of proteins which
contain helix-loop-helix, basic regions, and leucine zipper motifs; these
motifs are most frequently seen in transcription factors [115]. Myc protein
has been found to complex with another transcription factor, called max in
human cells and myn in the mouse. These products increase the complex's

affinity for a specific DNA hexamer, CACGTG (termed the E box) [116, 117]. To date, the function of this complex in transcription is unknown.

Products of the *myc* locus have long been suspected in deregulation of cell growth since many tumor cells have aberrant expression of this gene. Burkitt's lymphoma, an aggressive form of B-cell lymphoma, often is characterized by reciprocal translocations between the *c-myc* locus and one of the immunoglobulin gene loci. In fact, the translocation of a *c-myc* allele in close proximity with the μ-chain enhancer ($E\mu$) and into light chain loci is implicated in deregulation of *c-myc* mRNA expression and overexpression of myc protein, which may be involved in uncontrolled tumor growth [118–120].

Examination of *c-myc* in tumor cells such as Burkitt's lymphoma has led to the hypothesis that myc is involved in cell cycle progression. The *c-myc* gene is an early response gene; increased or decreased expression of *c-myc* has been seen as early as 30 min to 3 h when cells are stimulated with a variety of factors. These include a primary increase by 1 h and then a decrease to below background by 8 h of *c-myc* message in WEHI-231 B-lymphoma cells treated with anti-IgM [110, 111]. *C-myc* is also down-regulated by TGF-β in mink lung epithelial cells [103] and human keratinocytes [105], and by anti-CD40 antibodies in resting human tonsillar B lymphocytes [121].

More direct evidence for myc involvement in proliferation comes from utilization of oligonucleotides complimentary to the translation start sites in *c-myc* mRNA. Addition of antisense *c-myc* oligonucleotides to mitogen-activated human peripheral T lymphocytes blocked entry of these cells into S phase [122] and prevented DNA synthesis in hematopoietic cells [123]. Antisense *c-myc* oligos reversed anti-IgM and TGF-β induced growth inhibition and prevented apoptosis of the murine B-cell lymphomas, WEHI-231 and CH31 [109], but did not abrogate *c-myc* mRNA expression. In contrast, addition of antisense *c-myc* oligos prevented mitogen activation-induced apoptosis in T-cell hybridomas, and, significantly, prevented expression of myc protein [124]. Additionally, down-regulation of *c-myc* mRNA and myc protein was found when fibroblasts were serum- or mitogen-deprived [125, 126]. Serum-deprived rat fibroblasts with constitutively as well as increased *myc* expression were unable to growth arrest and were prone to apoptosis [127].

Thus, we consider it likely that inappropriate myc expression plays a role in regulating B-cell growth and may act via pRB. Clearly, the regulation of level of myc expression varies in different cell types. A clue as to how anti-IgM and TGF-β can growth inhibit murine B lymphomas came from examination of pRB phosphorylation in these cells. As noted above, anti-IgM or TGF-β treatment of WEHI-231 or CH31 led to accumulation of cells in G_1 and a block at the G_1/S border [75]. Importantly, anti-IgM treatment led to the accumulation of the underphosphorylated form of pRB, but phosphorylation of this protein occurred normally in the presence of antisense *c-myc* [75, 109].

Taken together, these data suggest that anti-IgM treatment leads to an early perturbation in myc expression and an accumulation of underphosphorylated pRB in these B-cell lymphomas with a growth block in G_1. It is tempting to speculate that increased myc protein synthesis without a concomitant second signal leads to arrest and apoptosis. We would suggest that the myc protein may bind to pRB in some manner, as reported by Rustgi et al. [128], or that myc alters the binding of other proteins to the pRB pocket. Alternatively, antisense *c-myc* may allow myc protein to function normally under conditions of growth regulation modified by anti-μ. However, the exact sequence of events and involvement of pRB, myc, cyclins, and cdc2/cdk kinases [76] in growth arrest must be further examined in the regulation of lymphoma proliferation, as well as in tolerance.

Analysis of Signal Transduction Pathways in B-Cell Tolerance

The Ig Complex and Second Messengers

Clearly, the signals for tolerance induction must be initiated by antigen binding to the Ig receptor complex. One of the earliest consequences of that binding in B-cell signaling is the rapid increase in the concentration of cytoplasmic free Ca^{2+}, PI turnover and the activation of PKC [39, 129, 130]. There is now compelling evidence that would allow one to place the activation of a tyrosine kinase as a proximal event to PI turnover, based on the observation that phospholipase C-γ (PLC-γ), which hydrolyzes membrane phospholipid to yield IP_3 and diacylglycerol, is phosphorylated on tyrosine residues after ligation of B-cell receptor. Indeed, a linear relationship between tyrosine phosphorylation of PLC-γ and its catalytic activity exists; furthermore, tyrosine-specific phosphatase treatment substantially decreases the catalytic activity of PLC-γ [131]. Of the two isoforms of PLC-γ (PLC-γ1 and PLC-γ2), there is evidence for the predominant expression of PLC-γ2 in splenic B cells [132]. In fact, stimulation of B-cell lymphoma lines, as well as normal splenic B cells, with anti-μ induced an increase in tyrosine phosphorylation of PLC-γ2, but no detectable tyrosine phosphorylation of PLC-γ1 in treated or untreated cells [131, 132].

How does the cross-linking of B-cell receptors activate PLC-γ? Since immunoglobulin receptors lack typical kinase domains, they must be directly or indirectly associated with nonreceptor tyrosine kinases. Clearly, stimulation of B cells with anti-Ig antibody induces rapid increases in a number of tyrosine phosphorylated intracellular proteins, including PLC-γ and the Ig-associated transmembrane signaling proteins, mb-1 and B29, as early as 30 s after adding anti-Ig and presisting for approximately 60 min [133–138]. Recently, the association of three kinases, p55^{blk}, p59^{fyn} and p53/56^{lyn}, was demonstrated by co-immunoprecipitation to mIg receptors in murine splenic B cells [139, 140]. Moreover, all three of these tyrosine kinases are activated upon cross-linking of mIgM or mIgD receptors. A fourth non-*src* kinase, called p72^{syk} [141], is also associated with mIgs [142].

Is there any evidence for the activity of these PTKs in positive or negative signaling? We have found that in the presence of the PTK inhibitor, herbimycin Anti-μ stimulated transition of resting B cells from G_0 to G_1 was effectively prevented as determined by ^{3}H-uridine incorporation [Yao and Scott, Cell Immunol 1993, vol 149, in press]. The inhibitor-treated cells showed impaired proliferation and differentiation in response to anti-μ or specific antigen stimulation, but they retained the ability to be activated by LPS. These cells also were less susceptible to tolerance [38]. Biochemical analysis showed that anti-μ stimulated tyrosine phosphorylation of intracellular proteins, including PLC-γ, was almost completely inhibited [137]. Taken together, these data support the hypothesis that tyrosine phosphorylation of intracellular substrates might be one of the mechanisms through which PTKs accomplish their role in signaling transduction.

Recent data suggest that antibody to IgMα (mb-1) and B29 (Igβ) co-immunoprecipitates the *blk*, *lyn* and *fyn* kinases in 1% NP-40; this treatment disrupts the association between mIg and mb-1/B29 heterodimers [143]. Since only a small fraction (1–3%) of the mb-1/B29 complex appears to be complex with PTKs, the heterodimer seems to function as a bridge between the Ig receptor and signal transducer proteins involved in regulation of B-cell activation [142, 143]. Secondly, the ability of multiple *src*-like PTKs to be associated with the heterodimer suggests the existence of conserved binding sites, such as SH2 domains located in NH$_2$-terminal half of these PTKs. Finally, each PTK may compete for binding to the heterodimers, depending on their relative abundance of expression. Thus, overexpression of particular PTK would be able to effectively compete for and saturate available binding sites.

These findings have significant implication in B-cell ontogeny and tolerance, because it has been thought that expression of PTKs are developmentally regulated. Indeed, we have found differential expression of *blk*, *lyn*, *fyn* and *lck* kinases in a panel of B-cell lymphomas which may represent different stages in B-cell maturation [137]. However, the differential expression of these kinases in CD5+ (B-1) and conventional (B-2) cells has not been observed so far [144]. Preferential expression and association with mIg receptors of PTKs may facilitate B cells choosing one signaling pathway or another. Thus, it is possible that each protein kinase may act on its own set of target of substrates. This arrangement allows one signal receptor to turn on a specific subset of downstream cellular enzymes in different B cells.

Finally, how do these pathways fit into models of B-cell tolerance? Based on studies in B-cell lymphomas, Page and DeFranco [145] proposed that PI turnover and rises in intracellular calcium were critical for negative signaling. This was rendered unlikely when it was found that growth inhibition could occur in the absence of detectable calcium signals and the PKC inhibitors were unable to prevent anti-μ-mediated growth inhibition [33, 146]. Indeed, activation of PKC with phorbol ester can actually prevent growth inhibitory signaling in synchronized lymphoma cells [33]. This implies that negative signaling in these cells may be delivered through divergent second messenger pathways.

More importantly, B cells treated with different PTK inhibitors (tyrphostin, genistein, and herbimycin) showed decreased phosphorylation of intracellular substrates when stimulated with anti-IgM [38, 135, 136]. Using this approach, Beckwith et al. [135] recently reported that herbimycin A was able to completely reverse anti-μ-mediated growth inhibition of human RL lymphoma cells. Interestingly, a tyrosine-phosphatase inhibitor, orthovanadate, enhanced tyrosine phosphorylation upon anti-μ stimulation and synergistically increased growth inhibition [135]. Since many of these inhibitors are not absolutely specific to PTKs, these results could be due to alterations in other signaling pathways, such as PKC, PKA and calmodulin-dependent kinases. Nonetheless, these studies suggest a PTK-dependent pathway in growth inhibition.

To avoid these limitations, we have utilized antisense oligonucleotides to 'delete' particular PTK activity in the anti-μ growth-inhibitable CH31 lymphoma cell line. These cells are IgM$^+$, IgD$^-$, CD5$^+$, and appear to represent an immature stage of development in the B-cell lineage. By Western blotting, we found that CH31 cells possess high levels of *blk*, *lyn*, and *lck* proteins, but did not contain detectable *fyn*. When exposed to antisense for *blk*, *lyn*, *lck* or *fyn*, the growth inhibition was significantly reduced by *blk* antisense, but not by oligos of *lyn*, *fyn* or *lck* antisense [137]. Transfection of this cell line with a *fyn* gene construct neither enhanced the sensitivity to anti-μ, nor did it prevent growth inhibition. These data imply a role of *blk* PTK in inhibitory signal pathway and, therefore, in B-cell tolerance. Indeed, as cited earlier, tyrphostin treatment of normal B cells can block the induction of anergy in adult B cells by anti-μ [see 38].

The events that lead to cellular growth arrest after ligation of the surface immunoglobulin molecule involve many factors. Data from our laboratory indicate that negative signaling can occur when the tyrosine kinase blk is present. The subsequent tyrosine phosphorylation of intracellular substrates would then presumably lead directly or indirectly to up-regulation of DNA-binding proteins active in transcription of genes important in cell cycle or growth regulation. Early events in transcription include transient up-regulation of *c-myc* by 1–2 h, followed by depletion of *c-myc* mRNA, and the subsequent prevention of pRB phosphorylation 12–14 h after anti-IgM stimulation. The exact sequence of events and the nature of all the factors remains to be determined. However, we believe that some of the key events have been elucidated. Thus, multiple events must take place in order for a negative growth signal to be delivered to a cell leading to in vitro growth arrest and apoptosis and, possibly, to in vivo tolerance induction.

Co-Stimulation: Why Do We Need T Cells?

It is clear from the above that B-cell tolerance does, indeed, exist despite the protestations of many T-cell immunologists. One might turn the tables and ask then 'why do we need T cells?' The answer is best expressed in a poem, originally from Winnie the Pooh, that has been adapted as an Ode to cellular collaboration[1]:

> So wherever I am, there's always T,
> There's always T and B.
> 'What would I do?', I said to T,
> 'If it wasn't for you,' and T said, 'True,
> It wasn't much fun for one, but two
> Can stick together'. Says T, says he,
> 'That's how it is,' says T.
>
> Us two (A.A. Milne)

Stated simply, B cells need T cells for help and T cells need B cells for antigen presentation, but T-cell help goes one step further in that it not only promotes differentiation to antibody synthesis, but it acts to override abortive signaling in B cells and to reverse negative signaling for tolerance. Thus, as has been repeatedly shown during the last several decades, the presence of optimal T-cell help when B cells are exposed to a tolerogenic epitope usually promotes positive activation and subsequent differentiation towards antibody synthesis, as well as the development of memory. It needs to be reiterated that in this mutual admiration society, T cells do not develop appropriately in the absence of B cells [147].

---

[1] With T = Pooh and B = me, and apologies to A.A. Milne!

Self-antigens are expressed at various times during development and vary in their chemical composition and persistence during the lifetime of an individual. Evidence has been presented that these self-antigens produce varying degrees of B-cell unresponsiveness depending on their chemical nature and tissue expression. Tolerance at the B-cell level is manifested either by deletion of such self-reactive clones or by the induction of a state of anergy in which they are incapable of reacting immediately to antigen even in the presence of T-cell help. In some cases, a state of immunologic ignorance may occur at the B-cell level, but the potential consequences of such ignorance is not bliss for the organism, should sufficient T-cell help, for example by a viral infection, be present to turn on such B cells. Moreover, the potential for somatic mutation during the maturation of immune response requires that some form of purging of the immunologic repertoire occurs throughout life before antigen, as well as after its introduction to the system. Thus, not only T-cell, but also B-cell tolerance is a requirement to maintain the integrity of the vertebrate organism.

Acknowledgements

The authors work summarized herein has been supported by grants from the NIH, ACS and CTR. This is publication No. 97 from the Immunology and Immunotherapy Division of the University of Rochester Cancer Center.

References

1 Ehrlich P, Morgenroth J: On hemolysins; 5th commun. (reprinted from Berl Klin Wochenschr 1901; No 10); in Ehrlich P: Collected Studies on Immunity. London, Wiley, 1906, chap VII, pp 71–87.
2 Owen RD: Immunogenetic consequences of vascular anastomoses between bovine twins. Science 1945;102:400–402.
3 Billingham RE, Brent L, Medawar PB: Actively acquired tolerance of foreign cells. Nature 1953;172:603–606.
4 Golub ES, Weigle WO: Studies on the induction of immunologic unresponsiveness. II. Kinetics. J Immunol 1967;99:624–628.
5 Chiller JM, Habicht GS, Weigle WO: Kinetic differences in unresponsiveness of thymus and bone marrow cells. Science 1971;171:813–815.
6 Werblin TP, Siskind GW: Effect of tolerance and immunity on antibody affinity. Transplant Rev 1972;8:104–136.
7 Sanfilippo AS, Scott DW: Cellular events in tolerance. III. Carrier tolerance as a model for T cell unresponsiveness. J Immunol 1974;113:1661–1667.
8 Scott DW, Venkataraman M, Jandinski JJ, Multiple pathways of B lymphocyte tolerance. Immunol Rev 1979;43:241–280.
9 Venkataraman M, Scott DW: Cellular events in tolerance. VII. Decrease in clonable precursors stimulatable in vitro by specific antigens or LPS. Cell Immunol 1979;47:323–331.

10 Linton PJ, Rudie A, Klinman NR: Tolerance susceptibility of newly generating memory B cells. J Immunol 1991;146:4099–4014.

11 Goodnow CC: Transgenic mice and analysis of B-cell tolerance. Annu Rev Immunol 1992;10:489–518.

12 Tiegs S, Russell D, Nemazee D: Receptor editing in self-reactive bone marrow B cells. J Exp Med 1993;177:1009–1020.

13 Diener E, Paetkau VH: Antigen recognition: Early surface-receptor phenomena induced by binding of tritium-labeled antigen. Proc Natl Acad Sci USA 1972;69:2364–2368.

14 Lawton AR III, Cooper MD; Modification of B lymphocyte differentiation by anti-immunoglobulin; in Cooper MD, Warner NL (eds): Contemporary Topics in Immunobiology. New York, Plenum Press, 1974, vol 3, pp 193–226.

15 Raff MC, Owen JJT, Cooper MD, Lawton AR III, Megson M, Gathings W: Differences in susceptibility of mature and immature mouse B lymphocytes to anti-immunoglobulin-induced immunoglobulin suppression in vitro. J Exp Med 1975;142:1052–1064.

16 Sidman CL, Unanue ER: Receptor-mediated inactivation of early B lymphocytes. Nature 1975;257:149–151.

17 Scott DW, Borrello M, Liou L-B, Yao X-R, Warner GL: B cell tolerance; in Singh B (ed): Advances in Molecular and Cellular Immunology. Greenwich, JAI Press, 1992, vol 1, p 119–143.

18 Brown D, Warner GL, Alés-Martínez J, Scott DW, Phipps RP: Prostaglandin E$_2$ induces apoptosis in normal and malignant B lymphocytes. Clin Immunol Immunopathol 1992;63:221–229.

19 Swat W, Ignatowicz L, Kisielow P: Detection of apoptosis of immature CD4+8+ thymocytes by flow cytometry. J Immunol 1991;137:79–87.

20 Nossal GJV, Pike B, Battye F: Mechanisms of clonal abortion tolerogenesis. II. Clonal behaviour of immature B cells following exposure to anti-μ chain antibody. Immunology 1979;37:203–215.

21 Metcalf E, Klinman N: In vitro tolerance induction of neonatal murine spleen cells. J Exp Med 1976;143:1327–1340.

22 Burnet FM: The Clonal Selection Theory of Acquired Immunity. Nashville, Vanderbilt University Press, 1959, pp 1–202.

23 Seyfert V, Sukhatme V, Monroe J: Differential expression of a zinc finger-encoding gene in response to positive versus negative signaling through receptor immunoglobulin in murine B lymphocytes. Mol Cell Biol 1989;9:2083–2088.

24 Monroe J, Yellen-Shaw A, Seyfert V: Molecular basis for unresponsiveness and tolerance induction in immature stage B lymphocytes; in Singh B (ed): Advances in Molecular and Cellular Immunology. Greenwich, JAI Press, 1992, vol 1, p 1–30.

25 Strasser A, Whittingham S, Vaux D, Bath ML, Adams JM, Cory S, Harris J: Enforced bcl-2 expression in B-lymphoid cells prolongs antibody responses and elicits autoimmune disease. Proc Natl Acad Sci USA 1991;88:8661–8665.

26 Scott DW: Cellular events in tolerance. V. Detection, isolation and fate of lymphoid cells which bind fluoresceinated antigens in vivo. Cell Immunol 1976;22:311–317.

27 Warner G, Scott DW: A polyclonal model for B cell tolerance. I. Fc-dependent and FC-independent induction of nonresponsiveness by pretreatment of normal splenic B cells with anti-Ig. J Immunol 1991;146:2185–2191.

28 Flahart R, Lawton A: Anti-μ antibody blocks LPS-driven B cell differentiation by suppressing specific RNAs. J Mol Cell Immunol 1987;3:61–67.

29 Yuan D: Molecular basis for the inhibition of LPS-induced differentiation by anti-Ig. Mol Cell Immunol 1987;3:133–141.

30 Chen U: Anti-IgM antibodies down-modulate mu-enhancer activity and OTF2 level in LPS-stimulated murine splenic B-cells. Nucleic Acids Res 1991;19:5981–5989.

31 Martenson I-L, Iglesias A, Leanderson T: Regulation of Ig gene expression in trans by phorbol esters. Eur J Immunol 1990;19:1497–1500.

32 Bijsterbosch M, Klaus GGB: Cross-linking of surface immunoglobulin on B lymphocytes inhibits stimulation of inositol phospholipid breakdown via the antigen receptors. J Exp Med 1988;162:1825–1836.

33 Warner GL, Scott DW: Cholera toxin-sensitive and insensitive signalling via surface Ig. J Immunol 1983;143:458–463.

34 Warner G, Gaur A, Scott DW: A polyclonal model for B cell tolerance. II. Linkage between abortive signaling of B cell egress from G_0, class II up-regulation and unresponsiveness. Cell Immunol 1991;138:404–412.

35 Klaus GGB, Hawrylowicz CM: Activation and proliferation signals in mouse B cells. II. Evidence for activation (G_0 to G_1) signals differing in sensitivity to cyclosporine. Eur J Immunol 1984;14:250–254.

36 Brines R, Klaus GGB: Effects of anti-immunoglobulin antibodies, interleukin-4, and second messenger agonists on B cells from neonatal mice. Int Immunol 1992;2:461–467.

37 Yellen-Shaw AJ, Monroe JG: Developmentally regulated association of a 56-kD member of the surface immunoglobulin M receptor complex. J Exp Med 1992;176:129–137.

38 Gaur A, Yao X-r, Scott DW: B-cell tolerance induction by cross-linking IgM but not IgD, and synergy by cross-linking both isotypes. J Immunol 1993;150:1663–1669.

39 Cambier JC, Campbell KS: Membrane immunoglobulin and its accomplices: New lessons from an old receptor. FASEB J 1992;6:3207–3211.

40 Campbell KS, Cambier JC: B lymphocyte receptors (mIg) are covalently associated with a disulfide-linked, inducibly phosphorylated glycoprotein complex. EMBO J 1990;9:441–448.

41 Brunswick M, Samelson LE, Mond JJ: Surface immunoglobulin cross-linking activates a tyrosine kinase pathway in B cells that is independent of protein kinase C. Proc Natl Acad Sci USA 1991;88:1311–1314.

42 Alés-Martínez J-E, Scott DW, Phipps RP, Casnellie J, Kroemer G, Martinez-AC, Pezzi L: Cross-linking of surface IgM or IgD causes differential biologic effects in spite of overlap in tyrosine (de)phosphorylation. Eur J Immunol 1992;22:845–850.

43 Hayakawa K, Hardy RR, Honda M, Herzenberg LA, Steinberg AD, Herzenberg LA: Ly-1 B cells: Functionally distinct lymphocytes that secrete IgM auto-antibodies. Proc Natl Acad Sci USA 1984;81:2494–2498.

44 Liou L-B, Warner GL, Scott DW: Can peritoneal B cells be rendered unresponsive? Int Immunol 1992;4:5–21.

45 Cohen DP, Rothstein TL: Elevated levels of protein kinase C and α-isoenzyme expression in murine peritoneal B cells. J Immunol 1991;146:2921–2927.

46 Cong Y, Rabin E, Wortis HH: Treatment of murine $CD5^-$ B cells with anti-Ig, but not LPS, induces surface CD5: Two B-cell activation pathways. Int Immunol 1991;3:467–476.

47 Yao X-r, Scott D: Effect of priming with a thymus-independent antigen on susceptibility to B-cell tolerance. Cell Immunol 1992;142:434–443.

48 Ishida H, Hastings R, Kearney J, Howard M: Continuous anti-interleukin 10 antibody administration depletes mice of Ly-1 B cells, but not conventional B cells. J Exp Med 1992;175:1213–1220.

49 Venkataraman M, Scott DW: Cellular events in tolerance. VI. Neonatal versus adult B cell tolerance: Differences in antigen-binding cell patterns and lipopolysaccaride stimulation. J Immunol 1977;119:1879–1886.

50 Scott DW: Cellular events in tolerance. I. Failure to demonstrate activation of lymphocytes, blocking factors or suppressor cells during the induction of tolerance to a soluble protein. J Immunol 1973;111:789–796.

51 Pillai PS, Scott DW: Cellular events in tolerance. IX. Maintenance of immunological tolerance in the presence of normal B cell precursors and in the absence of demonstrable suppression. Cell Immunol 1983;77:69–76.

52 Aldo-Benson M, Borel Y: The tolerant cell: Direct evidence for receptor blockage. J Immunol 1974;112:1793–1803.

53 Venkataraman M, Aldo-Benson M, Borel Y, Scott DW: Persistence of antigen-binding cells with surface tolerogen: Isologous versus heterologous immunoglobulin carriers. J Immunol 1977;119:1006–1009.

54 Chace JH, Scott DW: Activation events in hapten-specific B cells from tolerant mice. J Immunol 1988;141:3258–3262.

55 Chace JH, Scott DW: The tolerance defect: Properties and fate of tolerant B cells in adult mice; in Cruse JM, Lewis RE Jr (eds): The Year in Immunology. Basel, Karger, 1988, pp 181–192.

56 Goodnow CC, Crosbie J, Adelstein S, Lavoie T, Smith-Gill S, Brink A, Pritchard-Briscoe H, Wotherspoon J, Loblay H, Raphael K, Trent R, Basten A: A transgenic model of immunologic tolerance: Absence of secretion and altered surface immunoglobulin expression in self-reactive B lymphocytes. Nature 1988;334:676–682.

57 Nemazee D, Bürki K: Clonal deletion of B lymphocytes in a transgenic mouse bearing anti-MHC class I antibody genes. Nature 1989;337:562–566.

58 Goodnow CC, Crosbie J, Jorgensen H, Brink A, Basten A: Induction of self-tolerance in mature peripheral B lymphocytes. Nature 1989;342:385–390.

59 Basten A, Brink RA, Mason DY, Crosbie J, Goodnow CC: Self-tolerance in B cells from different lines of Lysozyme double transgenic mice.

60 Mason D, Jones C, Goodnow CC: Development and follicular localization of tolerant B lymphocytes in lysozyme/anti-lysozyme IgM/IgD transgenic mice. Int Immunol 1992;4:163–175.

61 Scott DW, Layton JE, Nossal GJV: Role of IgD in the immune response and tolerance. I. Anti-pretreatment facilitates tolerance induction in adult B cells in vitro. J Exp Med 1977;146:1473–1483.

62 Carsetti R, Köhler G, Lamers M: A role for immunogloulin D: Interference with tolerance induction. Eur J Immunol 1992;23:168–178.

63 Mitchell GM: Personal communication, 1977.

64 Nemazee D, Bürki K: Clonal deletion of autoreactive B lymphocytes in bone marrow chimeras. Proc Natl Acad Sci USA 1989;86:8039–8043.

65 Hartley SB, Crosbie J, Brink A, Kantor A, Basten A, Goodnow CC: Elimination from peripheral lymphoid tissues of self-reactive B lymphocytes recognizing membrane-bound antigens. Nature 1991;353:765–769.

66 Jerne N: The somatic generation of immune recognition. Eur J Immunol 1971;1:1–10.

67 Zöller M: 2,4,6-Trinitrophenyl (TNP) responsiveness of anti-TNP (Sp6) transgenic mice. Eur J Immunol 1991;21:1601–1610.

68 Scott DW: All self antigens are not created (tolerated) equally. Immunol Today 1984;5:68–71.

69 Theopold U, Köhler G: Partial tolerance in β-galactosidase-transgenic mice. Eur J Immunol 1990;20:1311–1316.

70 Zinkernagel R, Cooper S, Chambers J, Lazzarini RA, Hengartner H, Arnheiter H: Virus-induced autoantibody response to a transgenic viral antigen. Nature 1990;344:68–71.

71 Erikson J, Radic M, Camper S, Hardy R, Carmack C, Weigert M: Expression of

anti-DNA immunoglobulin transgenes in non-autoimmune mice. Nature 1991;349: 331–334.

72 Murakami M, Tsubata T, Okamoto M, Shimizu A, Kumagai S, Imura H, Honjo T: Antigen-induced apoptotic death of Ly-1 B cells responsible for autoimmune disease in transgenic mice. Nature 1992;357:77–80.

73 Pennell C, Scott D: Models and mechanisms for signal transduction in B cells. Surv Immunol Res 1986;5:61–70.

74 Scott DW, Livnat D, Pennell C, Keng P: Lymphoma models for B cell activation and tolerance. III. Cell cycle dependence for negative signalling of WEHI-231 B lymphoma cells by anti-μ. J Exp Med 1986;164:156–164.

75 Warner GL, Nelson D, Ludlow J, Scott DW: Anti-immunoglobulin treatment of murine B-cell lymphomas induces transforming growth factor β but pRB hypophosphorylation is transforming growth factor β independent. Cell Growth Differ 1992;3:175–181.

76 Draetta G: Cell cycle control in eucaryotes: Molecular mechanisms of cdc2 activation. Trends Biochem Sci 1990;15:378–383.

77 Hunter T, Pines J: Cyclins and cancer. Cell 1991;66:1071–1074.

78 Bandara LR, Adamczewski JP, Hunt T, La Thangue NB: Cyclin A and the retinoblastoma gene product complex with a common transcription factor. Nature 1991;352:249–251.

79 Milner J, Cook A, Mason J: p53 is associated with p34cdc2 in transformed cells. EMBO J 1990;9:2885–2889.

80 Friend SH, Horowitz JM, Gerber MR, Wang XF, Bogenmann E, Li FP, Weinberg RA: Deletions of a DNA sequence in retinoblastoma and mesenchymal tumors: Organization of the sequence and encoded protein. Proc Natl Acad Sci USA 1987;84:9059–9063.

81 Bookstein R, Shew J, Chen P, Scully P, Lee W: Suppression of tumorigenicity of human prostate carcinoma cells by replacing a mutated RB gene. Science 1990;247:712–715.

82 Huang H-J, Yee J-K, Shew Y-J, Chen P-L, Bookstein T, Friedmann E, Lee Y-H, Lee W-H: Suppression of the neoplastic phenotype by replacement of the RB gene in human cancer cells. Science 1988;242:1563–1566.

83 Lane DP, Benchimol S: p53: Oncogene or anti-oncogene? Genes Dev 1990;4:1–8.

84 Levine AJ, Momand J, Findlay CA: The p53 tumor suppressor gene. Nature 1991;351:453–456.

85 Yin Y, Tainsky MA, Bischoff FZ, Strong LC, Wahl GM: Wild-type p53 restores cell cycle control and inhibits gene amplification in cell with mutant p53 alleles. Cell 1992;70:937–948.

86 Livingstone LR, White A, Sprouse J, Livanos E, Jacks T, Tisty TD, Altered cell cycle arrest and gene amplification potential accompany loss of wild-type p53. Cell 1992;70:923–935.

87 DeCaprio JA, Ludlow JW, Lynch D, Furukawa Y, Griffin J, Piwnica-Worms H, Huang C-M, Livingstone DM: The product of the retinoblastoma susceptibility gene has properties of a cell cycle regulatory element. Cell 1989;58:1085–1095.

88 Buchkovich K, Duffy KA, Harlow E; The retinoblastoma protein is phosphorylated during specific phases of the cell cycle. Cell 1989;58:1097–1105.

89 Chen P-L, Scully P, Shew JY, Wang JYY, Lee W-H: Phosphorylation of the retinoblastoma gene product is modulated during the cell cycle and cellular differentiation. Cell 1989;58:1193–1198.

90 Mihara K, Cao X-R, Yen A, Chandler S, Driscoll B, Murphree AL, T'Ang A, Fung YT: Cell cycle-dependent regulation of phosphorylation of the human retinoblastoma gene product. Science 1990;246:1300–1303.

91 Goodrich DW, Wang NP, Qian YW, Lee EY-HP, Lee W-H: The retinoblastoma gene

product regulates progression through the G_1 phase of the cell cycle. Cell 1991;67:293–302.

92 Ludlow JW, Shon J, Pipas JM, Livingston DM, DeCaprio JA: The retinoblastoma susceptibility gene product undergoes cell cycle-dependent dephosphorylation binding to and release from SV40 large T antigen. Cell 1990;56:57–65.

93 Whyte PK, Buchkovich J, Harowitz JM, Friend SH, Raybuck M, Weinberg RA, Harlow E: Association between an oncogene and an anti-oncogene: The adenovirus E1A proteins bind to the retinoblastoma gene product. Nature 1988;334:124–129.

94 Dyson N, Buchkovich K, Whyte P, Harlow E: The cellular 107 kD protein that binds to adenovirus E1A also associates with the large T antigens of SV40 and JC virus. Cell 1989;58:249–255.

95 Mittnacht S, Weinberg RA: G1/S phosphorylation of the retinoblastoma protein is associated with an altered affinity for the nuclear compartment. Cell 1991;65:381–393.

96 Hinds PW, Mittnacht S, Dulic V, Arnold A, Reed SI, Weinberg RA: Regulation of retinoblastoma protein functions by ectopic expression of human cyclins. Cell 1990; 70:993–1006.

97 Chellapan SP, Heibert S, Mudryj M, Horowitz JM, Nevins JR: The E2F transcription factor is a cellular target for the RB protein. Cell 1990;65:1053–1061.

98 Chittenden T, Livingston DM, Kaelin W Jr: The T/E1A-binding domain of the retinoblastoma product can interact selectively with a sequence-specific DNA-binding protein. Cell 1991;65:1073–1082.

99 Cao L, Faha B, Dembski m, Tsai L-H, Harlow E, Dyson N: Independent binding of the retinoblastoma protein and p107 to the transcription factor E3F. Nature 1992;355:176–179.

100 Moses HL, Yang EY, Pietenpol JA: TGF-β stimulation and inhibition of cell proliferation: New mechanistic insights. Cell 1990;63:245–247.

101 Massagué J: The transforming growth factor β family. Annu Rev Cell Biol 1990;6:597–641.

102 Laiho M, DeCaprio JA, Ludlow JW, Livingston D, Massagué J: Growth inhibition by TGF-β linked to suppression of retinoblastoma protein phosphorylation. Cell 1990;62:175–185.

103 Zentella A, Weiss FBM, Ralph DA, Laiho M, Massagué J: Early gene responses to transforming growth factor-β in cell lacking growth-suppressive RB function. Mol Cell Biol 1991;11:4952–4958.

104 Pietenpol JA, Holt JT, Stein RW, Moses HL: Transforming growth factor-β1 suppression of c-myc gene transcription: Role in inhibition of keratinocyte proliferation. Proc Natl Acad Sci USA 1990;87:3758–3762.

105 Pietenpol JA, Stein RW, Moran E, Yaciuk P, Schlegel R, Lyons RM, Pittelkow M, Munger K, Howley P, Moses HL: TGF-β1 inhibition of c-myc transcription and growth in keratinocytes is abrogated by viral transforming proteins with pRB binding domains. Cell 1991;61:777–785.

106 Munger K, Pietenpol JA, Pittelkow MR, Holt JT, Moses HL: Transforming growth factor-β1 regulation of c-myc expression, pRB phosphorylation, and cell cycle progression in keratinocytes. Cell Growth Differ 1992;3:291–298.

107 Benhamou LE, Casenave P-A, Sarthou P: Anti-immunoglobulins induce death by apoptosis in WEHI-231 B lymphoma cells. Eur J Immunol 1990;20:1405–1407.

108 Hasbold J, Klaus GGB: Anti-immunoglobulin antibodies induce apoptosis in immature B-cell lymphomas. Eur J Immunol 1990;20:1685–1690.

109 Fischer G, Kent SC, Joseph L, Green DR, Scott DW: Lymphoma models for B-cell activation and tolerance. X. Anti-μ-mediated growth arrest and apoptosis of murine B-cell lymphomas is prevented by the stabilization of myc. 1993; submitted.

110 McCormick JE, Pepe VH, Kent BR, Dea M, Marshak-Rothstein A, Sonneshein G: Proc Natl Acad Sci USA 1984;81:5546–5550.
111 Mashesaran S, McCormick JE, Sonenshein G: Changes in phosphorylation of *myc* oncogene and *RB* anti-oncogene protein products during growth arrest of the murine lymphoma WEHI-231 cell line. Oncogene 1992;6:1965–1972.
112 Spencer CA, Groudine M: Control of *c-myc* regulation in normal and neoplastic cells. Adv Cancer Res 1991;56:1–48.
113 Pietenpol JA, Munger K, Howley PM, Stein R, Moses HL: Factor-binding element in the human *c-myc* promoter involved in transcriptional regulation by transforming growth factor-β1 and the retinoblastoma gene product. Proc Natl Acad Sci USA 1991;88:10227–10231.
114 Duyao MP, Buckler AJ, Sonenshein G: Interaction of an NF-κB-like factor with a site upstream of the *c-myc* promoter. Proc Natl Acad Sci USA 1990;87:4727–4731.
115 Luscher B, Eisenman RN: New light on myc and myb. Genes Dev 1990;4:2025–2035.
116 Blackwood EM, Eisenman RN: A helix-loop-helix-zipper protein that forms a sequence-specific DNA-binding complex with myc. Science 1991;251:1211–1217.
117 Prendergast GC, Lawe D, Ziff E: Association of myn, the murine homolog of max, with c-myc stimulates methylation-sensitive DNA binding and Ras cotransformation. Cell 1991;65:395–407.
118 Dalla-Favera R, Bregni M, Erickson J, Patterson D, Gallo RC, Croce CM: Human c-myc oncogene is located on the region of chromosome 8 that is translocated in Burkitt lymphoma cells. Proc Natl Acad Sci USA 1982;79:7824–7827.
119 Showe LC, Croce CM: The role of chromosomal translocations in B- and T-cell neoplasia. Annu Rev Immunol 1987;5:253–277.
120 Taub R, Kirsch I, Morton C, Lenoir G, Swan D, Tronick S, Aaronson S, Leder P: Translocation of the *c-myc* gene into the immunoglobulin heavy chain locus in human Burkitt lymphoma and murine plasmocytoma cells. Proc Natl Acad Sci USA 1982; 79:7837–7841.
121 Golay J, Cusamo G, Introna M: Independent regulation of *c-myc*, B-*myb*, and *c-myb* gene expression by inducers and inhibitors of proliferation in human B lymphocytes. J Immunol 1992;149:300–308.
122 Hekkila R, Schwab G, Wickstrom E, Loke SL, Pluznik DH, Watt R, Neckers LN: A *c-myc* antisense oligonucleotide inhibits entry into S phase but not progress from G_0 to G_1. Nature 1987;328:445–449.
123 Loke SL, Stein C, Zhang X, Avigan M, Cohen J, Neckers LM: Delivery of *c-myc* antisense phosphorothioate oligonucleotides to hematopoietic cells in culture by liposome fusion: Specific reduction in c-myc protein expression correlates with inhibition of cell growth and DNA synthesis. Curr Top Microbiol Immunol 1988;141:282–289.
124 Shi Y, Glynn JM, Guilbert LJ, Cotter TG, Bissonnette RP, Green DR: Role for *c-myc* in activation-induced apoptotic cell death in T cell hybridomas. Science 1992;257:212–214.
125 Freytag SO: Enforced expression of the *c-myc* oncogene inhibits cell differentiation by precluding entry into a distinct predifferentiation state in G_0/G_1. Mol Cell Biol 1988; 8:1614–1624.
126 Waters CT, Littlewood T, Handcock D, Moore J, Evan G: *C-myc* protein expression in untransformed fibroblasts. Oncogene 1991;6:101–109.
127 Evan GI, Wyllie AH, Gilberg CS, Littlewood TD, Land H, Brooks M: Induction of apoptosis in fibroblasts by c-myc protein. Cell 1992;69:119–128.
128 Rustgi AK, Dyson N, Bernards R: Amino-terminal domains of *c-myc* and N-*myc* proteins mediate binding to the retinoblastoma gene product. Nature 1991;352:541–544.
129 Cambier J, Ransom JT: Molecular mechanisms of transmembrane signaling in B lymphocytes. Annu Rev Immunol 1987;5:175–199.

130 Klaus GGB, Bijsterbosch M, O'Garra A, Harnett M, Rigley K: Receptor signaling and cross-talk in B lymphocytes. Immunol Rev 1987;99:19.

131 Hempel WM, Schatzman RC, DeFranco AL: Tyrosine phosphorylation of phospholipase-Cγ2 upon cross-linking of membrane immunoglobulin on murine B lymphocytes. J Immunol 1992;148:3021–3027.

132 Coggeshall KM, McHugh JC, Altman A: Predominant expression and activation-induced tyrosine phosphorylation of phospholipase C-γ2 in B lymphocytes. Proc Natl Acad Sci USA 1992;89:5660–5664.

133 Sefton BM, Campbell MA: The role of tyrosine protein phosphorylation in ly activation. Annu Rev Cell Biol 1991;7:257–274.

134 Gold M, Law D, DeFranco A: Stimulation of protein tyrosine phosphorylation by the B-lymphocyte antigen receptor. Nature 1990;345:810–813.

135 Beckwith M, Urba WJ, Ferris DK, Freter CE, Kuhns DB, Moratz CM, Longo DL: Anti-Ig-mediated growth inhibition of a human B lymphoma cell line is independent of phosphatidylinositol turnover and protein kinase C activation and involves tyrosine phosphorylation. J Immunol 1991;147:2411–2418.

136 Lane PJL, Ledbetter JA, McConnell FM, Draves K, Deans J, Schieven GL, Clark EA: The role of tyrosine phosphorylation in signal transduction through surface Ig in human B cells. J Immunol 1991;146:715–722.

137 Yao X-r, Scott DW: A role for the p55^{blk} kinase in the Ig complex signaling growth arrest in anti-μ-sensitive B-cell lymphomas. Immunol Rev 1993;132:163–187.

138 Burkhardt AL, Brunswick M, Bolen JB, Mond JJ: Anti-immunoglobulin stimulation of B lymphocytes activates src-related protein-tyrosine kinases. Proc Natl Acad Sci USA 1991;88:7410–7414.

139 Campbell M-A, Sefton BM: Association between B-lymphocyte membrane immunoglobulin and multiple members of the src family of protein tyrosine kinases. Mol Cell Biol 1992;12:2315–2321.

140 Yamanashi Y, Kakiuchi T, Mizuguchi J, Yamamoto T, Toyoshima K: Association of B-cell antigen receptor with protein tyrosine kinase lyn. Science 1991;251:192–194.

141 Hutchroft JE, Harrison ML, Geahlen RL: B lymphocyte activation is accompanied by phosphorylation of a 72-kD protein-tyrosine kinase. J Biol Chem 1991;266:14846–14849.

142 Clark M, Campbell KS, Kazlauskas A, Johnson SA, Hertz M, Potter TA, Pleiman C, Cambier JC: The B cell antigen receptor complex: Association of Ig-α and Ig-β with distinct cytoplasmic effectors. Science 1992;258:123–126.

143 Lin JJ, Justement LB. The MB-1/B29 heterodimer couples the B cell antigen receptor to multiple *src* family protein tyrosine kinases. J Immunol 1992;149:1548–1555.

144 Rothstein T: Personal communication, 1993.

145 Page D, DeFranco A: Role of phosphoinositide-derived second messengers in mediating anti-IgM-induced arrest of WEHI-231 B lymphoma cells. J Immunol 1984;140:3717–3726.

146 Scott DW, Livnat D, Whitin J, Dillon SB, Snyderman R, Pennell C: Lymphoma models for B cell activation and tolerance. V. Anti-Ig-mediated growth inhibition is reversed by phorbol myristate acetate but does not involve changes in cytosolic free calcium. J Mol Cell Immunol 1987;3:109–120.

147 Hayglass KT, Naides SJ, Scott CF, Benacerraf B, Sy M-N: T cell development in B cell-deficient mice. IV. The role of B cells as antigen-presenting cells in vivo. J Immunol 1986;136:823–829.

Dr. David W. Scott, Division of Immunology and Immunotherapy,
University of Rochester Cancer Center, 601 Elmwood Avenue,
Rochester, NY 14642 (USA)

Granstein RD (ed): Mechanisms of Immune Regulation.
Chem Immunol. Basel, Karger, 1994, vol 58, pp 67–91

Regulation of B Cell Tolerance and Triggering by Immune Complexes

Eric R. Fedyk[a], *Melinda A. Borrello*[a], *Deborah M. Brown*[a], *Richard P. Phipps*[a,b]

[a]Cancer Center, Departments of Microbiology and Immunology, and [b]Pediatrics, University of Rochester School of Medicine and Dentistry, Rochester, N.Y., USA

Introduction

Antigen-antibody complexes (i.e. immune complexes) are potent regulators of the immune system functioning as a critical component of normal antigen clearance mechanisms. However, accumulation of immune complexes may promote tissue injury and lead to numerous diseases. In principle, accumulation could result from decreased immune complex catabolism and/or increased immune complex formation. Current evidence indicates that in vivo catabolic rates are constant whereas the rate of immune complex formation fluctuates [1]. Formation of immune complexes requires optimal concentrations of antigen and antibody. Regulating the concentration of either component would influence formation. From a therapeutic perspective, most diseases are diagnosed after antigen has been introduced and after the immune system has reacted. Therefore, only the ongoing immune response is accessible to therapeutic manipulation. In this respect, it is of critical importance to understand how antigen-antibody complex interaction with B lymphocytes affects their differentiation into antibody-secreting plasma cells. Recently, several in vitro models have been developed to examine potential mechanisms whereby immune complexes regulate B cell responsiveness. In this article, data obtained with these systems is reviewed. Factors affecting the formation of immune complexes and the effect of immune complex components on B cell responsiveness will be explored. Next, the role of immune complexes, Fc receptors and prostaglandin E_2 in inducing B cell unresponsiveness will be reviewed. Finally, this article will analyze the role of immune complexes and prostaglandin E_2 in *promoting* B cell responsiveness. The reader will discover that the current data reinforces the concept that regulation of B cell responsiveness by immune complexes and associated factors is a complex process in which lymphocytes can be regulated at multiple levels.

Factors Which Influence the Development of Immune Complexes

During an immune response, antibodies are produced as part of the physiologic mechanism of eliminating antigenic material. Immune complexes form due to the noncovalent interaction of antigen with specific antibody. In addition, complement components may covalently associate with the antigen-antibody aggregate. These complexes play a role in normal immune regulation, but in many instances, they can also be mediators of pathogenesis. Interestingly, B lymphocytes are particularly sensitive to antigen-antibody regulation. This can occur when a B cell is a direct target of the immune complex, or in an indirect fashion, mediated by immune complex interaction with other cell types, such as macrophages or dendritic cells.

The properties of the antigen and responding antibody can profoundly affect the nature of the immune complex and the role it plays in controlling B lymphocytes. The nature of the antigen will determine its localization, and in part, the mechanism by which it will be eradicated. Antigens present on the surface of virally infected cells, tumor cells, or parasites will elicit immune complex formation wherever they are situated [2]. These complexes may remain tethered to that location. Bacteria or viruses in the interstitial or intravascular fluids can themselves be immunogenic, or they may release antigenic materials that result in formation of circulating immune complexes. Antigen-antibody complexes formed in the tissues or in the circulation are cleared by the mononuclear phagocyte system [3]. Regional macrophages (e.g. histiocytes) can remove low levels of immune complexes which are deposited in the tissues. Circulating complexes are removed by fixed Kupffer cells in the liver. Generalized tissue deposition, especially in the kidneys, occurs when the quantity of immune complexes exceeds the clearance capacity of mononuclear phagocytes, or when complexes are small. Size is a critical factor in determining whether or not an immune complex is effectively cleared. Large complexes are most easily phagocytized, while those that are small may not be as readily ingested [3]. The size and valence of the antigen contribute to the overall size of the resultant aggregate [4]. For example, a particulate antigen (such as an erythrocyte) with multiple unique determinants can result in formation of large immune complexes. Alternatively, although DNA is quite large, its identical, repetitive determinants favor bivalent, non-cross-linking antibody binding. This leads to formation of small complexes. Haptens and other monovalent antigens do not form lattices with their corresponding antibodies, but instead, form small complexes which remain in the circulation [4]. Finally, some antigens may exhibit tropism for certain tissues or their components. For example, DNA preferentially binds to collagen, resulting in enhanced connective tissue deposition of immune complexes containing the nucleic acid [5]. The charge of the antigen can also influence its tissue tropism, especially in the kidney, where the glomerular capillary wall is rich in negative charges, which serve to attract immune complexes containing cationic antigens [6].

The nature of the responding antibody is extremely important in dictating the physiologic properties of the immune complex. The isotype, determined by the heavy chain class, affects the ability of the antibody to induce the complement cascade. Complement is a collection of proteins in the serum which, once activated by the appropriate antibody, can interact to coat or lyse an antigen. In humans, IgM, IgG1, and IgG3 are the most effective at activating complement, while IgG2 is less effective [2]. Antibodies of the IgG4 subclass cannot fix complement. The binding of complement to the immune complex can either dissolve the antigen or it can recruit phagocytes to ingest the conglomerate. This chemotactic effect of complement results in improved clearance of the immune complexes, but can also cause inflammation. The immunoglobulin isotype also influences phagocytosis. IgG1 and IgG3 are effective initiators of phagocytosis whereas IgG2, IgG4, and IgM are quite poor [3]. The size and valence of the antibody, as for antigen, are additional factors which contribute to the size of the immune complex. Pentameric IgM forms large, precipitating immune complexes, whereas monomeric immunoglobulin forms smaller, soluble complexes. Antibody affinity and avidity also influence immune complex size. For example, high affinity antibodies tend to cross-link the antigen and form large immune complexes. In contrast, the interaction of a low avidity antibody with its corresponding antigen tends to get disrupted, and results in formation of smaller complexes. Lastly, anti-immunoglobulin antibodies such as rheumatoid factors (e.g., IgM anti-IgG antibodies) can become part of immune complexes, increasing their overall size, improving their ability to activate complement, and decreasing their abiltiy to induce phagocytosis [3]. As such, locally produced immune complexes in rheumatoid arthritis can augment the inflammatory process in the joints.

The ratio of antigen to antibody is a factor in determining the size of the resultant immune complex. Antigen excess, which can occur during chronic overexposure to viruses, bacteria, or tumor antigens, results in the formation of small, soluble complexes. Similarly, antibody excess, as occurs in disease states of deregulated immunoglobulin production, will also result in formation of small complexes. Immune complexes formed under these conditions are most commonly found in the circulation [4]. When there is an equivalence between antigen and antibody, large, cross-linked lattices are formed which are cleared at the highest rate.

When prolonged production of immune complexes occurs, the natural clearance systems become overloaded. In addition, levels of complement in the serum can be consumed. This is evident in a plethora of diseases where immune complexes play a role in pathogenesis. Some of the major disorders associated with circulating immune complexes are systemic lupus erythematosus, rheumatoid arthritis, and cancer. Indeed, assays for circulating immune complexes serve as a diagnostic and prognostic tool, especially in cancer, where patient sera contain tumor antigens in immune complexes, and their decreased levels may correlate with a favorable response to therapy [7]. Paradoxically, immune complexes can serve to protect the tumor, as it has been demonstrated that they are capable of suppressing cell-mediated tumor immunity [8, 9].

Immune Complex Interaction with B Lymphocytes

An immune complex, by virtue of its composition (antigen, antibody, complement) may interact with any cell which expresses an antigen receptor, Fc receptor, or complement receptor. B lymphocytes possess surface receptors for *all* of the constituents of an immune complex and are particularly susceptible to immune complex regulation. Surface immunoglobulin (sIg) is a specific antigen receptor which can transduce signals across the plasma membrane [10]. Extensive cross-linkage of sIg by so-called 'thymus (T)-independent' antigens can promote B cell activation. However, for the majority of antigens which are 'T-dependent', an additional signal is required with antigen binding. This is provided by lymphokines. Following antigen binding to the B lymphocyte, sIg initiates antigen internalization, which results in processing and presentation of fragments in association with class II MHC on the cell surface. The specificity of the interaction of an immune complex with a B lymphocyte is conferred by sIg. This is critical, because it allows the immune complex to enhance or down-regulate antibody production in an antigen-specific manner.

B lymphocytes have two types of receptors for the proteolytic cleavage products formed during the complement cascade. Most express CR1, a receptor for the initial cleavage fragments of C3 and C4, namely, C3b and C4b. Monoclonal antibody to human CR1 can enhance specific B cell differentiation [11]. Some B cells also express CR2, a receptor for the terminal cleavage products of C3 activation, C3dg, and C3g [12]. CR2 is exploited by the Epstein-Barr virus (EBV), which utilizes it as a portal of entry to infect human B lymphocytes. EBV and other CR2 ligands can induce B cells to proliferate. Expression of this receptor is restricted to a window of B cell maturation, from the immature stage to the point of activation. It disappears upon proliferation. Thus, immune complexes coated with fragments of C3 or C4 could have regulatory effects on B lymphocytes by acting as ligands for the complement receptors CR1 or CR2. In contrast, immune complexes composed of immunoglobulin isotypes which are not proficient at fixing complement, such as IgG4, would not be able to regulate B lymphocytes through the complement receptors.

Many cell types express receptors for the Fc portions of immunoglobulin, especially IgG and IgE. The FcγR, of which there are three types (FcγRI, FcγRII, FcγRIII), have been studied extensively on macrophages. These are distinguished based on their affinity for IgG. B lymphocytes constitutively express only the FcγRII [13]. The FcγRII expressed by B cells differs from the macrophage FcγRII, in that a cytoplasmic extension of the B cell receptor for IgG renders it incapable of mediating endocytosis [14]. This would eliminate the FcγRII from being involved in antigen presentation, making sIg unique in this aspect of B lymphocyte function. The only activity that has been well characterized for the FcγRII is its regulation of sIg-induced activation. This Fc-dependent regulation will be reviewed in more detail later in this article. B lymphocytes also express a low-affinity receptor for IgE, called CD23, which will not be considered in this review.

Immune complexes can regulate B lymphocytes, and depending on their composition, can suppress or enhance antibody production [15]. This regulation is antigen specific and involves binding of at least the antigenic component of the immune complex to sIg. Large immune complexes which contain IgG have been shown to be immunosuppressive, while large IgM-containing complexes actually enhance antibody production. It is hypothesized that the large IgG complex cross-links the FcγR and antigen receptor to such an extent that a 'negative signal' is delivered, which terminates the IgM and IgG primary response [16]. This is illustrated in figure 1. Occupancy of the complement receptor evidently cannot overcome this strong down-regulatory stimulus. On the other hand, as there are apparently no Fc receptors for IgM on the B cell, a negative signal is not delivered in association with sIg receptor occupancy by an immune complex composed of this isotype. Moreover, since IgM is an excellent inducer of the complement cascade, a stimulatory signal may be delivered through the B lymphocyte complement receptors. Indeed, the primary IgM and IgG

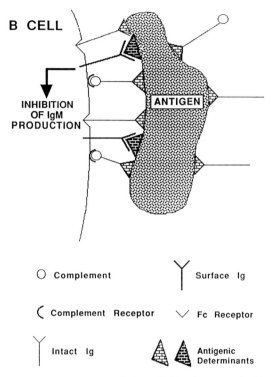

Fig. 1. All components of a large IgG-antigen complex bind the B cell surface resulting in a 'negative signal' which blocks differentiation to IgM-secreting plasma cells. Binding of complement cannot overcome this strong negative signal. If this complex were small, or if it were coated with antibodies of the IgM class, FcR cross-linkage would not occur. Instead, ligation of sIgM and the complement receptors would result in an enhanced IgM response.

responses to particulate antigens are enhanced by IgM antibodies [15, 17]. This is a potential mechanism by which IgM rheumatoid factors can deregulate the immune response. By recognizing IgG, rheumatoid factor may block a negative signal mediated by IgG-containing immune complexes. Furthermore, the IgM anti-IgG immune complex could bind complement and enhance B lymphocyte antibody production. Another level of complexity exists that depends on the size of the antigen-antibody complex. Small, soluble IgG immune complexes do not down-regulate antibody production, but instead, stimulate B lymphocytes [16]. Although this is not fully understood, it is postulated that a small complex does not contain enough antibody or antigen to cross-link the critical number of FcR and sIg required to transduce a negative signal. Stimuli below the threshold for inhibitory signaling may also allow complement to send an enhancing signal through the complement receptor that promotes antibody production.

Tolerance in the B Cell Repertoire Involves Two Mechanisms

Clonal Deletion versus Clonal Anergy

The importance of self tolerance in the immune system is to prohibit the recognition and subsequent destruction of endogenous or autoantigens. Discrimination between self and non-self structures was thought to be a mechanism which occurred developmentally early in lymphocyte ontogeny where an immature cell 'seeing' self antigen would be clonally deleted. While this mechanism exists, the central problem of this paradigm of tolerance is that not all self antigens can be encountered at sites of B cell maturation such as the bone marrow. The immune system must, therefore, have a second mechanism for inducing tolerance in peripheral and mature B cell populations.

The first mechanism by which B lymphocytes can be rendered tolerant to self antigens occurs in the bone marrow and is characterized by immature cells being clonally deleted after exposure to membrane-bound antigens in high concentrations. The seminal experiment to demonstrate this phenomenon employed transgenic mice that carried a gene encoding class I MHC antigens. Throughout ontogeny, B cells with receptors for this membrane-bound antigen considered it a 'self' molecule and were deleted in the periphery [18].

The second mechanism of tolerance affects mature B cells which encounter antigen in the periphery. In this case, persistent contact with antigen present in soluble form results in a functionally anergic B lymphocyte that has decreased expression of IgM, but continued expression of IgD [19]. Clonal anergy as a mechanism of B cell tolerance was elegantly demonstrated utilizing a double transgenic system in which mice expressing anti-hen egg lysozyme (HEL) immunoglobulin were crossed with mice that constitutively expressed HEL. Flow cytometric analysis showed that the anti-HEL-secreting B cells were present in the periphery but would not produce antibodies to challenge with a normally immunogenic form of HEL [20].

Interaction of immune complexes with B lymphocytes can also induce 'unresponsiveness'. In these cases, unresponsiveness resembles anergy since these cells are functionally silent to subsequent challenge with antigen. However, in contrast to circumstances that result in anergy, B cells encounter antigen in the form of antigen/antibody complexes. Another critical difference is that antigen may be present in lower concentrations and exposure may be of short duration. Immune complex-mediated inhibition is, therefore, distinct from the aforementioned mechanisms of tolerance, and has typically been defined (particularly in vitro) as the failure of B cells to differentiate into IgM-secreting plasma cells.

Models of B Cell Unresponsiveness Mediated by Immune Complexes and Fc Receptors

Immune Complex Induction of Unresponsiveness in vivo

As discussed earlier, immune complexes can form by interaction of circulating antibodies with antigen. These complexes can be potent immunoregulators in vivo, and have long been postulated as having a negative feedback effect on further production of antibody [1]. Previous work in this area has proved confusing in that immune complexes induce unresponsiveness in some cases, while in others the humoral response is enhanced.

One of the most powerful examples of immune complex-mediated induction of unresponsiveness is the prevention of an antibody response to the Rh antigen on erythrocytes. At parturition, Rh$^-$ mothers become exposed to Rh$^+$ fetal erythrocytes. Usually, an Rh$^-$ mother will mount an antibody response to these Rh$^+$ cells and with a subsequent Rh$^+$ fetus, the transplacental movement of IgG antibodies causes destruction of fetal red blood cells. Passively administered IgG antibodies specific for the D antigen of the Rh antigen complex given to women after giving birth effectively prevents hemolytic disease of subsequent Rh$^+$ newborns [15, 21]. The mechanism of this response was not attributed to simple clearance of the IgG-erythrocyte complex from the bloodstream, but was postulated as an antigen-specific, antibody-mediated mode of suppression.

A number of murine in vivo systems have been developed to further examine this phenomenon. Briefly, antitrinitrophenyl (TNP) monoclonal IgG antibodies were injected intravenously, followed by injection of TNP-conjugated sheep red blood cells (SRBC). Spleens from these mice were then removed, challenged with SRBC in vitro and IgM SRBC-specific plaque-forming cells (PFC) measured. Only 5 out of 9 monoclonal antibodies tested induced unresponsiveness. The important requirement for suppression was the ability of IgG antibodies to efficiently bind to SRBC with high affinity. There was no correlation between suppressive ability and antibody isotype, binding to protein A or ability to activate complement [16].

Another experimental in vivo system which demonstrated immune complex-mediated unresponsiveness involved injection of preformed antigen/antibody complexes consisting of PnC (a cell wall polysaccharide antigen extracted from *Streptococcus pneumoniae*) and TEPC-15 (a myeloma protein of BALB/c origin which binds PnC). On day 5, spleens were removed and PnC-specific PFC were determined. It was demonstrated that only injection of complexes formed in antibody excess could suppress the PnC-specific antibody response [22]. Furthermore, when an immunogenic dose of PnC was injected simultaneously or 1 day after injection of immune complexes (in antibody excess), the PnC-specific PFC response was suppressed. These findings suggest that the ratio of antigen to antibody in an immune complex determines whether the complex will be immunosuppressive or not, with those complexes formed in antibody excess inducing unresponsiveness, while complexes that are in antigen excess will enhance a response [22].

Immune Complex Induction of Unresponsiveness in vitro

In order to examine the mechanism of immune complex-mediated unresponsiveness, in vitro models must be developed to control for the influence of other effectors such as T lymphocytes and accessory cells. Not all B cells are sensitive to immune complex-mediated unresponsiveness. One model system indicating this, consisted of purified B cells cultured with Con A-stimulated spleen cell supernatant as a source of lymphokines, ovalbumin/anti-ovalbumin immune complexes (in antigen excess) and F(ab')$_2$ anti-μ as a source of antigen. As the amount of surface Ig cross-linked by F(ab')$_2$ anti-μ was increased, inhibition of the polyclonal PFC response also increased [23]. The antigen-specific response was also inhibited when B cells were cultured with immune complexes and TNP-bacterial lipopolysaccharide (LPS) as a source of antigen. TNP-specific PFC responses were inhibited 50–55% [23]. Inhibition of the antibody response was not complete, suggesting that a subset of B lymphocytes was resistant to immune compiex-mediated unresponsiveness. When these experiments were performed on different B cell populations, it was shown that resting B lymphocytes were more susceptible to an inhibitory signal than larger, activated B cells [24]. This is in agreement with our own findings in which only small resting B lymphocytes exhibited complete unresponsiveness after treatment with immune complexes as compared to only partial inhibition in the large activated fraction of B cells [25].

Further studies with another in vitro system demonstrated that intact IgG induced unresponsiveness to an antigen-specific challenge. B cells were first treated with anti-Ig (intact or F(ab')$_2$ fragment) for 20 h and subsequently challenged with either a polyclonal B cell activator, LPS, or the antigen, fluoresceinated *Brucella abortus* (FL-BA). Surprisingly, the LPS-induced responses were inhibited even with F(ab')$_2$ antibodies, suggesting that sIg cross-linking is all that is required for unresponsiveness. In contrast, however, only intact IgG can inhibit antigen-specific antibody responses. This suggests a role for the Fc receptor in mediating unresponsiveness [26]. To further investigate the role of the Fc receptor in inducing unresponsiveness, our laboratory utilized ligands which bind B cell sIg or FcR. Briefly, heat-aggregated ligands for sIg (FL-conjugated BSA) and FcR (IgG2b) were added to cultures of FL-specific B cells. After challenge with FL-BA, the anti-FL PFC response was reduced by 70% [27]. Thus, while independent cross-linking of sIg and FcR is inhibitory, some B cells still responded by differentiation to PFC. Complete inhibition of the antibody response was only attained when aggregated FL-conjugated 2.4G2 was used. This ligand consisted of a hapten-specific group (FL) and the monoclonal antibody to the FcγRII-B (2.4G2) and bound both the sIg and Fc receptor and induced maximum cross-linking [27]. These data clarify some of the reports mentioned earlier which utilized preformed immune complexes, where the composition of the immune complex or the ratio of antigen to antibody led to different results. Immune complex-mediated unresponsiveness requires maximal cross-linking of sIg and FcR. For example, an immune complex that is formed in vivo can be composed of an isotype that does not bind well to FcR [27, 28]. Alternatively, there may not be enough antibody present to sufficiently cross-link the sIg and Fc receptor [22, 24]. Enhancement may be observed in situations where maximal cross-linking is not achieved.

Mechanisms of Unresponsiveness

The Role of the Fc Receptor

The aforementioned studies suggest a central role for the FcR in mediating immune complex-induced unresponsiveness. Interestingly, the FcγRII and sIg co-cap on the B cell surface only when both receptors are cross-linked by their respective ligands [29, 30]. This model mimics antigen/antibody complexes since both antigen, which binds sIg, and antibody which binds FcR are present. Other data which favor the role of FcR in immune complex-mediated negative signaling utilize monoclonal antibodies to the Fc portion of mouse IgG. Goat antibody to the Fc portion of mouse IgG reconstitutes the response of splenic B cells to T-dependent antigens when present early in the response [31]. If intact IgG is then added to this system, a down-regulation of antibody-forming cells is observed [31]. Monoclonal antibodies to the Fc receptor itself also block immune complex-mediated unresponsiveness. The monoclonal antibody 2.4G2 specifically inhibits the suppression mediated by intact IgG in splenic cultures challenged with SRBC [32]. Another finding implicating FcR binding as the regulatory signal in immune complex-mediated unresponsiveness utilizes rheumatoid factors to block the Fc portion of IgG antibodies. When rheumatoid factors were incubated with immune complexes and then added to splenocyte cultures, the suppression mediated by immune complexes alone was abolished [33].

Signaling through sIg and the FcR

Cross-linking of sIg on mature B cells initiates an intracellular cascade of second messengers which ultimately leads to activation. The first step in this process is the hydrolysis of phosphoinositols into inositol 1,4,5-triphosphate (IP_3) (which releases Ca^{2+} from intracellular stores) and diacylglycerol which activates protein kinase C (fig. 2). Unfortunately, less is known about the biochemical events induced after FcR cross-linking.

Studies with $F(ab')_2$ fragments and intact anti-Ig reveal that both antibodies induce comparable inositol triphosphate (IP_3) release for the first 30 s following binding. However, while $F(ab')_2$ induces a sustained IP_3 signal, the increase observed with intact Ig quickly dissipates [34]. When both antibodies were mixed, $F(ab')_2$-induced IP breakdown was inhibited. This inhibition was attributed to signaling through the FcR since antibodies blocking FcR could block inhibition [34].

Similar results were obtained in Ca^{2+} signaling experiments. Both $F(ab')_2$ and intact anti-Ig induce comparable increases in intracellular Ca^{2+} levels. Ca^{2+} levels in $F(ab')_2$-stimulated cells peak at approximately 5 min and remain elevated for hours. Conversely, treatment with intact anti-Ig reached similar levels of elevation, but decreased to baseline within minutes [35]. Again, intact antibody and $F(ab')_2$ fragment added together decreased the calcium response and addition of FcR antibodies abrogated the inhibition of signal seen with intact anti-Ig. Transfection of an FcR-negative lymphoma with B cell FcγRII-B1 gave similar results when sIg and FcR were cross-linked; the initial Ca^{2+} response observed with sIg alone was decreased when FcR was cross-linked simultaneously with sIg. Addition of the FcR antibody 2.4G2 to intact Ig prevented the Ca^{2+} decay [14].

Antigen receptors are thought to be associated with a polyphosphoinositide-specific phosphodiesterase via a yet uncharacterized G protein termed Gp [36]. Stimulating B cells with intact anti-Ig uncouples sIg receptors from Gp [36] (fig. 2, dashed arrow). This finding provides a biochemical basis for the suppression of antibody responses mediated by immune complexes. Since occupancy of both receptors are required to induce a negative signal, the uncoupling of the G protein from sIg, by FcR occupancy, may be the necessary signal to inhibit phosphatidylinositol-4,5-bisphosphate (PIP_2) hydrolysis and Ca^{2+} elevation.

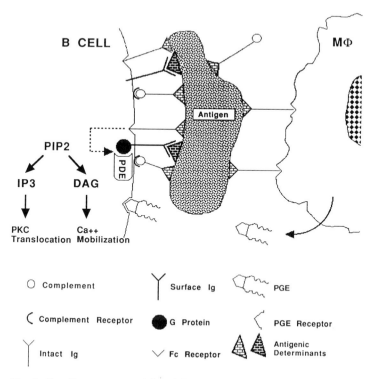

Fig. 2. Signaling events precipitated by sIg and Fc receptor cross-linking. Immune complexes can suppress IgM antibody responses by simultaneously cross-linking sIg and Fc receptors. Normally, engagement of sIg initiates the breakdown of inositol phosphates leading to activation. When FcR are concurrently cross-linked, sIg becomes uncoupled from its G protein, thus deactivating the polyphosphoinositide-specific phosphodiesterase (PDE) (dashed arrow). Engagement of FcR on macrophages stimulates production of PGE which potentiates immune complex-mediated unresponsiveness.

Effect of the Macrophage Product Prostaglandin E_2 on Immune-Complex-Induced Unresponsiveness

Treatment of human monocytes with immune complexes or Fc fragments stimulate a burst of prostaglandin E (PGE) production [37]. Murine splenic macrophages also release PGE after treatment with Fc fragments, but not $F(ab')_2$ fragments, in a dose-dependent manner [38]. Moreover, PGE synthase inhibitors such as indomethacin inhibit PGE production stimulated by Fc fragments [37, 38].

Addition of exogenous PGE_2 to splenic cultures treated with Fc fragments inhibited total immunoglobulin production [38]. Since splenic B cells contact immune complexes in areas that also contain macrophages, it was postulated that immune complexes would have a dual role in inducing unresponsiveness in vivo by (1) stimulating macrophages to secrete prostaglandins, and (2) inhibiting antibody production by cross-linking sIg and FcR (fig. 2). Our laboratory also developed an in vitro murine model system to investigate the role of PGE_2 in immune complex-mediated unresponsiveness. First, immune complexes consisting of FL-keyhole limpet hemocyanin (KLH) and anti-FL IgG2b antibodies (FL-KLH/anti-FL), formed at equivalence, were added to accessory cell monolayers. Spleen cells were then added for 24 h and subsequently challenged with FL-BA. Antibody-secreting cells were measured in a direct PFC assay. Interestingly, macrophages pulsed with immune complexes decreased the number of FL-specific antibody-secreting cells, while immune complex pulsed lymphoid dendritic cells (LDC) augmented the anti-FL antibody response. Moreover, the inclusion of indomethacin in cultures of macrophages pulsed with immune complexes alleviated the decreased antibody response and exogenously added PGE_2 restored it [39].

The inhibitory effect of PGE_2 was further analyzed by using a defined population of hapten-specific B lymphocytes which were cultured in an accessory cell-free system that included immune complexes and PGE_2. Immune complexes at high concentrations blocked differentiation of B cells into IgM-secreting plasma cells. However, with the addition of PGE_2, IgM-secreting cells were blocked at concentrations of immune complexes that are normally not inhibitory [39]. Thus, PGE_2 can sensitize B cells to the inhibitory effects of immune complexes.

A model of events which occur in vivo can be proposed from these data in which immune complexes contact B cells either in soluble form of bound to macrophages. In germinal centers, 'tingible body' macrophages and B cells are in close proximity and the immune complex can act on both of these cell types either directly by simultaneously cross-linking sIg and FcR, or indirectly by stimulating macrophages to produce PGE which can synergize with antigen/antibody complexes to block B cell differentiation to IgM plasma cells (fig. 2).

Immune Complexes Inhibit Only Some Aspects of B Cell Responsiveness

In the previous section, unresponsiveness was typically measured by the failure of a B cell to differentiate into an IgM-secreting plasma cell. Although immune complex-treated B cells are unresponsive in this respect, these B cells may respond via alternate pathways. With this in mind, we investigated whether B cell activation to antigen-specific or polyclonal stimuli was affected. Purified small dense resting splenic murine B cells pulsed with immune complexes (monoclonal IgG2b anti-FL and FL-KLH) for 24 h were challenged with either FL-BA or LPS. In both the antigen-specific system (FL-BA) and the polyclonal system (LPS), pulsing with immune complex had no affect on B cell proliferation [25]. This indicates B cells are activated by antigen despite immune complex treatment and reinforces the theory that B cells may respond via a non-IgM antibody pathway. We speculated that immune complex treatment promotes memory B cell formation [25]. Memory B cells are responsible for secondary responses which are characterized by immunoglobulin class switching. Therefore, analysis of these results suggested that immune complexes inhibit differentiation to IgM plasma cells as a prelude to an immunoglobulin isotype switch.

Induction of Antibody Secretion by Immune Complexes

In vivo, the primary antibody response initiates during the period immediately following antigen introduction. Virgin B cells bind free antigen directly and activation occurs in the extrafollicular areas of lymphoid organs. Some of these activated B cells remain in the periphery and differentiate into plasma cells, whereas others migrate to the follicles. Within the follicles, some B cells physically interact with follicular dendritic cells. Follicular dendritic cells trap immune complexes on the surface of their cell membranes and present these complexes to B cells [40]. B cells which associate with follicular dendritic cells have been observed to differentiate into memory B cells [41]. The primary response culminates with antigen-specific IgM secretion approximately 1 week following antigen introduction [42]. In contrast, secondary responses are characterized by secreted antibody of the IgG, IgE, or IgA isotypes. Secondary responses initiate with the activation of memory B cells. Memory B cells bind immune complexes, internalize them and present processed peptide fragments, in association with class II MHC, to T helper (T_H) cells. Binding of these processed fragments by T_H cells results in their secretion of cytokines. In turn, binding of immune complexes, physical interaction with T_H cells and binding of cytokines all contribute to B cell activation and differentiation. Activated B cells enter the marginal zones, recirculate in the lymph and differentiate into plasma cells. Secondary responses occur from 1 week after antigen introduction throughout the lifetime of the organism [43].

Immune Complexes and IFN-γ Promote
Hapten-Specific IgG2a Secretion

As mentioned, treatment of B cells with immune complexes can block differentiation to antigen-specific IgM-secreting plasma cells, yet these B cells are activated by the antigen. These activated B cells may differentiate into memory cells, providing a foundation for secondary responses. Once restimulated with antigen and appropriate accessory signals, these memory B cells may differentiate into plasma cells secreting antigen-specific antibody of a non-IgM isotype. One hypothesis to explain the decrease in IgM plasma cells mediated by immune complexes was that this was a critical event in the process of immunoglobulin class switching. Using in vitro systems developed to examine isotype switching, cytokines were shown to specify the immunoglobulin isotype which was subsequently expressed. The system which best exemplifies cytokine-specific selection of isotypes is murine splenic B cells stimulated with LPS and cytokines. LPS alone induces IgM and IgG3 expression, whereas LPS and IL-4 induce B cells to switch to IgG1 and IgE while suppressing expression of IgM, IgG2a, IgG2b, and IgG3. In contrast, LPS and IFN-γ act in an antagonistic manner to IL-4 and induce synthesis of IgG2a, while suppressing IgM, IgG1, in addition to IgG2b, IgG3, and IgE [44]. Furthermore, LPS and TGF-β induce B cells to switch to IgA [45]. In analogy with this murine system, we reasoned that addition of IFN-γ to splenic B cells treated with an immune complex would result in antigen-specific IgG2a secretion. Small dense resting splenic B cells were first treated with IFN-γ for 24 h and then pulsed with immune complexes (monoclonal IgG2b anti-FL and FL-KLH). These B cells were then challenged with an optimal dose of FL-BA. Culture supernatant was removed after 6 days and the anti-FL IgM and anti-FL IgG2a responses determined via enzyme linked immunoabsorbent assay (ELISA). B cells incubated with immune complexes and challenged with FL-BA secreted 5-fold less IgM (as compared to FL-BA-treated cells (−)) and undetectable levels of IgG2a (table 1). Interestingly, the combination of immune complex, FL-BA, and IFN-γ increased anti-FL IgG2a production at least 10-fold whereas IgM production was comparable to those treated with immune complex alone (table 1). Furthermore, in PFC assays, immune complexes and IFN-γ treatment suppressed IgM anti-FL PFC responses more than immune complex treatment alone, implying that IFN-γ sensitizes B cells to the effects of immune complex and promotes secretion of another isotype. In sum, culturing B cells with immune complexes inhibits B cell differentiation to anti-FL IgM-producing plasma cells and this inhibition appears to be part of a response promoting antibody secretion of another isotype [46]. Independent investigations using

Table 1. Treatment with immune complexes and IFN-γ or IFN-γ and PGE$_2$ promote anti-FL IgG2a antibody production

Treatment[1]	Supernatant levels of anti-FL antibody, μg/ml	
	IgM	IgG2a
Not pulsed[2]	5.5	<0.1
Immune complex (1 μg/ml)	0.9*	<0.1
IFN-γ (50 U/ml)	6.1	<0.1
Immune complex + IFN-γ	0.8*	1.1*
PGE$_2$ (0.1 μM)	5.3	<0.1
IFN-γ + PGE$_2$	0.9*	1.4*
Immune complex + PGE$_2$	0.7*	<0.1
Immune complex + IFN-γ + PGE$_2$	1.5*	0.9*

* Values indicate anti-FL responses which were statistically different ($p < 0.05$) from the corresponding values obtained with untreated (buffer) controls.

[1] Small resting B cells were incubated for 24 h with IFN-γ, PGE$_2$ or media and then pulsed with immune complexes (a monoclonal IgG2b anti-FL antibody and FL-KLH constructed at equivalence), for an additional 24 h.

[2] Treatment was not pulsed with immune complexes. B cells were then challenged with FL-BA. After 6 days' incubation, culture supernatants were removed and anti-FL IgM or IgG2a concentrations were determined by ELISA. In the absence of FL-BA, no IgM or IgG2a anti-FL responses were detected.

LPS and IFN-γ implicated that IFN-γ does induce a class switch to IgG2a [47, 48]. However, controversy surrounds these conclusions and the converse has also been proposed; IFN-γ does not induce a class switch but enhances IgG2a secretion from B cells already committed to IgG2a [49]. Our current data does not distinguish whether immune complexes trigger a class-switching mechanism or promote differentiation of B cells already committed to IgG2a secretion.

PGE$_2$ Potentiates IFN-γ, but not Immune Complex and IFN-γ Induction of Hapten-Specific IgG2a Production

As mentioned earlier, PGE$_2$ potentiates the decrease in anti-FL IgM responses induced by immune complexes [39]. A decrease in anti-FL IgM responses appears indicative of a potential class switch. Moreover, PGE$_2$ promotes isotype switching in polyclonal systems [44, 50]. Therefore, it was hypothesized that addition of PGE$_2$ to immune complex and IFN-γ-treated cultures would increase IgG2a production more than immune complex and

IFN-γ treatment alone. Likewise, anti-FL IgM responses induced by immune complex and IFN-γ should be decreased more in the presence of PGE$_2$. In order to test this, B cells were first treated with IFN-γ and/or PGE$_2$ for 24 h and subsequently treated as above (after 24 h, B cells were pulsed with immune complexes (monoclonal IgG2b anti-FL and FL-KLH) and then challenged with an optimal dose of FL-BA). Responses were determined by PFC and by ELISA. Curiously, immune complex, IFN-γ, and PGE$_2$ treatment did increase anti-FL IgG2a secretion, however not more than immune complex and IFN-γ treatment alone (table 1). Furthermore, immune complex, IFN-γ, and PGE$_2$ treatment diminished anti-FL IgM responses less than immune complex and PGE$_2$ treatment (table 1). It is clear PGE$_2$ does not enhance immune complex and IFN-γ-induced IgG2a production. However, treatment with IFN-γ and PGE$_2$ did enhance anti-FL IgG2a secretion while decreasing anti-FL IgM responses (table 1). Hence, IFN-γ and PGE$_2$ are also potent stimuli for promoting FL-specific IgG2a production at the expense of IgM production. Since PGE$_2$ promotes anti-FL IgG2a responses independently of immune complexes, it appears two pathways for promoting anti-FL IgG2a responses exist [46].

cAMP-Dependent and cAMP-Independent Mechanisms Promoting Immune Complex-Induced FL-Specific IgG2a Production

Small dense resting murine B cells treated with PGE$_2$ or cholera toxin demonstrate elevated levels of intracellular cAMP [46]. Therefore, it was hypothesized that a rise in cAMP potentiates IgG2a synthesis by B cells. Cholera toxin and dibutyryl cAMP also elevate cAMP and potentiate IgG2a production in B cells treated with IFN-γ and FL-BA, while further decreasing IgM responses. In contrast, PGF2α, which does not elevate cAMP, had no effect [46]. Moreover, agents which block cAMP production inhibit IgG2a promotion induced by IFN-γ and PGE$_2$. The cAMP antagonist, RpcAMP, is a competitive inhibitor of protein kinase A which cannot be hydrolyzed by common cyclic phosphodiesterases [51, 52]. Pretreatment with RpcAMP inhibited IgG2a production by at least 80% [25]. These results clearly indicate that a cAMP signal occupies a central role in promoting IgG2a production in response to IFN-γ and PGE$_2$. In another class-switching system, treating splenic murine B cells with LPS and IL-4, PGE$_2$ potentiates the class switch from IgM to IgE [53]. At the mRNA level, it was demonstrated that PGE$_2$ increases the frequency of switching events from the μ to the ε C$_H$ gene via a cAMP-dependent mechanism [54]. In contrast, little is known of the intracellular signals induced by immune complex and/or IFN-γ binding to receptors on B cells. We have not observed changes in intracellular cAMP levels following immune complex and/or IFN-γ binding to B cells [25]. These results agree with those of Cambier and Ransom [55], who also did not observe changes in cAMP following binding of IFN-γ to murine splenic B cells. It is not known precisely which intracellular signaling pathways transmit IFN-γ-specific signals promoting IgG2a production nor immune complex-specific signals which inhibit IgM production. However, we know two pathways exist for inducing anti-FL IgG2a murine splenic B cell responses. The first involves IFN-γ and PGE$_2$ induction of IgG2a responses and requires cAMP as a second messenger. In the second pathway, immune complexes and IFN-γ act in a cAMP-independent manner.

Interestingly, it appears both cAMP-dependent and cAMP-independent pathways could occur in vivo. Numerous studies have indicated that differentiation of B cells into IgG2a-secreting plasma cells in vivo requires physical (cognate) interaction between T and B cells [56]. It has also been demonstrated that cognate interaction between B and T$_H$ cells enhances class switching to IgE by B cells treated with LPS and IL-4 [57]. Cognate interaction between B and T$_H$ cells increases intracellular cAMP levels within the B cells [58]. Likewise, binding of B cell class II MHC by antibody induces a rise in intracellular cAMP [59]. We have demonstrated that anti-class II antibody potentiates IgG2a production by B cells treated with LPS and IFN-γ via a cAMP-dependent mechanism [60]. Thus, when a B cell presents processed immune complex fragments in the context of class II MHC to T$_H$ cells in vivo, a rise in the level of B cell intracellular cAMP would occur. If IFN-γ is concomitantly bound by the B cell (secretion of IFN-γ by the T$_H$ cell would promote this), increased levels of intracellular cAMP would potentiate synthesis of IgG2a (fig. 3a). This cAMP-dependent pathway could serve as a common mechanism for enhancing isotype switching events in vivo due to the requirement for cognate interaction between B cells and T$_H$ cells. Which isotype is selected would depend on the cytokine secretion profile of the T$_H$ cell with which the B cell is interacting. Factors which increase B cell intracellular cAMP levels could also contribute to isotype switching in vivo. For example, PGE is secreted by accessory cells such as macrophages, follicular dendritic cells, fibroblasts, and epithelial cells in response to a variety of stimuli including cross-linking of Fc receptors and complement components [61–65]. B cell association with these PGE-secreting accessory cells in the spleen and lymph nodes provides an opportunity for B cell exposure to high local concentrations of PGE$_2$. High PGE$_2$ titers in humans have been correlated with a number of disorders associated with excessive antibody production and subsequent immune complex deposition such as, Hodgkin's disease [66, 67], rheumatoid arthritis [68], hyper-IgE syndrome [69] and periodon-

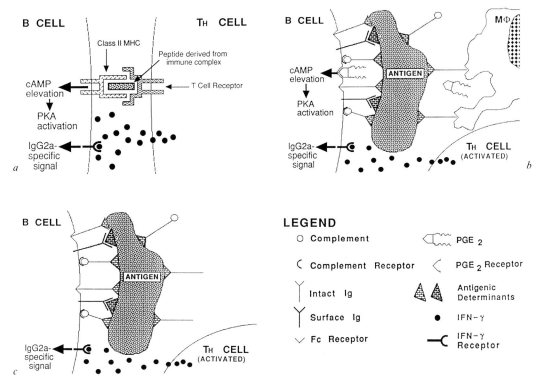

Fig. 3. Models for cAMP-dependent and cAMP-independent promotion of IgG2a secretion in vivo by B cells stimulated with immune complexes and IFN-γ. *a* cAMP-dependent promotion by interaction of B cell peptide-bound class II MHC with the T cell receptor. *b* cAMP-dependent promotion by B cell binding of macrophage-derived PGE₂. *c* cAMP-independent promotion.

tal disease [70, 71]. Therefore, in the absence (fig. 3b) or in conjunction with cognate interaction between B cells and T_H cells, binding of PGE_2 by B cells elevates intracellular cAMP levels which could potentiate class switching induced by cytokines. In contrast, if an activated B cell binds only immune complex and IFN-γ (the B cell does not bind factors which elevate cAMP such as T_H cells and PGE_2), then a cAMP-independent class switching pathway could occur (fig. 3c). From an evolutionary perspective, one can rationalize the maintenance of both pathways as a need for complimentary pathways each of which are required to ensure IgG2a antibody production in sufficient quantities to ward off infection during the lifetime of an organism.

Conclusion

Experimental and clinical observations reviewed in this article demonstrate that immune complexes are potent regulators of the humoral immune response. Review of the literature reveals that diverse systems have been employed to study the effects of immune complexes on B cell responsiveness. Before comparing results obtained in these systems, it is important to note that a multitude of factors influence the effect of immune complexes on B cell responsiveness. Differences between any of these factors may limit meaningful comparisons. For example, physiologic features of immune complexes themselves are important. These features include the size and valence of the antigen, size and valence of the responding immunoglobulin, and whether complement associates with the complex. Moreover, whether immune complexes form under conditions of antibody excess, antigen excess, or equivalence is influential. A rewarding reductionist approach is under way which is aimed at understanding signaling pathways and gene expression induced by binding of surface Ig, FcR, or complement receptors individually. Much has been learned about binding of membrane Ig, however more investigation of FcR or complement receptor binding is required. Only after understanding pathways induced by binding of individual receptors will we fully appreciate events triggered by immune complex binding of all three receptor types.

Accessory cells and their secreted products also influence B cell responsiveness. We presented an experimental system which uses a single type of immune complex in effort to avoid variability in the factors mentioned above. In this system, immune complexes can inhibit antigen-specific primary IgM responses and render B cells unresponsive in this respect. PGE_2, secreted by macrophages as well as other accessory cell types, enhances inhibition via a cAMP-dependent mechanism. Although these B cells do not synthesize IgM, they are activated by antigen and can proliferate. Moreover, addition of cytokines can induce these B cells to secrete antibody of a non-IgM isotype. It appears that the immune complex-mediated decrease in antigen-specific IgM production may be an integral component in an immunoglobulin class-switching pathway. However, this pathway needs to be further investigated. Since B cells encounter antigen as an immune complex in vivo, possible relationships between immune complex binding and secondary immune response events, such as isotype switching and affinity maturation, should be investigated. Moreover, accessory cell factors which elevate intracellular cAMP, such as PGE_2, may potentiate immune complex-induced secondary immune response events. Future investigation requires additional well-defined experimental systems which definitively examine these issues. Understanding what signaling pathways and genes are expressed in response to an immune complex will allow us to better predict the nature of the subsequent B cell response. More importantly, comprehension of these events could provide the rationale for insightful therapies designed to manipulate B cell responsiveness to immune complexes. Such therapies could be used to more effectively treat individuals with immune complex-associated disease.

Acknowledgements

The authors wish to thank Dr. S. Derdak and D. Phipps for reviewing the manuscript. This research was supported by the US Public Health Service Grants CA42739, CA55305, CA11198, and T32-AI07285. This publication is number 96 of the Immunology and Immunotherapy Division of the Cancer Center.

References

1   Tew, J.G.; Phipps, R.P.: Cyclic antibody production: Role of antigen retaining follicular dendritic cells and antibody feedback regulation; in DeLisi, C.; Hiernaux, J.R.J. (eds): Regulation of Immune Response Dynamics. Boca Raton, CRC Press, 1982, vol. 1, pp. 27–42.

2   Salinas, A.S.; Wee, K.H.; Silver, H.K: Clinical relevance of immune complexes, associated antigen, and antibody in cancer; in Salinas, FA.; Hanna, G.H. Jr. (eds): Immune Complexes in Human Cancer. Contemporary Topics in Immunobiology. New York, Plenum Press, 1985, vol. 15, pp. 55–109.

3   Lamers, M.C.: Factors influencing the development of immune-complex diseases. Allergy 36: 527–535 (1981).

4   Espinoza, L.R.: Assays for circulating immune complexes; in Espinoza, L.R.; Osterland, C.K. (eds): Circulating Immune Complexes: Their Clinical Significance. Mount Kisco, Futura, 1983, pp. 21–49.

5   Izui, S.; Lambert, P.S.; Miescher, P.A.: In vitro demonstration of a particular affinity of glomerular basement membrane and collagen for DNA. A possible basis for local formation of DNA-anti-DNA complexes in systemic lupus erythematosus. J. Exp. Med. 144: 428–443 (1976).

6   Koyama, A.; Inage, H.; Kobayashi, M.; Nakamura, H.; Narita, M.; Tojo, S.: Effect of chemical modification of antigen on characteristics of immune complexes and their glomerular localization in the murine renal tissues. Immunology 58:535–540 (1986).

7   Dorval, G.; Pross, H.: Immune complexes in cancer; in Espinoza, L.R.; Osterland, C.K. (eds): Circulating Immune Complexes: Their Clinical Significance. Mount Kisco, Futura, 1983, pp. 161–177.

8   Hellstrom, I.; Sjogren, H.O.; Warner, G.; Hellstrom, K.E.: Blocking of cell-mediated tumor immunity by sera from patients with growing neoplasms. Int. J. Cancer 7: 226–237 (1971).

9   Heppner, G.H.; Stolbach, L.; Byrne, M.; Cummings, F.J.; McDonough, E.; Calabresi, P.: Cell-mediated and serum blocking reactivity to tumor antigens in patients with malignant melanoma. Int. J. Cancer 11: 245–260 (1973).

10 Cambier, J.C.; Campbell, K.S.: Membrane immunoglobulin and its accomplices: New lessons from an old receptor. FASEB J. 6: 3207–3217 (1992).

11 Weiss, L.; Delfraissy, J.F.; Vazquez, A.; Wallon, C.; Galanaud, P.; Kazatchkine, M.D.: Monoclonal antibodies to the human C3b/C4b receptor (CR1) enhance specific B cell differentiation. J. Immunol. 138: 2988–2993 (1987).

12 Cooper, N.R.; Moore, M.D.; Nemerow, G.R.: Immunobiology of CR2, the B lymphocyte receptor for Epstein-Barr virus and the C3d complement fragment. Annu. Rev. Immunol. 6: 85–113 (1988).

13 Gergely, J.; Sarmay, G.: B-cell activation-induced phosphorylation of FcγRII: A possible prerequisite of proteolytic receptor release. Immunol. Rev. 125: 5–19 (1992).

14 Amigorena, S.; Bonnerot, C.; Drake, J.R.; Choquet, D.; Hunziker, W.; Guillet, J.G.; Webster, P.; Sautes, C.; Mellman, I.; Fridman, W.H.: Cytoplasmic domain heterogeneity and functions of IgG Fc receptors in B lymphocytes. Science 256: 1808–1812 (1992).

15 Heyman, B.: The immune complex: Possible ways of regulating the antibody response. Immunol. Today 11: 310–313 (1990).

16 Wiersma, E.J.; Coulie, P.G.; Heyman, B.: Dual immunoregulatory effects of monoclonal IgG antibodies: Suppression and enhancement of the antibody response. Scand. J. Immunol. 29: 439–448 (1989).

17 Henry, C.; Jerne, N.K.: Competition of 19S and 7S antigen receptors in the regulation of the primary immune response. J. Exp. Med. 128: 133–152 (1968).

18 Nemazee, D.A.; Buerki, K.: Clonal deletion of B lymphocytes in a transgenic mouse bearing anti-MHC class I antibody genes. Nature 337: 562–566 (1989).

19 De Franco, A.L.: Tolerance: a second mechanism. Nature 342: 340–341 (1989).

20 Goodnow, C.C.; Crosbie, J.; Jorgensen, H.; Brink, R.A.; Basten, A.: Induction of self-tolerance in mature peripheral B lymphocytes. Nature 342: 385–391 (1989).

21 Pollack, W.; Ascari, W.Q.; Crispin, J.F.; O'Connor, R.R.; Ho, T.Y.: Studies on Rh prophylaxis. II. Rh immune prophylaxis after transfusion with Rh-positive blood. Transfusion 11: 340–344 (1971).

22 Caulfield, M.J.; Shaffer, D.: Immunoregulation by antigen/antibody complexes. I. Specific immunosuppression induced in vivo with immune complexes formed in antibody excess. J. Immunol. 138: 3680–3683 (1987).

23 Uher, F.; Dickler, H.B.: Cooperativity between B lymphocyte membrane molecules: Independent ligand occupancy and cross-linking of antigen receptors down-regulates B lymphocyte function. J. Immunol. 137: 3124–3129 (1986).

24 Uher, F.; Lamers, M.C.; Dickler, H.B.: Antigen/antibody complexes bound to Fcγ receptors regulate B lymphocyte differentiation. Cell Immunol. 95: 368–379 (1985).

25 Stein, S.H.; Phipps, R.P.: Macrophage-secreted prostaglandin E$_2$ potentiates immune complex-induced B cell unresponsiveness. Eur. J. Immunol. 20: 403–407 (1990).

26 Warner, G.L.; Scott, D.W.: A polyclonal model for B cell tolerance. I. Fc-dependent and Fc-independent induction of nonresponsiveness by pretreatment of normal splenic B cells with anti-Ig. J. Immunol. 146: 2185–2191 (1991).

27 Schad, V.C.; Phipps, R.P.: Prostaglandin E$_2$-dependent induction of B cell unresponsiveness role of surface Ig and Fc receptors. J. Immunol. 143: 2127–2132 (1989).

28 Phillips, N.E.; Parker, D.C.: Subclass specificity of Fcγ receptor-mediated inhibition of mouse B cell activation. J. Immunol. 134: 2835–2838 (1985).

29 Dickler, H.B.; Kubicek, M.T.: Interactions between lymphocyte membrane molecules. I. Interaction between B lymphocyte surface IgM and Fc receptors requires occupancy of both receptors. J. Exp. Med. 153: 1329–1343 (1981).

30 Tsokos, G.C.; Kinoshita, T.; Typhronitis, G.; Patel, A.D.; Dickler, H.D.; Finkelman, F.D.: Loaded but not free receptors for complement and the Fc portion of IgG co-cap independently with cross-linked surface Ig. J. Immunol. 144: 239–243 (1990).

31 Sinclair, N.R.; Panoskaltsis, A.: Rheumatoid factor and Fc signaling: A tale of two Cinderellas. Clin. Immunol. Immunopathol. *52:* 133–146 (1989).

32 Heyman, B.: Inhibition of IgG-mediated immunosuppression by a monoclonal anti-Fc receptor antibody. Scand. J. Immunol. *29:* 121–126 (1989).

33 Panoskaltsis, A.; Sinclair, N.R.: Rheumatoid factor blocks regulatory Fc signals. Cell Immunol. *123:* 177–188 (1989).

34 Klaus, G.G.B.; Bjisterbosch, M.K.; O'Garra, A.; Harnett, M.M.; Rigley, K.P.: Receptor signaling and cross-talk in B lymphocytes. Immunol. Rev. *99:* 19–38 (1987).

35 Wilson, H.A.; Greenblatt, D.; Taylor, C.W.; Putney, J.W.; Tsien, R.Y.; Findelman, F.D.; Chused, T.M.: The B lymphocyte calcium response to anti-Ig is diminished by membrane immunoglobulin cross-linkage to the Fcγ receptor. J. Immunol. *138:* 1712–1718 (1987).

36 Rigley, K.P.; Harnett, M.M.; Klaus, G.G.B.: Co-cross-linking of surface immunoglobulin Fcγ receptors on B lymphocytes uncouples the antigen receptors from their associated G protein. Eur. J. Immunol. *19:* 481–485 (1989).

37 Passwell, J.; Rosen, FS; Merler, E.: The effect of Fc fragments of IgG on human mononuclear cell responses. Cell Immunol. *53:* 395–403 (1980).

38 Morgan, E.L.; Hobbs, M.V.; Wiegle, W.O.: Lymphocyte activation by the Fc region of immunoglobulin. I. Role of prostaglandins in the down-regulation of Fc fragment-induced polyclonal antibody production. J. Immunol. *134:* 2247–2253 (1985).

39 Stein, S.H.; Phipps, R.P.: Elevated levels of intracellular cAMP sensitize resting B lymphocytes to immune-complex-induced unresponsiveness. Eur. J. Immunol. *21:* 313–318 (1991).

40 Tew, J.G.; Koscow, M.H.; Szakal, A.K.: The alternate antigen pathway. Immunol. Today *7:* 229–232 (1989).

41 Tsiagbe, C.K.; Linton, P.; Thorbecke, G.J.: The path of memory B-cell development. Immunol. Rev. *126:* 113–141 (1992).

42 MacLennan, I.C.M.; Gray, D.: Antigen-driven selection of virgin and memory B cells. Immunol. Rev. *91:* 61–85 (1986).

43 MacLennan, I.C.M.; Liu, Y.; Johnson, G.D.: Maturation and dispersal of B-cell clones during T cell-dependent antibody responses. Immunol. Rev. *126:* 143–161 (1992).

44 Phipps, R.P.; Stein, S.H.; Roper, R.L.: A new view of prostaglandin E regulation of the immune response. Immunol. Today *12:* 349–352 (1991).

45 Lebman, D.A.; Nomura, D.Y.; Coffman, R.L.; Lee, F.D.: Molecular characterization of germ-line immunoglobulin A transcripts produced during transforming growth factor type β-induced isotype switching. Proc. Natl Acad. Sci. USA *87:* 3962–3969 (1990).

46 Stein, S.H.; Phipps, R.P.: Antigen-specific IgG2a production in response to prostaglandin E_2, immune complexes, and IFN-γ. J. Immunol. *147:* 2500–2506 (1991).

47 Severinson, E.; Fernandez, C.; Stavnezer, J.: Induction of germ-line immunoglobulin heavy chain transcripts by mitogens and interleukins prior to switch recombination. Eur. J. Immunol. *20:* 1079–1084 (1990).

48 Snapper, C.M.; Peschel, C.; Paul, W.E.: IFN-γ stimulates IgG2a secretion by murine B cells stimulated with bacterial lipopolysaccharide. J. Immunol. *140:* 2121–2127 (1988).

49 Bossie, A.; Vitetta, E.S.: IFN-γ enhances secretion of IgG2a from IgG2a-committed LPS-stimulated murine B cells: Implications for the role of IFNγ in class switching. Cell Immunol. *135:* 95–104 (1991).

50 Phipps, R.P.; Stein, S.H.; Roper, R.L.: Regulation of B cell triggering by macrophages and lymphoid dendritic cells. Immunol. Rev. *117:* 135–158 (1990).

51 Rothermel, J.D.; Parker-Botehlo, L.H.: A mechanistic and kinetic analysis of the interactions of the diastereomers of adenosine 3′,5′-(cyclic) phosphorothioate. Biochem. J. *251:* 757–762 (1988).

52 Parker-Botehlo, L.H.; Rothermel, J.D.; Coombs, R.V.; Jastroff, B.: cAMP analog antagonists of cAMP action. Methods Enzymol. *159:* 159–172 (1988).

53 Roper, R.; Conrad, D.H.; Brown, D.M.; Warner, G.L.; Phipps, R.P.: Prostglandin E_2 promotes IL-4-induced IgE and IgG1 synthesis. J. Immunol. *145:* 2644–2651 (1990).

54 Roper, R.; Phipps, R.P.: Prostaglandin E enhances class switching to IgE and IgG1. 8th Int Congr Immunol, 37 (W-9): abstr 20 (1992).

55 Cambier, J.C.; Ransom, J.T.: Molecular mechanisms of transmembrane signaling in B lymphocytes. Annu. Rev. Immunol. *5:* 175 (1987).

56 Vitetta, E.S.; Fernandez-Botran, R.; Meyers, C.D.; Sanders, V.M.: Cellular interactions in the humoral immune response. Adv. Immunol. *45:* 1–105 (1989).

57 Berton, M.T.; Vitetta, E.S.: IL-4-induced expression of germline γ1 transcripts in B cells following cognate interactions with T helper cells. Int. Immunol. *4:* 387–396 (1992).

58 Pollock, K.E.; O'Brian, V.; Marshall, L.; Olson, J.W.; Noelle, R.J.; Snow, E.C.: The development of competence in resting B cells. The induction of ornithine decarboxylase activity after direct contact between B and T helper cells. J. Immunol. *146:* 1633–1639 (1991).

59 Cambier, J.C.; Newell, M.K.; Justement, L.B.; McGuire, J.C.; Leach, K.L.; Chen, Z.Z.: Ia binding ligands and cAMP stimulate nuclear translocation of PKC in B lymphocytes. Nature *327:* 629–632 (1987).

60 Stein, S.H.; Phipps, R.P.: Anti-class II antibodies potentiate IgG2a production by lipopolysaccharide-stimulated B lymphocytes treated with PGE_2 and IFN-gamma. J. Immunol. *148:* 3943–3949 (1992).

61 Frey, J.; Janes, M.; Engelhardt, W.; Afting, E.G.; Geerds, C.; Muller, B.: Fc gamma-receptor-mediated changes in the plasma membrane potential induce prostaglandin release from human fibroblasts. Eur. J. Biochem. *158:* 85–89 (1986).

62 Heinen, E.; Cormann, N.; Braun, M.; Kinet-Denoel, C.; Vanderschelden, K.; Simar, L.J.: Isolation of follicular dendritic cells from human tonsils and adenoids. Ann. Inst. Pasteur *137D:* 369–382 (1986).

63 Kurland, J.I.; Bockman, R.: Prostaglandin E production by human blood monocytes and mouse peritoneal macrophages. J. Exp. Med. *147:* 952–957 (1978).

64 Hsueh, W.; Arroyave, C.M.; Jordan, R.L.: Identification of C3b as the major serum protein that stimulates prostaglandin and thromboxane synthesis by macrophages. Prostaglandins *28:* 889–904 (1984).

65 Ferreri, N.R.; Howland, W.C.; Spiegelberg, H.L.: Release of leukotrienes C4 and B4 and prostaglandin E_2 from human monocytes stimulated with aggregated IgG, IgA, and IgE. J. Immunol. *136:* 4188–4193 (1986).

66 Waldmann, T.A.; Bull, J.M.; Bruce, R.M.; Broder, S.; Jost, M.C.; Balestra, S.T.; Suer, M.E.: Serum immunoglobulin E levels in patients with neoplastic disease. J. Immunol. *113:* 379–386 (1974).

67 Goodwin, J.S.; Messner, R.; Bankhurst, A.; Peake, G.T.; Saiki, J.H.; Williams, R.C.: Prostaglandin-producing suppressor cells in Hodgkin's disease. N. Engl. J. Med. *297:* 963–968 (1977).

68 Higgs, G.A.; Vane, J.R.; Hart, F.D.; Wojtulewski, J.A.: Effects of anti-inflammatory drugs on prostaglandins in rheumatoid arthritis; in Robinson, H.J.; Vane, J.R. (eds): Prostaglandin Synthetase Inhibitors. New York, Raven Press, 1987, pp. 165–173.

69 Leung, D.; Key, L.; Steinberg, J.; Young, M.C.; Von Deck, M.; Wilkinson, R.; Geha, R.S.: In vitro bone reabsorption by monocytes in the hyper-immunoglobulin E syndrome. J. Immunol. *140:* 84–88 (1988).

70 Offenbacher, S.; Odle, B.M.; Van Dyke, T.E.: The use of cervicular fluid prostaglandin E$_2$ levels as a predictor of periodontal attachment loss. J. Periodont. Res. *21:* 101–112 (1986).
71 Stern, M.A.; Dreizen, S.; Mackler, B.F.; Levy, B.M.: Antibody-producing cells in human periapical granulomas and cysts. J. Endodont. *7:* 447–452 (1981).

Richard P. Phipps, University of Rochester Cancer Center, Box 704,
601 Elmwood Ave., University of Rochester School of Medicine and Dentistry,
Rochester, NY 14642 (USA)

Granstein RD (ed): Mechanisms of Immune Regulation.
Chem Immunol. Basel, Karger, 1994, vol 58, pp 92–116

T Cell Clonal Anergy

Kathleen M. Gilbert

The Scripps Research Institute, Department of Immunology, La Jolla, Calif., USA

Introduction

Immunologic self-tolerance is achieved largely by clonal deletion of self-reactive CD4+/CD8+ cells in the thymus (reviewed on pp 1–33). However, it has been proposed that two other nondeletional mechanisms exist to functionally silence mature autoreactive T cells that escape negative selection in the thymus or that are specific for extrathymic host antigens. These mechanisms include active suppression and clonal anergy. T cell clonal anergy will be discussed here.

The existence of a nondeletional mechanism involved in maintaining peripheral self-tolerance was proposed in response to the fact that clonal deletion is not absolute. For example, healthy (nonautoimmune) individuals have been shown to contain both T cells and B cells that recognize self-antigens [1, 2]. However, the functional significance of these self-reactive lymphocytes was not clear since they were often generated in vitro in response to a mitogen, and are presumed to be present in low numbers in vivo, and to have a low avidity for self-antigens. Thanks to the recent development of transgenic mice whose germlines contain rearranged T cell receptors (TCR) or immunoglobulin genes, it is now easier to examine the fate of large numbers of self-reactive lymphocytes. It has been shown that double transgenic mice which have T cells with anti-MHC antigen TCR, and which bear MHC antigens on pancreatic cells, developed a state of T cell unresponsiveness to these antigens [3]. Although the mechanism surrounding the induction of this unresponsiveness is not yet clear, T cell inactivation in this system was not mediated by clonal deletion or suppressor cell activity [4]. Similarly, transplantation tolerance to minor histocompatibility antigens has been found to occur in mice despite the continued presence of normal numbers of the Mls-reactive $V\beta6$ CD4+ T cells [5, 6]. Tolerance induction to MHC antigens induced by administrating antibody (Ab) against CD4 at the time of antigen exposure [7, 8], was also found to be associated with the inactivation, rather than deletion, of the potentially MHC-reactive TCR Vβ phenotype [7].

A nondeletional mechanism of T cell inactivation has also been implicated in tolerance induction to soluble antigens. Mice injected with monomeric (deaggregated) human γ-globulin (HGG) are unable to mount an immunological response upon subsequent challenge with immunogenic forms of this antigen, such as aggregated HGG or soluble HGG emulsified in complete Freund's adjuvant [9, 10]. T cell inactivation in mice treated with monomeric HGG (mHGG) was demonstrated by the inability of peripheral lymph node T cells from these mice to proliferate in response to stimulation with HGG in vitro, or to provide help for primed B cells in adoptive transfer experiments [10, 11]. The fact that tolerance in Th cells in the HGG mouse model can be achieved in athymic mice, is independent of CD8+ suppressor cells, and is dependent upon the persistence of circulating tolerogen, suggests that tolerance in this model is the result of T cell clonal anergy rather than clonal deletion or active suppression. Similarly, it has been shown that clonal deletion of CD4+ cells specific for bovine γ-globulin did not occur during induction of tolerance to this antigen in adult mice [12].

Although clonal anergy appears to be involved in several models of experimental tolerance, not all examples of peripheral T cell inactivation can be ascribed to such a mechanism. For example, it was shown that proliferative T cell unresponsiveness to MIs-1a in one system correlated with the development of memory cells rather than tolerance [13]. In addition, active suppression, rather than clonal anergy, has been demonstrated in systems in which (a) deletion of regulatory subsets from normal cell populations resulted in autoimmune responses [14, 15], or (b) tolerance was transferred by suppressor or 'tolerizing' cells [16]. However, taken together, the findings make it likely that a nondeletional mechanism such as clonal anergy can play an important role in maintaining peripheral self-tolerance. Furthermore, although deletion appears to be the main mechanism that mediates tolerance to MHC or MIs antigens in the thymus, thymus epithelial cells can generate a nondeletional form of tolerance [6, 17, 19].

Inducing Tolerance of Th Cells in vitro

Although in vivo models are important for demonstrating the relevance of clonal anergy in the maintenance of peripheral tolerance, the mechanisms involved in the induction and effector functions of T cell clonal anergy are often more easily studied in vitro. Such studies have benefitted greatly from the generation of Th cell clone subsets as a source of antigen-specific T cells. Cloned Th1 and Th2 cells can be distinguished on the basis of their functions and lymphokine production [20]. Th1 cells secrete IL-2, lymphotoxin and IFN-γ, are efficient mediators of delayed-type hypersensitivity, but are poor providers of B cell help, promoting only IgM or IgG2a Ab responses. Th2 cells secrete IL-4 and IL-5, do not transfer delayed-type hypersensitivity, but are efficient providers of B cell help, promoting IgM, IgG1, IgA and IgE antibody responses.

It has been shown that Th1 cells exposed to tolerogen in vitro lose their ability to proliferate or produce IL-2 upon subsequent antigen challenge [21–24]. Th1 cell tolerogens exist in several forms, but consist primarily of antigen presented in a way that permits T cell recognition of the epitope in the absence of T cell activation. This entails presentation of antigen by an antigen-presenting cell (APC) which (a) expresses MHC class II molecules, (b) lacks certain costimulator molecules necessary for T cell activation, and (c) in the case of intact antigen, is capable of antigen uptake and processing. Thus, murine Th1 clones have been inactivated by exposing them to peptide presented by chemically fixed APC [23] or planar lipid membrane [24], or to fixed APC that had been previously pulsed with antigen [21]. Anergy has also been induced by treating Th1 cells with Con A or immobilized anti-CD3 mAb in the absence of stimulatory APC [25–27]. If used in high concentrations, or for a long enough time period, anti-CD3 Ab induced Th1 cell unresponsiveness even in the presence of accessory cells [25, 28].

Studies of Th tolerance in vitro have been confined primarily to Th1 clones. However, this does not mean that Th2 cells are not susceptible to tolerance induction. It was initially demonstrated that HGG-specific Th clones treated with monomeric HGG in vitro retained their proliferative capacity, but lost their ability to promote antibody production by HGG-primed B cells [29]. These clones are subsequently found to be of the Th2 type. Similarly, it has since been shown that HGG-specific Th2 clones exposed to HGG-pulsed fixed spleen cells lost their ability to supply helper activity to B cells in HGG-stimulated secondary cultures [21, 22]. Similar to the requirements for Th1 tolerance, Th2 tolerance requires exposure to both antigen and MHC class II molecules. HGG-specific Th2 cells were incubated with one of three different populations of HGG-pulsed fixed APC. The populations consisted of unseparated spleen cells, splenic B cells or adherent peritoneal exudate cells (PEC). The adherent PEC population was comprised primarily of macrophages which were phagocytic, but expressed very low levels of MHC class II molecules

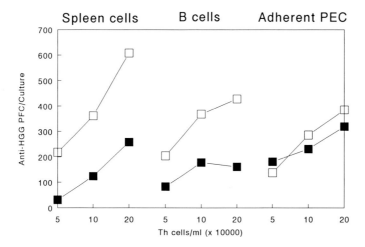

Fig. 1. The induction of Th2 cell unresponsiveness depends upon recognition of antigen in the presence of MHC class II molecules. HGG-specific Th2 cells were incubated for 2 days in primary cultures with APC which had first been preincubated with (■–■) or without (□–□) HGG and then fixed with paraformaldehyde. The Th2 cells from the primary cultures were then isolated, and added in various concentrations to HGG-stimulated secondary cultures containing HGG-primed B cells. The results represent the number of IgG anti-HGG plaque-forming cells measured subsequently in the secondary cultures by means of an indirect hemolytic plaque assay. The APC used in the primary cultures consisted of unseparated spleen cells, resting splenic B cells or PEC. The PEC population used consisted of cells which remained adherent after overnight culture. This population was highly phagocytic, expressed low levels of MHC class II molecules and was presumed to contain primarily macrophages.

[Gilbert et al., unpubl. data]. Unlike Th2 cells exposed to HGG-pulsed fixed spleen cells or HGG-pulsed fixed B cells, Th2 cells exposed to HGG-pulsed fixed adherent PEC did not experience a decrease in helper activity (fig. 1). None of the APC populations induced Th2 unresponsiveness in the absence of HGG.

Unlike murine Th cells which are inactivated only when exposed to antigen presented by class II-bearing tolerogenic APC, human peptide-specific T cell clones have been inactivated in vitro by exposure to high concentrations of peptide alone [30, 31]. This apparent disregard for the need of tolerogenic APC is most probably due to the fact that human T cells, unlike murine T cells, express MHC class II molecules and can therefore present peptide to themselves. Lacking the costimulator molecules expressed by 'professional' APC, human T cells present antigen in a tolerogenic fashion to other T cells. In support of this theory is the recent finding that human T cell clones were unable to act as autonomous APC, such that in the absence of third-party APC they induced unresponsiveness in other antigen-specific or alloreactive T cells [32]. Antigen-specific unresponsiveness has also been induced in human T cell clones [33] and resting peripheral T cells [34] by treatment with anti-CD3 Ab in the absence of accessory cells.

Inducing Th Cell Tolerance with Naturally Tolerogenic APC

In addition to fixed APC, nonfixed cells which can present antigen, but naturally lack costimulator molecules on their surface, are also capable of serving as tolerogenic APC. For example, murine Th cells have been inactivated in vitro by exposure to keratinocytes or endocrine cells which have been treated with IFN-γ to induce expression of MHC class II molecules [35, 36]. There is evidence to suggest that unlike nonlymphoid cells which must be stimulated to express MHC class II in order to become tolerogenic, resting B cells exist naturally in a tolerogenic state. Lightly irradiated resting splenic B cells are very inefficient stimulators of Th1 proliferation to HGG [22]. Other investigators have similarly noted that resting B cells are ineffective or inferior APCs for the initiation of primary Th1 responses to alloantigens and foreign antigens [37–42]. In the HGG system it was found that this B cell inefficiency was associated with antigen-specific inactivation: Th1 exposed to HGG and lightly irradiated resting B cells in primary cultures were unable to proliferate to HGG and immunogenic APC in secondary cultures [22]. Similar findings were described in a system which showed that heavily irradiated B cells together with peptide could tolerize peptide-specific Th1 cells [43]. The tolerogenic effects of lightly irradiated B cells on HGG-specific Th1 cells were mimicked by mitomycin C-treated B cells [22], confirming that essentially intact B cells are capable of presenting antigen in a tolerogenic fashion to Th1 cells. The fact that HGG-pulsed fixed B cell lymphomas also tolerized Th1 cells [44] make it unlikely that Th1 anergy induced by populations of splenic B cells was due to contaminating adherent cells. In fact, irradiated adherent cells do not appear to be tolerogenic for Th1 cells. Unseparated spleen cells, resting splenic B cells, and a population of PEC enriched for dendritic cells were compared for their ability to present antigen in an immunogenic rather than tolerogenic fashion to Th cells (fig. 2). When irradiated, the unseparated spleen cells and dendritic cells, unlike the splenic B cells, stimulated Th1 proliferation in HGG-stimulated primary cultures [Gilbert et al., unpubl. data]. The Th1 cells from the primary cultures were subsequently tested in HGG-stimulated secondary cultures,

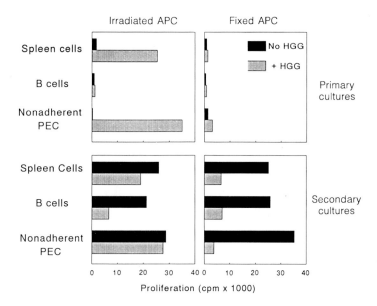

Fig. 2. Unlike B cells, adherent cells must be fixed in order to induce Th1 cell unresponsiveness. HGG-specific Th1 cells were incubated for 2 days in primary culture with APC which had first been preincubated with or without HGG and then irradiated or fixed with paraformaldehyde. Some of the Th1 cells from the primary cultures were then isolated and reincubated in secondary cultures stimulated with HGG and irradiated spleen cells. Proliferation of the Th1 cells in both the primary and secondary cultures were measured as a function of ^{3}H-TdR uptake. The APC used in the primary cultures consisted of unseparated spleen cells, resting splenic B cells or PEC. The PEC population used consisted of cells which although initially adherent, became nonadherent after overnight culture. This population was only marginally phagocytic, expressed high levels of MHC class II molecules, and was presumed to contain primarily dendritic cells.

and only those Th1 cells which had been exposed to HGG in the presence of irradiated B cells lost their antigen-specific proliferative activity. In contrast, when fixed, none of the three populations of APC stimulated Th1 proliferation in primary cultures, and all induced Th1 inactivation as measured in antigen-stimulated secondary cultures. Taken together, it appears that while chemical fixation may convert most immunogenic APC such as dendritic cells and splenic macrophages into tolerogenic APC, resting B cells, unlike these immunogenic APC, do not require fixation to present antigen in a tolerogenic fashion to Th cells. Presumably, this is because resting B cells can present antigen in conjunction with MHC class II molecules, but lack the costimulatory signals required to fully activate Th cells.

In vivo, the role of B cells as tolerogenic APC has been demonstrated in mice by using ultracentrifuged Fab fragments of rabbit anti-mouse IgD to induce tolerance specific for nonimmune rabbit Fab [5]. Tolerance in this system apparently required antigen presentation by B cells since intravenous injection of deaggregated nonimmune rabbit Fab, which is not targeted to B cells, induced a lower degree of tolerance. It has also been demonstrated that injection of a population of spleen cells consisting primarily of small resting B cells induced hyporesponsiveness to MHC alloantigens [46, 47], and to the male-specific antigen H-Y [48] in mice. Thus it appears that under conditions in which antigen is presented by resting B cells in a way that precludes presentation by stimulatory APC such as dendritic cells and macrophages, resting B cells induce antigen-specific unresponsiveness in CD4+ cells.

Tolerance of Th1 and Th2 Cells in vivo

While there is evidence that Th1-like and Th2-like CD4+ cells exist in vivo [49, 51], the effect of experimental tolerance on the two subsets in vivo is somewhat controversial. Attempts have been made to study the responsiveness of Th1 and Th2 cells in vivo by taking advantage of the fact that the Th cell subsets can be distinguished on the basis of which Ig isotype they stimulate. Using this approach, De Wit et al. [52] observed that in vivo challenge of monomeric HGG-treated mice with hapten-coupled HGG revealed a selective deficiency in the IgG2a subclass of anti-hapten Abs, and concluded that monomeric HGG inactivates Th1, but not Th2 lymphocytes in vivo. In apparent contradiction of this study are the results from early adoptive transfer experiments demonstrating that the serum levels of total anti-HGG Abs were severely suppressed in mice given T cells from the spleen or bone marrow of monomeric HGG-treated mice together with B cells from nontolerized mice [10, 53]. Although it was not possible at that time to distinguish between the Th1 and Th2 cells from the tolerized mice, it can be inferred from the virtually complete lack of T cell-mediated Ab response that both Th cell subsets were inactivated in terms of helper activity. In addition, it was reported that the high level of Ab unresponsiveness (as measured by IgG plaque-forming cell responses to HGG challenge) induced by treatment with mHGG remained at the same high level (>95%) at time periods (90 days) when B cell responsiveness had been restored to normal levels or higher [54]. This finding would also seem to suggest that anergy encompassed both Th1 and Th2 cells. One study which examined this question more directly found that mice tolerized to rabbit Fab exhibited decreased levels of both

IgG2a and IgG1 isotopes of anti-rabbit Fab Abs [45], implying that both Th1 and Th2 cell subsets were inactivated in these mice. These findings, taken together with our observations concerning the loss of Th2 helper activity following exposure to tolerogens in vitro, suggest that Th2 cells are susceptible to tolerance induction, and that both Th1 and Th2 cells can be inactivated in vivo.

Examining the Mechanism for T Cell Tolerance

Attempts have been made to determine how tolerance induction affects the mechanisms that regulate normal T cell proliferation and differentiation. These mechanisms involve a series of events, some of which are well defined. For example, it has been shown that ligand binding to the TCR complex, $\alpha\beta$-CD3-$\zeta\eta$, activates two related second messenger pathways [for reviews, see 55, 56]. The ζ chain of the complex is phosphorylated on a tyrosine residue after antigen triggering due to action of the CD4-associated protein tyrosine kinase, lck, and the $\alpha\beta$-associated protein tyrosine kinase, fyn [for reviews, see 57, 58]. The dephosphorylated, tyrosine kinases in turn activate phospholipase C, resulting in hydrolysis of polyphosphoinositides to yield diacylglycerol and inositol 1,4,5-triphosphate. These second messengers mediate mobilization of intracellular and extracellular Ca^{2+} and activation of protein kinase C. By a mechanism yet undefined these second messenger pathways are integrated in G_1 to promote Th cell cycle progression [for reviews, see 59]. This promotion involves the activation of a series of genes which appear to be related to T cell growth and maturation [for reviews, see 60, 61]. These genes include the somewhat functionally obscure oncogenes, some of which may serve as transactivating factors that bind to DNA sequences to activate genes later. Also activated during T cell growth and maturation are genes whose mRNAs are ultimately translated into the various proteins and enzymes required for mitosis (e.g. cyclins and lymphokines such as IL-2).

The finding that normal resting T cells proliferate in the presence of a combination of a calcium ionophore and a PKC-activating phorbol ester suggested that increased $[Ca^{2+}]_i$ and PKC activation were sufficient to stimulate IL-2 production [67]. However, it has subsequently been shown that an accessory cell-generated costimulatory signal, which was delivered via a cell-cell interaction, and which acted independently of the rise in $[Ca^{2+}]_i$ and PKC activation, was required during the response to ionomycin and phorbol ester [63]. At high cell densities the T cells themselves apparently provided this signal (presumably acquired as a result of intense activation), since at limiting cell numbers, the T cells did not proliferate. A requirement for cell contact-mediated costimulation has also been reported for T cell proliferation in response to immobilized anti-CD3 Mab [25, 63], and for antigen-induced proliferation of Th1 cells [64–66]. Thus, these results led to the prediction that all methods of IL-2-dependent promotion of T cell proliferation in response to TCR ligation requires the delivery of a distinct accessory cell-derived costimulatory signal.

Attempts to delineate the mechanism(s) by which T cells are tolerized have been concentrated in several areas. These include examining the affects of tolerance induction on normal T cell functional activities and signal transduction, and attempting to define the costimulatory signal whose absence is associated with APC tolerogenicity.

Role of Costimulator Signals in T Cell Anergy

It has been demonstrated that the crucial APC costimulatory signal, the absence of which leads to T cell anergy, can be delivered to Th1 cells by nonfixed low density splenocytes [67], but cannot be mimicked by any lymphokine yet tested, including IL-1 or IL-6 [24, 25, 67]. There is some controversy over whether the costimulator signal must appear on the same APC that presents antigen. The addition of allogeneic spleen cells (which are incapable of presenting antigen, but which can present costimulator signals) to cultures containing cloned Th1 cells and antigen-pulsed fixed APC stimulated proliferation and prevented Th1 inactivation [26, 68]. In contrast, clonal expansion of splenic CD4+ T cells in response to anti-CD3 Ab was most efficiently costimulated when both costimulator signals and FcR (required for cross-linking of the anti-CD3 Ab) were present on the same cell [69]. It was suggested that the apparently conflicting results reflected the different costimulator requirements of naive versus cloned CD4+ T cells.

Although many different accessory molecules on APC appear to be involved in various aspects of T cell activation [for reviews, see 70, 71] most of these molecules do not meet the criteria for the costimulator signal whose absence is associated with APC tolerogenicity. In some cases, ligating the T cell receptors for these accessory molecules bypasses the need for TCR cross-linking, or merely augments the biochemical signals provided by the TCR. Such accessory molecules are probably not linked to tolerance induction. Of the molecules which do appear to meet the criteria of a true costimulator signal, the B7/BB-1 antigen and ICAM-1 have been studied in the most detail.

B7/BB-1 antigen is a member of the Ig superfamily that is found on activated, but not resting, B cells [72], thymic stromal cells [73], splenic dendritic cells [74], and IFN-γ-stimulated macrophages [75]. ICAM-1, another member of the Ig superfamily, is expressed constitutively on resting B cells, dendritic cells and macrophages [reviewed in 76, 77]. Expression of both B7 and ICAM-1 can be unregulated by APC contact with CD4+ cells or lymphokines [72, 76, 78]. B7/BB-1 interacts with its T cell ligand CD28 on resting T cells, and two receptors, CD28 and CTLA-4, on activated T cells [72]. ICAM-1 interacts with CD11a (LFA-1) on T cells [79]. There is evidence that both LFA-1 and CD28 are important in cell contact-dependent activation of CD4+ T cells by APC. For example, it has been shown that otherwise non-costimulatory APC can stimulate antigen-specific or alloreactive T cell proliferation if transfected with the B7 molecule or ICAM-1 molecule [80–82]. Anti-CD28 stimulated proliferation of human tetanus toxoid-specific T cells cultured with antigen-pulsed costimulation-deficient APC [83]. The function of the CD28 receptor appears to be distinct from the TCR during T cell activation since CD28 stimulation can enhance lymphokine production even in the presence of maximal phorbol ester and calcium ionophore stimulation [84], and, unlike TCR-mediated IL-2 stimulation, is resistant to the immunosuppressant cyclosporine [85]. ICAM-1 has been shown to provide a costimulatory signal for TCR-mediated activation of resting human and murine T cells [86–88]. Thus, it appears that in addition to its adhesive properties, ICAM-1 can provide an activation signal that is distinct from, and yet synergistic with, TCR-transduced signals.

Especially relevant here are the effects of B7 and ICAM-1 on Th cell tolerance induction. Alloantigen-specific T cell hyporesponsiveness has been induced in a human mixed leukocyte culture by blocking interaction of B7 with its ligand CD28 [89]. Hyporesponsiveness was achieved through the use of a soluble analogue of the receptor for B7, or by the addition of the monovalent Fab fragments of anti-CD28 mAb which did not activate T cells. In addition, a stimulatory preparation of anti-CD28 Ab has been shown to interfere with the inactivation of λ-bacteriophage repressor-specific Th1 cells exposed to antigen-pulsed fixed spleen cells [90]. Antibodies to ICAM-1 and LFA-1 have been shown to act synergistically to prolong cardiac allografts in mice in the absence of CD4+ cell depletion [91]. Although the mechanism for the unresponsiveness in this system has yet to be established, it seems likely that the antibody treatment led to the induction of specific tolerance though the blockade of ICAM-1 mediated signaling events. Similarly, antibodies to ICAM-1 or LFA-1 have been used individually to induce tolerance in vivo to soluble antigens [92, 93].

Another candidate for a costimulatory molecule which confers immunogenicity on APC is IL-1. IL-1 has been described both as a molecule which costimulates Th2 cell proliferation by fixed APC [94], and as a molecule which induces costimulatory signals retained on APC following fixation [95]. In response to certain stimuli, IL-1 can be produced by macrophages, B cells, memory T cells, and cloned Th cells [96–98]. IL-1 has been shown to interfere with mHGG-induced tolerance induction of Th2 cells in vitro [99]. In addition, the injection of IL-1 shortly after the administration of mHGG inhibited tolerance induction at both the T and B cell level in vivo [110]. mHGG did not stimulate IL-1 production in vivo or in vitro, and was not degraded by peritoneal macrophages. In contrast, heat-aggregated HGG, which is immunogenic rather than tolerogenic, stimulated IL-1 production and was degraded by peritoneal macrophages. These findings are consistent with the suggestion that tolerance induction resulted from T cell recognition of HGG in the absence of IL-1.

Also apparently possessing costimulatory capacity is heat-stable antigen, a molecule that is variably expressed on B cells at different functional stages [68]. When expressed on transfected CHO cells, heat-stable antigen promoted proliferation of CD4+ T cells to anti-CD3 Ab [101]. The addition of blocking antibody to heat-stable antigen to primary cultures of anti-CD3-treated CD4+ T cells and LPS-activated B cells inhibited T cell proliferation. The T cells from the primary cultures were found to be unresponsive when restimulated with anti-CD3 and fresh APC in secondary cultures [101]. This T cell unresponsiveness was attributed to the fact that blocking heat-stable antigen on the activated B cells from immunogenic APC to tolerogenic APC.

Despite these promising beginnings, some questions remain to be answered in order to clarify the role of the aforementioned costimulatory molecules in preventing tolerance induction. For example, do Th1 and Th2 cells require distinct costimulator molecules? The likelihood of this possibility is supported by the finding that the Th cell subsets prefer different populations of APC, a preference that does not correlate with MHC-resistricting elements or involvement of CD4 [22, 39, 94]. Examining the costimulator requirements of Th2 cells is complicated however by the fact that Th2 cells reportedly have the capacity to produce one of their own costimulator signals, IL-1 [98]. In addition, Th2 cells have been shown to have accessory cell capacity of their own, i.e. can promote anti-CD3-mediated proliferation of resting splenic T cells [102]. The capacity of Th2 cells to make their own IL-1, and perhaps serve as accessory cells for themselves, could explain why Th2 cells, but not Th1 cells, proliferate in response to antigen presented by apparently costimulator-deficient B cells [22, 39]. In clarifying the role of APC costimulator molecules, it would also be useful to know what effect, direct or indirect, IL-1 has on tolerance induction in Th1 cells. Does the ability of IL-1 to block tolerance induction to mHGG in vivo, and thus maintain anti-HGG antibody production, pertain to Th1- as well as Th2-induced anti-HGG isotypes? Although Th1 cells can respond to IL-1 [103], they do not require this molecule to costimulate proliferation [104, 105], and IL-1 does not appear to block tolerance induction of Th1 cells exposed to peptide and planar membranes [24]. Also requiring clarification is the observation that activated B cells, shown to express a variety of costimulatory molecules, are still tolerogenic rather than immunogenic for Th1 cells. Fuchs and Matzinger [48] recently showed that LPS treatment did not inhibit the ability of B cells expressing H-Y to induce transplantation tolerance in mice: T cells from these animals exhibited a decrease in H-Y directed cytotoxicity. Examining B cell antigen presentation in vitro, it was found that LPS-treated B cells, although better than resting B cells at promoting antigen-specific proliferation of HGG-specific Th1 cells, were still inferior to unseparated spleen cells in this regard [Gilbert and Weigle, unpubl. data]. This was despite the fact that the same B cells efficiently promoted Th1 proliferation in a mixed lymphocyte reaction or in response to anti-CD3 stimulation. Preliminary studies showed that LPS-treated B cells were at least as efficient as resting B cells in tolerizing Th1 cells to HGG. It is possible that although activation with LPS may increase the ability of B cells to promote T cell proliferation in response to mitogens or anti-CD3 Ab, the effects of this treatment on the ability of B cells to present nominal antigen to Th1 cell clones may be less dramatic, and may not prevent the B cells from presenting antigen in a tolerogenic fashion. We are presently examining this possibility.

Taken together, it appears that the relationship between an APC's tolerogenicity and its expression of a particular costimulatory molecule may not be as straightforward as previously predicted. It seems likely that T cells receive several costimulatory signals, some of which may not yet be identified. Activation versus inactivation of Th cells may depend upon the summation of signals received. The relative importance of the costimulatory signals may vary for Th1 and Th2 cells, and for activated versus resting CD4+ cells. Along these lines, it was shown that B7 and ICAM-1 exerted contrasting regulatory effects on the proliferation of CD4+ T cells depending on their state of activation [89]. TCR-induced proliferation of resting CD4+ T cells was costimulated efficiently by ICAM-1 and inefficiently by B7, while similar proliferation of antigen-primed CD4+ T cells was promoted efficiently by B7 and poorly by ICAM-1. Also potentially relevant when speculating about the role of costimulatory molecules in T cell activation is the fact that the functional effects of the various molecules (or of antibodies to the molecules) is often monitored by examining T cell proliferation in an MLR or in the presence of anti-CD3 Ab. Th cell requirements for APC costimulatory signals in these systems may not be as stringent as those needed to promote proliferation in response to presentation of a nominal antigen by APC. Frank et al. [106] showed that cytoplasmic truncation of the ζ molecule of the TCR inhibited antigen-induced IL-2 secretion, but did not inhibit IL-2 production in response to anti-CD3 Ab. These investigators proposed that whatever function ζ performs during activation of the TCR by a physiologic ligand is bypassed by direct external cross-linking of the TCR by antibodies to the TCR. It is possible that the results from experiments examining T cell costimulatory requirements using anti-CD3 Ab to ligate the TCR may therefore not accurately reflect the complexity of costimulatory signals required for T cell responses to a nominal antigen.

Functional Consequences of T Cell Anergy

Perhaps the most interesting conclusion to be drawn from examining the functional consequences of T cell tolerance induction is that the loss of Th cell activity associated with T cell anergy is selective rather than absolute. For example, although tolerized Th1 cells lose their ability to produce IL-2 (and to some extent IL-3 and IFN-γ), when stimulated with antigen and immunogenic APC, they retain their capacity to proliferate in response to exogenous IL-2 [22, 24, 66, 107]. This split responsiveness is explained by the fact that tolerized Th1 cells do not suffer a significant loss in expression of IL-2R or in their ability to respond to IL-2 [66, 108]. TCR and IL-2R appear to be coupled to different signal transduction pathways [109], so it is not unlikely that tolerance induction can suppress one pathway without affecting the other pathway.

The effect of tolerance induction on the ability of Th1 cells to provide B cell help also appears to be selective. Activated Th1 cells and Th2 cells can induce antigen-unrestricted polyclonal B cell proliferation that is in part mediated by Th cell-secreted lymphokines, but requires B cell-Th cell contact for an optimal response [110, 111]. Antigen-activated Th1 cells are less efficient than antigen-activated Th2 cells in inducing polyclonal B cell proliferation [112]. Apparently Th1 cells are similar to Th2 cells in providing a cell contact-mediated signal to B cells, but are deficient in producing the lymphokines (e.g. IL-4) required for optimal B cell proliferation. Exposure to tolerogen was found to suppress the modest ability of antigen-stimulated Th1 cells to induce B cell DNA synthesis, but did not inhibit the ability of Th1 cells to enhance MHC class II expression on the B cells [107] (fig. 3). These results were interpreted to mean that Th1 cell anergy does not interfere with the capacity of Th1 cells to interact with B cells, and stimulate early events such as up-regulation of class II molecules. but does suppress some later signaling event in Th1 cell contact-mediated B cell help.

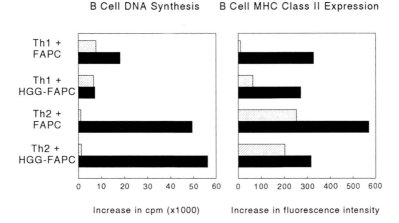

B Cell DNA Synthesis   B Cell MHC Class II Expression

Fig. 3. Th1 cells, but not Th2 cells, exposed to tolerogen lose their ability to promote polyclonal B cell proliferation. HGG-specific Th1 cells and Th2 cells were incubated in primary cultures together with paraformaldehyde-fixed APC which had been preincubated in the absence of HGG (FAPC) or the presence of HGG (HGG-FAPC). The Th cells from the primary cultures were isolated, and reincubated with (■) or without (□) HGG in secondary cultures of resting splenic B cells. After 48 h, the B cells from the secondary cultures were examined for DNA synthesis (as a measurement of ³H-TdR uptake) and for expression of MHC class II molecules (as a measurement of two-color immunofluorescent staining with flow cytofluorometric analysis). The results are represented as the increase in activity above that of unstimulated B cell controls.

It was recently reported that in addition to their ability to respond to IL-2, and to interact with B cells, anergized Th1 cells also retained their ability to lyse target cells with an efficiency comparable to nontolerized Th1 cells [113]. Th1 cells recently exposed to tolerogen were able to lyse target cells in an antigen/MHC-unrestricted manner. After the anergized Th1 cells were rested in culture, their cytolytic activity regained its antigen and MHC specificity. Taken together, these results underline the selective effect of tolerance induction on Th1 cell functional activities.

Similar to Th1 cell activities, Th2 cells activities also appear to be differentially affected by exposure to tolerogen. For example, we have shown that while cloned Th2 cells exposed to tolerogen lost their ability to promote antigen-specific secondary antibody production, they retained their antigen-specific proliferative capacity [21]. In fact, exposure of Th2 cells to HGG-pulsed fixed APC or HGG-pulsed resting B cells in primary cultures stimulated DNA synthesis. Tolerogen-treated Th2 cells also retained their ability to promote polyclonal B cell proliferation [107]. It is tempting to propose that alterations in Th2 lymphokine production may be involved in their response to tolerogens. For example, it is possible that exposure to tolerogen inhibits secretion of a Th2 lymphokine (e.g. IL-5) required for Ab production, but activates production of a Th2 lymphokine (e.g. IL-4) required to promote Th2 proliferation and Th2-induced poly-clonal B cell proliferation. Such selectivity in lymphokine inactivation has been reported in systems using Th0 clones that make both IL-2 and IL-4. Following exposure to tolerogen, Th0 cells lost their ability to make IL-2, but do not exhibit a decrease in IL-4 production [114, 115]. The capacity of Th2 cells for selective responsiveness has also been demonstrated by Evavold and Allen [116] who showed that Th2 proliferation could be separated from Th2 helper activity by an alteration in the TCR ligand. Obviously a detailed examination of Th2 lymphokine production following exposure to tolerogen would be useful in understanding the different aspects of Th2 cell hyporesponsiveness.

Molecular and Biochemical Consequences of T Cell Anergy

It is clear that one of the most crucial functional consequences of Th1 cell anergy is a defect in antigen-induced IL-2 production. Consequently, several studies have examined the effects of T cell tolerance on the molecular events that mediate IL-2 synthesis. Quill et al. [117] showed that anergic Th1 cells expressed constitutively reduced amounts of the protein tyrosine kinase $p56^{lck}$ and constitutively elevated levels of the protein tyrosine $p59^{fyn}$. This observation led to the suggestion that T cell anergy may be maintained by an alteration in the tyrosine phosphorylation signaling events important for normal IL-2 synthesis. In examining this possibility it would be useful to know whether the altered levels of the fyn and lck kinases are still demonstrable when the anergized Th1 cells are restimulated with antigen. Go and Miller [118] reported that, unlike Th1 cells exposed to antigen presented by immunogenic APC, Th1 cells exposed to antigen-pulsed fixed APC failed to express NF-AT and NF-κB binding factor, two transcription factors previously shown to regulate IL-2 expression. However, when restimulated with antigen and immunogenic APC, the anergized Th1 cells exhibited the same pattern of transcription factors as nontolerized Th1 cells, even though the anergized Th1 cells were still unable to proliferate. In contrast to NF-AT, another transcription factor, AP-1 has been shown to be moderately decreased in anergized Th1 cells at early time points after stimulation with antigen and immunogenic APC [119]. The fact that other transcription factors were not diminished in antigen-activated anergized Th1 cells supported the theory that a specific regulatory event, rather than a generalized signal transduction defect, occurs during anergy induction.

There is evidence to suggest that the specific regulatory mechanism associated with T cell anergy involves the production of an inhibitory substance by the unresponsive T cells. Quill and Schwartz [24] showed that Th cell unresponsiveness was blocked by cycloheximide, implying that this phenomenon was dependent upon the generation of new proteins. In addition, the existence of an inhibitory substance generated in response to T cell tolerance induction has been proposed to explain why several days of IL-2 induced growth reverses Th1 cell unresponsiveness and its gene regulatory phenotype [118–120]. A short exposure to exogenous IL-2 does not prevent Th1 tolerance induction [24, 26], and in some cases (e.g. when Th1 anergy is induced by immobilized anti-CD3 Ab) Th1 tolerance induction is accompanied by limited IL-2 production. Thus, it seems that IL-2-induced cell division rather than an IL-2 induced signaling event is responsible for the ability of this lymphokine to reverse Th1 cell tolerance. It has been proposed that normal proliferation of the anergized T cell may dilute out some negative regulatory factor induced as a consequence of tolerance induction, and allow the IL-2 gene to return to an activatable state.

T cell generation of a negative regulatory substance would most likely be linked to T cell recognition of costimulatory molecules. In the absence of this recognition, T cells still experience TCR-mediated increases in $[Ca^{2+}]_i$ [24, 26], but are unable to, however, mediate a second signal transduction pathway that is independent but related to that mediated by inositol phospholipid hydrolysis and intracellular calcium. It has been demonstrated that IL-2 secretion is more susceptible to a protein tyrosine kinase inhibitor than is Th1 cell anergy [121]. This led the investigators to suggest that TCR-mediated increases in inositol phospholipid hydrolysis and intracellular calcium lead to T cell inactivation, which is only avoided when costimulators mediate, via tyrosine kinase activation, a second pathway which results in IL-2-induced Th cell activation. Presumably, successful completion of this second pathway would prevent induction of the putative inhibitory substance, while unsuccessful completion of the second pathway would trigger the generation of this substance. This theory is supported by the finding that Th cell anergy can be induced under otherwise stimulatory conditions (e.g. in the presence of antigen-presenting immunogenic APC) if IL-2-induced effects are suppressed by the addition of anti-IL-2 Ab [28] or a protein kinase inhibitor which prevents IL-2 synthesis [121]. Removal of IL-2 and other factors from 16-hour cultures of Th1 cells and antigen-presenting immunogenic APC also induced Th1 cell tolerance [120]. There is evidence to suggest that Th cell unresponsiveness

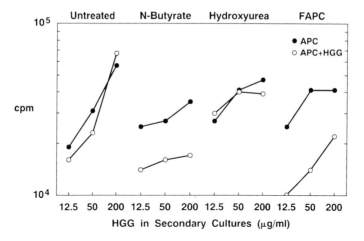

Fig. 4. Exposure of antigen-activated Th1 cells to a G_{1a} blocker induces antigen-specific unresponsiveness. HGG-specific Th1 cells were incubated in primary cultures with or without HGG together with irradiated spleen cells as APC. Some primary cultures also contained the G_{1a} blocker *n*-butyrate or the G_{1b} blocker hydroxyurea. Other primary cultures contained, instead of irradiated APC, fixed APC (FAPC) which had been prepared in the presence or absence of HGG. The Th1 cells from the primary cultures were isolated and reincubated in secondary cultures containing irradiated APC and various concentrations of HGG. The results represent Th1 proliferation in the secondary cultures.

is not induced by all events that inhibit Th cell cycle progression. We have found that treatment of Th1 cells with the G_{1a} blocker *n*-butyrate in the presence of HGG-presenting APC inhibited the ability of the Th1 cells to proliferate to a subsequent challenge with HGG [122] (fig. 4). In contrast, treatment of Th1 cells with drugs that blocked cell cycle progression in G_0, G_{1b}, or G_2/S phases did not induce subsequent antigen-specific anergy. The inhibitory effects induced by n-butyrate were antigen-specific in their induction and effector phases.

The importance of the G_{1a} phase of the cell cycle in regulating Th cell inactivation/activation mitosis is underlined by the fact that mitosis of eukaryotic cells is largely controlled in early G_1 phase [for review, see 59]. Almost all physiological conditions that inhibit mitosis, such as lack of essential amino acids, or exposure to calcitriol, transforming growth factor-β and T suppressor cells, act during G_{1a} [59, 123–125]. In view of these findings it is tempting to speculate that Th cell anergy may be regulated by a cell cycle checkpoint mechanism that controls the restriction point in mid-G_1 where eukaryotic cells become committed to a new round of division. This checkpoint mechanism would be initiated if normal cell cycle progression is prevented (e.g. by a lack of IL-2 production) or interrupted in G_{1a} phase. The consequences of initiating this mechanism may very likely involve the generation of a negative inhibitory substance that suppresses certain Th cell events such as antigen-induced activation of the IL-2 gene.

Concluding Remarks

Based on the observations described here it is possible to speculate that both the induction of experimental tolerance and the maintenance of self-tolerance represent dynamic states in which potentially reactive antigen-specific Th cells can be inactivated by exposure to antigen presented by resting B cells or nonlymphoid cells which have been stimulated to express MHC class II molecules. Stimulation of MHC class II molecules on nonlymphoid cells may occur naturally during the course of certain immune responses via IFN-γ production. T cell inactivation would most likely occur when CD4+ cells are exposed to antigen presented by tolerogenic APC in a manner which precludes antigen uptake and presentation to T cells by such immunogenic APC as splenic dendritic cells and macrophages. Several factors such as the route of antigen administration, the size, form and concentration of the antigen, and the general immune status of the host could influence which APC encounter the antigen, and the activation state of the APC when it presents antigen to CD4+ T cells. This in turn could determine whether the APC presents the antigen to T cells in an immunogenic or tolerogenic fashion. The activation state of the T cell when it encounters antigen could also play a role in determining the outcome of the response: naive T cells presented with antigen by tolerogenic APC may be inactivated, while memory T cells presented with antigen by the same APC may be activated. The subset to which the T cell belongs (e.g. IL-2 or IL-4 producing) may also affect its response to antigen presentation. Obviously more work is required to examine the integration of these various factors into a regulatory network designed to maintain peripheral tolerance.

References

1   Cohen, I.: Regulation of autoimmune disease – physiological and therapeutic. Immunol. Rev. *94:* 5 (1986).
2   Guilbert, B.; Dighiero, G.; Avrameas, S.: Naturally occurring antibodies against nine common antigens in human sera. I. Detector isolation and characterization. J. Immunol. *128:* 2779–2787 (1982).

3   Lo, D.; Burkly, L.C.; Widera, G.: Diabetes and tolerance in transgenic mice expressing class II MHC molecules in pancreatic beta cells. Cell 53: 159 (1988).

4   Miller, J.F.: Post-thymic tolerance to self antigens. J. Autoimmun. 5: suppl A:27 (1992).

5   Rammensee, H.-G.; Kroschewski, R.; Frangoulis, B.: Clonal anergy induced in mature Vβ6+ T lymphocytes on immunizing Mls-1ᵇ mice with Mls-1ᵃ expressing cells. Nature 339: 541 (1989).

6   Qin, S.; Cobbold, S.; Benjamin, R.; Waldmann, H.: Induction of classical transplantation tolerance in the adult. J. Exp. Med. 169: 779 (1989).

7   Alters, S.E.; Shizuru, J. A.; Ackerman, J.; Grossman, D.; Seydel, K. B.; Fathman, C.G.: Anti-CD4 mediates clonal anergy during transplantation tolerance induction. J. Exp. Med. 491: 4 (1991).

8   Benjamin, R.J.; Qin, S.; Wise, M.P.; Cobbold, S. P.; Waldmann, H.: Mechanisms of monoclonal antibody-facilitated tolerance induction: A possible role for the CD4 (L3T4) and CD11a (LFA-1) molecules in self-non-self discrimination. Eur. J. Immunol. 18: 1079 (1988).

9   Dresser, D. W.: Specific inhibition of antibody production. II. Paralysis induced in adult mice by small quantities of protein antigen. Immunology 5: 378 (1962).

10 Chiller, J.M.; Habicht, G.S.; Weigle, W.O.: Kinetic differences in unresponsiveness of thymus and bone marrow cells. Science 171: 813 (1971).

11 Gahring, L.C.; Weigle, W.O.: The induction of peripheral T cell unresponsiveness in adult mice by monomeric human gamma-globulin. J. Immunol. 143: 2094 (1989).

12 Burtles, S.S.; Taylor, R.B.; Hooper, D.C.: Bovine gamma globulin-specific CD4+ T cells are retained by bovine gamma globulin-tolerant mice. Eur. J. Immunol. 20: 1273 (1990).

13 Bandeira, A.; Mengel, J.; Burlen-Defranoux, O.; Coutinho, A.: Proliferative T cell anergy to Mls-1a does not correlate with in vivo tolerance. Int. Immunol. 3: 923 (1991).

14 Miller, R.D.; Calkins, C.E.: Development of self-tolerance in normal mice. J. Immunol. 141: 2206 (1988).

15 Taguchi, O.; Nishizuka, Y.: Self-tolerance and localized autoimmunity: Mouse models of autoimmune disease that suggest tissue-specific suppressor T cells are involved in self-tolerance. J. Exp. Med. 165: 146 (1987).

16 Miller S.D.; Wetzig, R.P.; Claman, H.N.: The induction of cell-mediated immunity and tolerance with protein. J. Exp. Med. 149: 758 (1979).

17 Ramsdell, F.; Lantz, T.; Fowlkes, J.: A non-deletional mechanism of thymic self-tolerance. Science 246: 1038 (1989).

18 Roberts, J.L.; Sharrow, S.O.; Singer, A.: Clonal deletion and clonal anergy in the thymus induced by cellular elements with different radiation sensitivities. J. Exp. Med. 171: 935 (1990).

19 Houssaint, E.; Flajnik, M.: The role of thymic epithelium in the acquisition of tolerance. Immunol. Today 11: 357 (1990).

20 Mosmann, T.R.; Coffman, R.L.: Th1 and Th2 cells: Different patterns of lymphokine secretion lead to different functional properties. Annu. Rev. Immunol. 7: 145 (1989).

21 Gilbert, K.M.; Hoang, K.D.; Weigle, W.O.: Th1 and Th2 clones differ in their response to a tolerogenic signal. J. Immunol. 144: 2063 (1990).

22 Gilbert, K.M.; Weigle, W.O.: B cell presentation of a tolerogenic signal to Th clones. Cell. Immunol. 139: 58 (1982).

23 Jenkins, M.K.; Schwartz, R.H.: Antigen presentation by chemically modified splenocytes induces antigen-specific T cell unresponsiveness in vitro and in vivo. J. Exp. Med. 165: 302 (1987).

24   Quill, H.; Schwartz, R.H.: Stimulation of normal inducer T cell clones with antigen
     presented by purified Ia molecules in planar lipid membranes: Specific induction of a
     long-lived state of proliferative nonresponsiveness. J. Immunol. *138:* 3704 (1987).

25   Williams, I.R.; Unanue, E.R.: Costimulatory requirements of murine Th1 clones. J.
     Immunol. *145:* 85 (1990).

26   Mueller, D.L.; Jenkins, M.K.; Schwartz, R. H.: An accessory cell-derived costimulatory
     signal acts independently of protein kinase C activation to allow T cell proliferation
     and prevent the induction of unresponsiveness. J. Immunol. *142:* 2617 (1989).

27   Jenkins, M.K.; Chen, C.; Jung, G.; Mueller, D.L.; Schwartz, R.H.: Inhibition of
     antigen-specific proliferation of type 1 murine T cell clones after stimulation with
     immobilized anti-CD3 monoclonal antibody. J. Immunol. *144:* 16 (1990).

28   DeSilva, D.R.; Urdahl, K. B.; Jenkins, M.K.: Clonal anergy is induced in vitro by
     T cell receptor occupancy in the absence of proliferation. J. Immunol. *147:* 3261
     (1991).

29   Levich, J.D.; Parks, D.E.; Weigle, W.O.: Tolerance induction in antigen-specific T cell
     clones and lines in vitro. J. Immunol. *135:* 873 (1985).

30   Feldmann, M.; Zanders, E.D.; Lamb, J.R.: Tolerance in T-cell clones. Immunol.
     Today *6:* 58 (1985).

31   LaSalle, J.M.; Tolentino, P.J.; Freeman, G.J.; Nadler, L.M.; Hafler, D.A.: Early
     signaling defects in human T cells anergized by T cell presentation of autoantigen. J.
     Exp. Med. *176:* 177 (1992).

32   Sidhu, S.; Deacock, S.; Bal, V.; Batchelor, J.R.; Lombardi, G.; Lechler, R.I.: Human T
     cells cannot act as autonomous antigen-presenting cells, but induce tolerance in anti-
     gen-specific and alloreactive responder cells. J. Exp. Med. *176:* 875 (1992).

33   Lamb, J.R.; Zanders, E.D.; Sewell, W.; Crumpton, M.J.; Feldmann, M.; Owen, M.J.:
     Antigen-specific T cell unresponsiveness in cloned helper T cells mediated via the CD2
     or CD3/T cell receptor pathways. Eur. J. Immunol. *17:* 1641 (1987).

34   Davis, L.S.; Wacholtz, M.C.; Lipsky, P.E.: The induction of T cell unresponsiveness by
     rapidly modulating CD3. J. Immunol. *142:* 1084 (1989).

35   Iwatani, Y.; Amino, N.; Miyai, K.: Peripheral self-tolerance and autoimmunity: The
     protective role of expression of class II major histocompatibility antigens on non-
     lymphoid cells. Biomed. Pharmacother. *43:* 593 (1989).

36   Gaspari, A.A.; Jenkins, M.K.; Katz, S.I.: Class II MHC-bearing keratinocytes induce
     antigen-specific unresponsiveness in hapten-specific Th1 clones. J. Immunol. *141:* 2216
     (1988).

37   Chestnut, R.W.; Colon, S.M.; Grey, H.M.: Antigen presentation by normal B cells, B
     cell tumors, and macrophages: Functional and biochemical comparison. J. Immunol.
     128: 1764 (1982).

38   Cowing, C.; Chapdelaine, J.M.: T cells discriminate between Ia antigens expressed on
     allogeneic accessory cells and B cells: A potential function for carbohydrate side chains
     on Ia molecules. Proc. Natl Acad. Sci. USA *80:* 6000 (1983).

39   Gajewski, T.F.; Pinnas, M.; Wong, T.; Fitch, F.W.: Murine Th1 and Th2 clones
     proliferate optimally in response to distinct antigen-presenting cell populations. J.
     Immunol. *146:* 1750 (1991).

40   Baumhuter, S.; Bron, C.; Corradin, G.J.: Different antigen-presenting cells differ in
     their capacity to induce lymphokine production and proliferation of an apo-cy-
     tochrome c-specific T cell clone. J. Immunol. *135:* 989 (1985).

41   McKenzie, J.L.; Prickett, T.C.R.; Hart, D.N.J.: Human dendritic cells stimulate allo-
     geneic T cells in the absence of IL-1. J. Immunol. *67:* 290 (1989).

42 Webb, S.R.; Hu Li, J.; Wilson, D.B.; Sprent, J.: Capacity of small B cell enriched populations to stimulate mixed lymphocyte reactions: Marked differences between irradiated vs. mitomycin c-treated stimulators. Eur. J. Immunol. *15:* 92 (1985).

43 Ashwell, J.D.; Jenkins, M.K.; Schwartz, R.H.: Effect of gamms radiation on resting B lymphocytes. II. Functional characterization of the antigen-presentation defect. J. Immunol. *141:* 2536 (1988).

44 Gilbert, K.M.; Gahring, L.C.; Weigle, W.O.: Tolerance induced by soluble antigens; in Thomson, A.W. (ed.): The Molecular Biology of Immunosuppression. New York, Wiley, 1992, pp. 1–105.

45 Eynon, E.E.; Parker, D.C.: Small B cells as antigen-presenting cells in the induction of tolerance to soluble protein antigens. J. Exp. Med. *175:* 131 (1992).

46 Ryan, J.J.; Gress, R.E.; Hathcock, K.S.; Hodes, R.J.: Recognition and response to alloantigens in vivo. II. Primary with accessory cell-depleted donor allogeneic splenocytes: Induction of specific unresponsiveness to foreign major histocompatibility complex determinants. J. Immunol. *133:* 2343 (1984).

47 Hori, S.; Sato, S.; Kitagawa, S.; Azuma, T.; Kukudo, S.; Hamaoka, T.; Fujiwara, H.: Tolerance induction of allo class II H-2 antigen-reactive L3T4+ helper T cells and prolonged survival of the corresponding class H-2 disparate skin graft. J. Immunol. *143:* 1447 (1989).

48 Fuchs, E.J.; Matzinger, P.: B cells turn off virgin but not memory T cells. Science *258:* 1156 (1992).

49 Bass, H.; Mosmann, T.; Strober, S.: Evidence for mouse Th1- and Th2-like helper T cells in vivo. J. Exp. Med. *170:* 1495 (1989).

50 Romagnani, S.: Human Th1 and Th2 subsets: Doubt no more. Immunol. Today *12:* 256 (1991).

51 Swain, S.L.; McKenzie, D.T.; Weinberg, A.D.; Hancock, W.: Characterization of T helper 1 and 2 subsets in normal mice. J. Immunol. *141:* 3445 (1988).

52 De Wit, D.; Van Mechelen, M.; Ryelandt, M.; Figueiredo, A.C.; Abramowicz, D.; Goldman, M.; Bazin, H.; Urbain, J.; Leo, O.: The injection of deaggregated gamma globulins in adult mice induces antigen-specific unresponsiveness of T helper type 1 but not type 2 lymphocytes. J. Exp. Med. *175:* 9 (1992).

53 Chiller, J.M.; Weigle, W.O.: Cellular events during induction of immunologic unresponsiveness in adult mice. J. Immunol. *106:* 1647 (1971).

54 Weigle, W.O.: Analysis of autoimmunity through experimental models of thryoiditis and allergic encephalomyelitis; in Advances in Immunology, 30th ed. New York, Academic Press, 1980, p. 159.

55 Finkel, T.H.; Kubo, R.T.; Cambier, J.C.: T-cell development and transmembrane signaling: Changing biological responses through an unchanging receptor. Immunol Today *12:* 79 (1991).

56 June, C.H.: Signal transduction in T cells. Curr. Opin. Immunol. *3:* 287 (1991).

57 Rudd, C.E.: CD4, CD8 and the TCR-CD3 complex: A novel class of protein-tyrosine kinase receptor. Immunol Today *11:* 400 (1990).

58 Klausner, R.D.; Samelson, L.E.: T cell antigen receptor activation pathways: The tyrosine kinase connection. Cell *64:* 875 (1991).

59 Pardee, A.B.: G_1 events and regulation of cell proliferation. Science *246:* 603 (1989).

60 Feder, J.N.; Guidos, C.J.; Kusler, B.; Carswell, C.; Lewis, D.; Schimke, R.T.: A cell cycle analysis of growth-related genes expressed during T lymphocyte maturation. J. Cell Biol. *111:* 2693 (1990).

61 Crabtree, G.R.: Contingent genetic regulatory events in T lymphocyte activation. Science *243:* 355 (1989).

62 Trunch, A.; Albert, F.; Golstein, P.; Schmitt-Verhulst, A. M.: Early steps of lymphocyte activation bypassed by synergy between calcium ionophores and phorbol ester. Nature *313:* 318 (1985).

63 Mueller, DL.; Jenkins, M.K.; Chiodetti, L.; Schwartz, R.H.: An intracellular calcium increase and protein kinase C activation fail to initiate T cell proliferation in the absence of a costimulatory signal. J. Immunol. *144:* 3701 (1990).

64 Roska, A.K.; Lipsky, P.E.: Dissection of the function of antigen presenting cells in the induction of T cell activation. J. Immunol. *135:* 2953 (1985).

65 Nisbet-Brown, E.R.; Lee, J.W.W..; Cheung, R.K.; Gelfand, E.W.: Antigen-specific and -nonspecific mitogenic signals in the activation of human T cell clones. J. Immunol. *138:* 3713 (1987).

66 Jenkins, M.K.; Pardell, D.M.; Mizuguchi, J.; Chused, T.M.: Molecular events in the induction of a nonresponsive state in interleukin-1 producing helper lymphocyte clones. Proc. Natl Acad. Sci. USA *84:* 5409 (1987).

67 Jenkins, M.K.; Aswell, J.D.; Schwartz, R.H.: Allogeneic non-T spleen cells restore the responsiveness of normal T cell clones stimulated with antigen and chemically modified antigen-presenting cells. J. Immunol. *140:* 3324 (1988).

68 Bruce, J.; Symington, F.W.; Mckearn, T.J.; Sprent, J.: A monoclonal antibody discriminate between subsets of T and B cells. J. Immunol. *127:* 2496 (1981).

69 Liu, Y.; Janeway, C. A., Jr.: Cells that present both specific ligand and costimulatory activity are the most efficient inducers of clonal expansion of normal CD4 T cells. Proc. Natl Acad. Sci. USA *89:* 3845 (1992).

70 Steinman, R.M.; Young, J.W.: Signals arising from antigen-presenting cells. Curr. Opin. Cell Biol. *3:* 361 (1991).

71 Geppert, T.D.; David, L.S.; Gur, H.; Wacholtz, M.C.; Lipsky, P.E.: Accessory cell signals involved in T-cell activation. Immunol. Rev. *117:* 5 (1990).

72 Freeman, G.J.; Freedman, A.S.; Segil, J.M.; Lee, G.; Whitman, J.F.; Nadler, L.M.: B7, a new member of the Ig superfamily with unique expression on activated and neoplastic B cells. J. Immunol. *143:* 2714 (1989).

73 Turka, L.A.; Linsley, P.S.; Paine, R.; Schieven, G.L.; Thompson, G.B.; Ledbetter, J.A.: Signal transduction via CD4, CD8, and CD28 in mature and immature thymocytes. Implications for thymic selection. J. Immunol. *146:* 1428 (1991).

74 Larsen, C.P.; Ritchie, S.C.; Pearson, T.C.; Linsley, P.S.; Lowry, R.P.: Functional expression of the costimulatory molecule, B7/BB1, on murine dendritic cell populations, J. Exp. Med. *176:* 1215 (1992).

75 Freedman, A.S.; Freeman, G.J.; Rhynhart, K.; Nadler, L.M..: Selective induction of B7/BB-1 of interferon-gamma stimulated monocytes: A potential mechanism for amplification of T cell activation through the CD28 pathway. Cell. Immunol. *137:* 429 (1991).

76 Makgoba, M.W.; Sanders, M.E.; Shaw, S.: The CD2-LFA-3 and LFA-1-ICAM pathways: Relevance to T-cell recognition. Immunol. Today *10:* 417 (1989).

77 Albelda, S.M.; Buck, C.A.: Integrins and other cell adhesion molecules. FASEB J. *4:* 2868 (1990).

78 June, C.H.; Ledbetter, J.A.; Linsley, P.S.; Thompson, C.B.: Role of the CD28 receptor in T-cell activation. Immunol. Today *11:* 211 (1990).

79 Wawryk, S.D.; Novotny, J.R.; Wicks, I.P.; Wilkinson, D.; Maher, D.; Salvaris, E.; Welch, K.; Ferondo, J.; Boyd, A.W.: The role of LFA-1/ICAM-1 interaction in human leukocyte homing and adhesion. Immunol. Rev. *108:* 135 (1989).

80 Altmann, D.M.; Hogg, N.; Trowsdale, J.; Wilkinson, D.: Cotransfection of ICAM-1 and HLA-DR reconstitutes human antigen-presenting cell function in mouse L cells. Nature *338:* 512 (1989).

81   Azuma, M.A.; Cayabyab, M.; Buck, D.; Phillips, J.H.; Lanier, L.L.: CD28 interaction
      with B7 costimulates primary allogeneic proliferative responses and cytotoxicity medi-
      ated by small, resting T lymphocytes. J. Exp. Med. *175:* 353 (1992).

82   Galvin, F.; Freeman, G.J.; Razi-Wolfe, Z.; Hall, W.; Benacerraf, B.; Nadler, L.; Reiser,
      H.: Murine B7 antigen provides a sufficient costimulatory signal for antigen-specific and
      MHC-restricted T cell activation. J. Immunol. *149:* 3802 (1992).

83   Jenkins, M.K.; Taylor, P.S.; Norton, S.D.; Urdahl, K.B.: CD28 delivers a costimulatory
      signal involved in antigen-specific IL-2 production by human T cells. J. Immunol. *147:*
      2461 (1991).

84   June, C.H.; Ledbetter, J.A.; Lindsten, T.; Thompson, C.B.: Evidence for the involve-
      ment of three distinct signals in the induction of IL-2 gene expression in human T
      lymphocytes. J. Immunol. *143:* 153 (1989).

85   Thompson, C.B.; Lindsten, T.; Ledbetter, J.A.; Kunkel, S.L.; Young, H.A.; Emerson,
      S.G; Leiden, J.M.; June, C.H.: CD28 activation pathways regulates the production of
      multiple T-cell-derived lymphokines/cytokines. Proc. Natl Acad. Sci. USA *86:* 1333
      (1989).

86   Wacholtz, M.C.; Patel, S.S.; Lipsky, P.E.: Leukocyte function-associated antigen 1 is an
      activation molecule for human T cells. J. Exp. Med. *170:* 431 (1989).

87   Van Seventer, G.A.; Shimizu, Y.; Horgan, K.J.; Shaw, S.: The LFA-1 ligand ICAM-1
      provides an important costimulatory signal for T cell receptor-mediated activation of
      resting T cells. J. Immunol. *144:* 4579 (1990).

88   Damle, N.K.; Klussman, K.; Linsley, P.S.; Aruffo A..; Ledbetter, J.A.; Differential
      regulatory effects of intercellular adhesion molecule-1 or costimulation by the CD28
      counter-receptor B7. J. Immunol. *149:* 2541 (1992).

89   Tan, P.; Anasetti, C.; Hansen, J.A.; Melrose, J.; Brunvand, M.; Bradshaw, J.; Ledbetter
      J.A.; Linsley, P.S.: Induction of alloantigen-specific hyporesponsiveness in human T
      lymphocytes by blocking interaction of CD28 with its natural ligand B7/BB1. J. Exp.
      Med. *177:* 165 (1993).

90   Harding, F.A.; McArthur, J.G.; Gross, J.A.; Raultet, D.H.; Allison, J.P.: CD28-medi-
      ated signalling co-stimulates murine T cells and prevents induction of anergy in T-cell
      clones. Nature *356:* 607 (1992).

91   Isobe, M.; Yagita, H.; Okumura, K.; Ihara, A.: Specific acceptance of cardiac allograft
      after treatment with antibodies to ICAM-1 and LFA-1. Science *255:* 1125 (1992).

92   Charlton, B.; Guymer, R.H.; Slattery, R.M.; Mandel, T.E.: Intercellular adhesion
      molecule (ICAM-1) inhibition can induce tolerance in vivo. Immunol. Cell Biol. *69:* 89
      (1991).

93   Springer, T.A.; Dustin, M.L.: Kishimoto, T.K.; Marlin, S.D.: The lymphocyte function
      associated LFA-1, CD2, and LFA-3 molecules: Cell adhesion receptors of the immune
      system. Annu. Rev. Immunol. *5:* 223 (1987).

94   Weaver, C.T.; Unanue, E.R.: The costimulatory function of antigen-presenting cells.
      Immunol Today *11:* 49 (1990).

95   Aiello, F.B.; Longo, D.L.; Overton, R.; Takacs, L.; Durum, S.K.: A role for cytokines
      in antigen presentation: IL-1 and IL-4 induce accessory functions of antigen-presenting
      cells. J. Immunol. *144:* 2572 (1990).

96   Bonnefoy, J.-Y.; Denoroy. M.-C.; Guillot, O.; Martens, C.L.; Banchereau, J.: Activation
      of normal human B cells through their antigen receptor induces membrane expression of
      IL-1α and secretion of IL-1β. J. Immunol. *143:* 864 (1989).

97   Van Kooten, C.; Rensink, I.; Pascual-Salcedo, D.; van Oers, R.; Aarden, L.: Monokine
      production by human T cells; IL-1α production restricted to memory T cells. J.
      Immunol. *146:* 2654 (1991).

98 Tartakovsky, B.: Finnegan, A.; Muegge, K.; Brody, D.T.; Kovacs, E.J.; Smith, M.R.; Berzofsky, J.A.; Young, H.A.; Durum, S.K.: IL-1 is an autocrine growth factor for T cell clones. J. Immunol. *141:* 3863 (1988).

99 Levich, J.S.; Signorella, A.P.; Wittenberg, G.; Weigle, W.O.: Macrophage handling of a tolerogen and the role of IL-1 in tolerance induction in a helper T cell clone in vitro. J. Immunol. *138:* 3675 (1987).

100 Weigle, W.O.; Scheuer, W.V.; Hobbs, M.V.; Morgan, E.L.; Parks, D.E.: Modulation of the induction and circumvention of immunological tolerance to human gamma-globulin by interleukin. J. Immunol. *138:* 2069 (1987).

101 Liu, Y.; Jones, B.; Aruffo.; A, Sullivan, K.M.; Linsley, P.S.; Janeway, C.A.: Heat-stable antigen in a costimulatory molecule for CD4 T cell growth. J. Exp. Med. *175:* 437 (1992).

102 Evavold, B.D.; Quintans, J.: Accessory cell function of Th2 clones. J. Immunol. *143:* 1784 (1989).

103 Quill, H.; Gaur, A.; Brown, P.; Infante, A.J.; Phipps, R.P.: Synergistic activation of granulocyte-macrophage-colony-stimulating factor production by IL-1 and IL-2 in murine Th1 cells. J. Immunol. *143:* 2242 (1989).

104 Weaver, C.T.; Hawrylowicz, C.M.; Unanue, E.R.: T helper cell subsets require the expression of distinct costimulatory signals by antigen-presenting cells. J. Immunol. *25:* 8181 (1988).

105 Lichtman, A.H.; Chin, J.; Schmidt, J.A.; Abbas, A.K.: Role of interluekin-1 in the activation of T lymphocytes. Proc. Natl Acad. Sci. USA *85:* 9699 (1988).

106 Frank, S.J.; Niklinska, B.B.; Orloff, D.G.; Mercep, M.; Ashwell, J.D.; Klausner, R.D.: Structural mutations of the T cell receptor zeta-chain and its role in T cell activation. Science *249:* 174 (1990).

107 Gilbert, K.M.; Rothermel, A.L.; Ernst, D.N.; Hobbs, M.V.; Weigle, W. O.: Ability of tolerized Th1 and Th2 clones to stimulate B cell activation and cell cycle progression. Cell Immunol. *142:* 1 (1992).

108 Gilbert, K.M.; Ernst, D.N.; Hobbs, M.V.; Weigle, W.O.: Effects of tolerance induction on early cell cycle progression by Th1 clones. Cell Immunol. *141:* 362 (1992).

109 Saltzman, E.M.; White, K.; Casnellie, J.E.: Stimulation of the antigen and interleukin-2 receptors on T lymphocytes activates distinct tyrosine protein kinases. J. Biol. Chem. *256:* 10138 (1990).

110 Tony, H.P.; Parker, D.C.: Major histocompatibility complex restructured polyclonal B cells resulting from helper T cell recognition of anti-immunoglobulin presented by small B lymphocytes. Exp. Med. *161:* 223 (1985).

111 Noelle, R.J.; McCann, J.: Cognate interactions between helper T cells and B cells-I-cloning and helper activity of a lymphokine-dependent helper T cell clone. J. Mol. Cell. Immunol. *4:* 161 (1989).

112 Rothermel, A.; Gilbert, K.M.; Weigle, W.O.: Differential abilities of Th1 and Th2 to induce polyclonal B cell proliferation. Cell Immunol. *135:* 1 (1991).

113 Go, C.; Lancki, D.W.; Fitch, F.W.; Miller, J.: Anergized T cell clones retain their cytolytic ability. J. Immunol. *150:* 367 (1993).

114 Torbett. BE.; Glasebrook, A.L.: Induction of antigen unresponsiveness in T helper clones inhibits IL-2 but not IL-4 secretion. FASEB J. *3:* A1269 (1989).

115 Mueller, D.L.; Chiodetti, L.; Bacon, P.A.; Schwartz, R.H.: Clonal anergy blocks the response to IL-4, as well as the production of IL-2, in dual-producing T helper cell clones. J. Immunol. *147:* 4118 (1991).

116 Evavold, B.D.; Allen, P.M.: Separation of IL-4 production from Th cell proliferation by an altered T cell receptor ligand. Science *252:* 1308 (1991).

117   Quill, H.; Riley, M.P.; Cho, E.A.; Casnellie, J.E.; Reed, J.C.; Torigae, T.: Anergic Th1
      cells expressed altered levels of the protein kinases p56[lck] and p59[fyn] J. Immunol. *149*:
      2887 (1992).

118   Go, C.; Miller, J.: Differential induction of transcription factors that regulate the
      interleukin-2 gene during anergy induction and restimulation. J. Exp. Med. *175*: 1327
      (1992).

119   Kang, S.-M.; Beverly, B.; Tran, A.-C..; Brorson, K.; Schwartz, R.H.; Lenardo, M.J.:
      Transactivation by AP-1 is a molecular target of T cell clonal anergy. Science *257*: 1134
      (1992).

120   Beverly, B.; Kang, S.M.; Lenardo, M.J.; Schwartz, R.H.: Reversal of in vitro T cell
      clonal anergy by IL-2 stimulation. Int. Immunol. *4*: 661 (1992).

121   Norton, S.D.; Hovinen, D.E.; Jenkins, M.K.: IL-2 secretion and T cell clonal anergy
      are induced by distinct biochemical pathways. J. Immunol. *146*: 1125 (1991).

122   Gilbert, K.M; Weigle, W.O.: Th1 cell anergy and blockade in G_{1a} phase of the cell cycle.
      J. Immunol, in press.

123   Morris, D.R.; Keupfer, C.A.; Ellingsworth, L.R.; Ogawa, Y.; Rabinovitch, P.S.: Trans-
      forming growth factor-β blocks proliferation but not early mitogenic signalling events in
      T-lymphocytes. Exp. Cell. Res. *185*: 529 (1989).

124   Loertscher, R.; Leichtman, A.B.; Williams, J.M.; Strom, T.B.: Blocking of T cell
      activation at the G_{1a}/G_{1b} interface of the cell cycle and prevention of expression of
      IL-2R by alloactivated suppressor T lymphocytes. Transplant *45*: 194 (1988).

125   Rigby, W.F.C.; Hamilton, B.J.; Waugh, M.G.: 1,25-Dihydroxyvitamin D_3 modulates
      the effects of interleukin-2 independent of IL-2 receptor binding. Cell. Immunol. *125*:
      396 (1990).

Kathleen M. Gilbert, The Scripps Research Institute, Department of Immunology,
10666 North Torrey Pines Road, La Jolla, CA 92037 (USA)

Granstein RD (ed): Mechanisms of Immune Regulation.
Chem Immunol. Basel, Karger, 1994, vol 58, pp 117–145

Regulation of the Immune Response within Privileged Sites

Bruce R. Ksander[a], *J. Wayne Streilein*[b]

[a]Department of Microbiology and Immunology, University of Miami Medical School;
[b]Department of Ophthalmology, Bascom Palmer Eye Institute, Miami, Fla., USA

Introduction

An immune privileged site is defined as an anatomical site in which immunogenic tissue survives for an extended period of time in an immunocompetent host. Privileged sites have been identified by experiments that compare the survival of allografts placed within the privileged site with the rejection of identical grafts placed in conventional nonprivileged sites such as the subcutaneous tissue of the skin. While many investigators have confirmed the existence of privileged sites, the literature contains many reports which refute the concept of immune privilege, primarily because not all foreign grafts are accepted at privileged sites for indefinite intervals of time. In fact, survival of allografts within privileged sites is usually prolonged, but often varies considerably. Although these inconsistent results have been used in the past as evidence against the existence of immune privilege, they most likely indicate that the extent to which immune privileged sites can protect allografts from immune-mediated destruction varies greatly. On the one hand, immune privilege can be absolute and provide indefinite protection; on the other hand, privilege may be transient and provide only brief protection. The extent to which immune privilege protects allografts depends upon (i) features of the immune privileged site; (ii) the immunogenicity of the graft, and (iii) the type of tissue used as an allograft (skin, tumor, or endocrine tissue). In general, the survival of allografts within immune privileged sites is inversely related to the immunogenicity of the graft; weakly immunogenic grafts survive longer than highly immunogenic grafts.

Immune privilege is believed to have developed evolutionarily as a protective mechanism for highly specialized organs in which a local immune response can disrupt normal function. For example, the anterior segment of the eye is responsible for focusing light on the retina so that normal vision can occur. A local inflammatory response within the anterior segment would disrupt vision by preventing this function. Since loss of vision is a major threat to host survival, immunologic privilege is maintained within the anterior chamber of the eye in order to preserve vision by preventing local immunogenic inflammation. Creation of immune privilege may make these sites more susceptible to invading pathogens. It is believed that immune privilege develops only within organs in which maintaining normal function is paramount to survival of the host, and in which disruption would be more threatening to the host than the threat of invading pathogens. Thus immune privilege is bestowed on sites that maintain neurological functions (brain and eye) and reproductive function (fetus and testis). In this review it will become obvious that immune privilege does not leave these sites completely without immune protection.

The original and simplest explanation of how immune privilege is maintained was that allografts within privileged sites are hidden from the immune system and are unable to induce an immune response within the host. Thus, immune privilege is maintained by sequestering antigens so that the host is unaware of their presence. Three common properties of immune privileged sites suggest they are isolated physically from the immune system. Most immune privileged sites possess the following features: (i) an abnormal, or deficient lymphatic drainage pathway; (ii) a blood:tissue barrier that limits the ingress of blood-borne molecules and cells, and (iii) an immunosuppressive local microenvironment that contains atypical antigen-presenting cells. The absence of conventional antigen-presenting cells and lymphatic drainage was believed to disrupt the afferent phase of the immune response, while the presence of a blood-tissue barrier was thought to prevent the efferent arm of immunity by blocking the infiltration of immune effectors into the site. The concept that immune privileged sites are physically isolated was also supported by experimental evidence. Artificial immune privileged sites were created by disrupting the lymphatic drainage to an isolated flap of skin. By surgically removing a skin flap, the vascular supply was kept intact, but the lymphatic drainage was prevented. Allogenetic skin grafts placed on isolated skin flaps survived indefinitely. This experimental evidence seemed to support the concept that immune privilege was established by preventing the escape of antigens. Current experimental evidence indicates that this model of immune privilege applies only to the artificial skin flap and the hamster cheek pouch, which is the only naturally occurring immune privileged site that truly lacks a lymph draining system, thus preventing immune recognition of antigens contained within this site.

Considerable evidence has accumulated over the past 20 years that indicates immune privileged sites are not isolated from the immune system of the host. It is now clear that antigens placed within immune privileged sites are capable of inducing an immune response. Although it is true that immune privileged sites often lack direct lymphatic drainage, antigens or antigens carried by antigen presenting cells escape by direct vascular routes and induce an immune response in the host. The contemporary view is that immune privilege is maintained by regulation of immune responses induced by antigens that escape privileged sites and it is now apparent that the immune response is regulated, at least in part, by the existence of a local immunosuppressive microenvironment that alters both afferent and efferent phases of immunity.

In this review we will also consider the concept that immune privilege may be established at sites where spontaneous tumors develop. Since immune privilege is a mechanism that allows immunogenic tissue to escape immune-mediated elimination, and since spontaneous tumors express immunogenic antigens, we believe that tumors may use similar mechanisms to those developed by privileged sites to escape immune-mediated destruction.

The Fetus as an Immune Privileged Site

The mating of two genetically disparate individuals results in a fetus that inherits and expresses paternal transplantation antigens while residing in the mother's uterus. In this way a fetus represents a semi-allogeneic graft that survives and escapes from a destructive alloimune response. The initial experiments that identified the fetus as an immune privileged site were conducted by Woodruff [1] in which fetal tissue was removed from the uterus and grafted on the thigh. The transplanted fetal tussue was rapidly rejected from the thigh, but the embryos remaining in the uterus were unaffected by the allogeneic response. Thus the fetal tissue was susceptible to immune-mediated destruction at a heterotopic site. Within the gravid uterus the privileged nature of the site was identified by grafting experiments that demonstrated that allogeneic grafts placed at the maternal-fetal junction (hemochorial junction) survived, but identical grafts placed on endometrium distant to the interface were rapidly rejected. These experiments identify the maternal-fetal junction as conveying privilege to the fetus.

Development of the Maternal-Fetal Junction

The maternal-fetal junction is formed when the conceptus embeds in the decidual lining. The conceptus consists of an inner and outer cell mass, the inner cell mass forming the fetus and the outer cell mass forming the trophoblastic tissue. Under the influence of the decidua basalis (side closest to the implantation site) trophoblasts differentiate into cells that comprise the placenta. The maternal-fetal junction is formed as the placenta becomes

embedded in the maternal sinuses in the decidua. At the histological level
the barrier between the mother and fetus occurs at the placental villi that
insert into the decidual sinuses. The placenta contains three types of
trophoblasts-villous trophoblasts consisting of the syncytiotrophoblasts
and the cytotrophoblasts, and extravillous trophoblasts. The syncytiotrophoblasts are in direct contact with the maternal blood and form the outer
most barrier of the placental villi. The cytotrophoblasts are directly be-
neath these cells and the extravillous trophoblasts are further removed
from the surface of the placenta. The trophoblasts are believed to play a
prominent role in the resistance of the fetus to 'normal' allograft immunity
directed against paternal transplantation antigens, because these fetal cells
form the barrier between the fetus and mother. Grafting experiments
indicate that (a) the fetus does not normally prime maternal mice to
paternal transplantation antigens, and (b) if the mice are previously primed
to paternal alloantigens, the allospecific cells fail to reject or eliminate the
fetus. The mechanisms behind this resistance to classical allograft immunity
are believed to involve the trophoblasts and are due to (a) altered expression of class I and II molecules on the cell surface, and (b) resistance of
trophoblasts to lysis by effector cells. Although the fetus is resistant to
classical allograft immunity the evidence suggests that it may, under certain
conditions, be susceptible to nonspecific immune responses mediated by
natural killer (NK)-like effector cells.

Expression of Class I and II Antigens on Trophoblasts

Since paternal transplantation antigens would be recognized by maternal lymphocytes if expressed at the interface between fetus and mother, the
expression of MHC on trophoblasts is obviously of extreme importance
and has been studied extensively in humans and rodents. In humans,
trophoblasts at all layers of the placenta fail to express HLA class II
molecules and treatment with IFN-γ fails to up-regulate class II expression
on the cell surface [2–4]. By contrast, the expression of class I molecules on
trophoblasts is highly regulated and varies depending upon the type of
trophoblast and the proximity with maternal decidual tissue. In the area
where exchange occurs between maternal blood cells and the fetus, villo
trophoblasts (syncytiotrophoblasts and cytotrophoblasts) fail to express
cell surface class I molecules [5–7]. Hunt et al. [8] used in situ hybridization for class I mRNA to examine HLA expression within placental tissue.
Their results indicate that syncytiotrophoblasts express neither cytoplasmic
mRNA for class I, nor class I cell-surface proteins. By contrast, cytotrophoblasts possess class I mRNA, but fail to express the protein on the cell
surface. The extravillous trophoblasts, which are not in contact with the
exchange area, express both mRNA and cell-surface class I molecules.

However, the expression of class I is abnormal on these trophoblasts. They express nonclassical HLA-G cell-surface proteins, and a limited amount of HLA-C can be detected, but no HLA-A or HLA-B is found [9-11]. HLA class I expression can be up regulated by IFN-γ treatment, however, it is not known which class I (HLA-G, or C) is expressed on IFN-γ treated trophoblastic cells [12]. These results indicate that expression of class I is regulated at the cellular level and varies among subpopulations of tro-phoblasts. This has given rise to the belief that the failure of trophoblasts to express classical HLA molecules contributes to the failure of allogeneic fetal tissue to sensitize the mother.

What is interesting about HLA-G expression on the extravillous trophoblasts is that HLA-G is only detected on embryonic tissue in the placenta. Moreover, expression appears to be linked to the development of the placenta and HLA-G levels are reduced on third trimester trophoblasts. HLA-G is unusual and differs from HLA-A, B, and C molecules in that it has (i) a truncated cytoplasmic region, and (ii) a smaller, nonpolymorphic 39 kD α-chain. In addition, a smaller 37-kD isoform of HLA-G has been detected that appears to be secreted and not expressed on the trophoblast cell surface. Thus, soluble HLA-G class I molecules may be secreted by trophoblasts within the placenta. Many important questions remain unan-swered about the expression of HLA-G on trophoblasts (i) what peptides might be presented by HLA-G class I; (ii) which maternal cells recognize peptides in the context of HLA-G, and (iii) what is the function of these cells. Regarding the latter, Loke and King [13] suggest that HLA-G may be involved in the recognition of trophoblasts by maternal NK large granular lymphocytes (LGL) cells in the decidua. These cells may function in limiting the invasion of trophoblasts into the decidua. Whether the nonclassical MHC expression is involved in nonspecific immune responses is discussed later.

It is interesting to speculate on the possible role of HLA-G in the failure of the fetus to induce classical allograft immunity. It is possible that cells responding to HLA-G may function in down-regulating normal alloimmune responses.

Resistance of Trophoblasts to Lysis by Cytotoxic Effector Cells
Since human trophoblasts express HLA-G and only low levels of HLA-C, these cells could only be lysed by maternal cytotoxic T cells that recognize these paternal MHC molecules. By contrast, murine trophoblasts express all of the classical MHC class I molecules on their cell surface [14]. However, even though these cells express the relevant paternal class I alloantigens, they are not lysed in vitro by either maternal allospecific cytotoxic T cells, or NK cells [15, 16]. The inability of maternal cells to lyse

trophoblasts was not due to insufficient expression of class I, or to the failure of cytotoxic T cells (CTL) to bind trophoblast target cells. Moreover, trophoblasts do not secrete soluble factors that prevent CTL from lysing other allogeneic target cells. These results suggest that trophoblasts are resistant to the lytic factors that CTL normally secrete to lyse target cells. This resistance to lysis was confirmed for one lytic factor, tumor necrosis factor-α (TNFα)-trophoblasts are resistant to lysis by TNFα [17]. The resistance of fetal trophoblasts to lysis, even though they express paternal class I alloantigens, may contribute to the resistance of the fetus to normal alloimmune cell-mediated immunity. If mice are primed to paternal class I alloantigens prior to, or during pregnancy, they fail to mount an effective immune response against the fetus [18, 19]. The failure of classical alloimmunity in fetal rejection is further confirmed by the fact that it is rare, even in aborting fetuses, to detect maternal T cells infiltrating into the decidua [20]. Together these results indicate that T cell-mediated immunity is unable to mount an effective allodestructive immune response against the fetus under normal circumstances, and therefore is unlikely to play a role in the induction of fetal abortions. However, alloimmune rejection may play a role in habitual abortion. To the contrary, there is evidence that primed T cells specific for paternal alloantigens can secrete cytokines that increase placental and fetal weight, resulting in increased fetal survival. T cells that secrete interleukin 3 (IL-3), granulocyte-macrophage colony-stimulating factor, (GM-CSF), and colony-stimulating factor-1 (CSF-1) serve as growth factors for placental trophoblasts [21, 22].

The resistance of trophoblasts to lysis by lytic factors is not complete. Lymphokine-activated killer (LAK) cells and activated NK cells can readily lyse trophoblast target cells [23]. This is significant because unlike T cells, NK cells or LGL are often detected within the decidua adjacent to the fetal trophoblasts [24, 25].

Identification of Unique NK-Like Cells within the Decidua

As fetal trophoblasts infiltrate into the decidua, maternal mononuclear cells can be detected within the placenta adjacent to the extravillous trophoblasts in both humans and rodents. Macrophage/monocytes and CD3 + cells comprise only a small percentage of these cells. The majority of the cells are NK-like LGL that appear to be derived from the NK cell lineage, but are phenotypically and functionally distinct from NK cells found among peripheral blood lymphocytes. Human LGL in the decidua express high levels of CD56 (a neural cell adhesion molecule marker) that is normally found on NK cells in the peripheral blood, but these cells fail to express CD16, CD3, CD4, or CD8 [26, 27]. They also express the early T cell markers CD2 and CD7. These LGL are distinct from the subpopulation of NK cells in the peripheral blood that are CD56+, CD16−, CD3−

in that the expression of CD56 is elevated on decidual LGL. Moreover, CD56+ NK cells in the PBL are normally small agranulated cells and do not possess the large granules found in decidual LGL. King and Loke [28] have demonstrated that decidual LGL can lyse NK-sensitive K562 target cells, but fail to lyse trophoblasts. By contrast, if these same LGL are cultured in the presence of IL-2 they proliferate and can now kill both K562 and extravillous trophoblasts. Loke and King [13] hypothesize that uterine-derived CD56+ LGL are uterine-resident leukocytes that migrate from the bone marrow and proliferate in response to progesterone in the uterus. These cells may function in limiting trophoblast migration into the decidua. There is considerable speculation on the possible role of HLA-G as a restricting element for LGL trophoblast recognition, however this is still under investigation.

Rodents also possess unique NK-like cells within the decidua that are found within the metrial gland – a small glandular structure within the myometrium [29]. These cells are called granulated metrial gland cells, (GMG) and appear to be similiar to human decidual LGL cells (although glands are not formed around human LGL cells). As in human LGL cells, murine GMG are similar to NK cells and express asialo-GM1 (ASGM1) plus Fc receptors, but are morphologically and functionally distinct from NK cells in the peripheral blood. GMG cells are found in the areas adjacent to degenerating trophoblasts in the region of the placenta where there is exchange between fetus and mother [30]. Evidence suggests that these NK-like cells may play an important role in fetal absorption. The frequency of GMG cells in the placenta increases during allogeneic pregnancies [31]. A direct role for NK-like cells in fetal absorption has been demonstrated in experiments conducted by Baines and colleagues [32, 33]. Increased fetal absorption is induced in CBA × DBA/2 mice treated with poly I:C (polyinosinic cytidilic acid), a potent inducer of IFN that activates NK cells. Fetal absorption can also be induced by the adoptive transfer of lymphocytes from poly I:C treated mice. The ability of transferred lymphocytes to induce absorption is eliminated if they are first treated with anti-ASGM1 antibody and complement. From these series of experiments it is currently believed that activated NK cells induce absorption by lysing fetal trophoblasts.

Role of the Local Microenvironment in Maintaining Immune Privilege

The previous evidence suggests that pregnancy failure can be mediated by NK cells and thus represents a condition where immune privilege for the fetus is terminated. If the normal function of these NK-like cells within the decidua is to control the growth of invasive trophoblasts, it is possible that NK cells normally are capable of limiting trophoblast growth without affecting fetal survival or terminating immune privilege. However, under conditions where NK cell activation is up-regulated (as in the poly I:C model) these effector cells terminate immune privilege and initiate absorption. To maintain fetal viability and immune privilege, there is evidence that a local microenvironment is established at the maternal fetal junction that down-regulates NK cells at this site. Clark and colleagues [34, 35] have identified a non-T/non-B suppressor cell within the decidua that can be detected 4–5 days post implantation. These suppressor cells release a soluble factor similar to transforming growth factor B_2 (TGFβ_2). As mentioned above, although trophoblasts are resistant to lysis by CTL and TNFα, they can be lysed by activated ASGM-1 + cells in the decidua. Lysis of trophoblasts can be blocked by anti-ASGM-1 antibody and complement treatment, or by treatment with TGFβ_2. Moreover, treatment of mice with anti-TGFβ_2 antibodies in vivo *increases* the absorption rate of fetuses. Thus treatments like poly I:C or intrauterine injection of lipopolysaccharide (LPS), that increase the abortion rate may block or overcome TGFβ_2-mediated suppression of NK cells. There is also evidence that the increased fetal survival observed after immunization with paternal alloantigens results from increased secretion of TGFβ_2 by cells in the decidua.

It is tempting to speculate that this local microenvironment also influences maternal antigen presenting cells within the decidua. Redline et al. [36] demonstrated that decidual cells inhibit macrophage adhesion and migration. Thus, it is possible that the microenvironment within the decidua prevents the induction of an effective T cell response to paternal alloantigens by altering the ability of antigen presenting cells to process and present antigens in a conventional manner.

In summary, the maternal-fetal interface is highly specialized immunologically, and as a consequence this region functions as an immune privileged site. The most obvious example of privilege at this site is the successful development to term of an allogeneic fetus. The factors which contribute to immune privilege at the maternal-fetal interface are similar, but not identical with, the three features mentioned at the beginning of this chapter as important in creating an immune privileged site. Clearly, the uterine microenvironment which bathes trophoblastic tissue is unique, and immunosuppressive. Moreover, a kind of blood:tissue barrier exists that limits the transfer of blood-borne maternal cells and molecules into the fetal circulation. Moreover, fetal cells and molecules that break off from the placenta are readily transported into the maternal venous circulation. However, the pregnant uterus displays a luxurious lymphatic drainage. Two other features of the maternal-fetal interface undoubtedly contribute to locally altered cell-mediated immune responses. First, expression of molecules encoded by the major histocompatibility complex is severely reduced, compared to other tissues, rendering the conceptus less accessible to MHC-restricted T immune effector cells. Second, NK cells within the uterine decidua are regulated by factors in the local microenvironment, preventing pernicious invasion and destruction of the developing placenta. The first of these may also contribute to immune privilege in the eye and in the brain.

Anterior Chamber of the Eye

It has been known for over 100 years that the anterior chamber (AC) of the eye is an immune privileged site. This early demonstration of immune privilege was probably due to the fact that the AC was easily accessible and grafts could be readily placed within the AC and their survival monitored visually through the transparent cornea. Early experiments reportedly observed that allogeneic and even xenogeneic tissue survived for a prolonged period of time within the AC [37]. As in other immune privileged sites it was believed that antigens placed within the AC failed to escape and therefore failed to induce an immune response within the host. This was supported by the fact that the AC lacks direct access to lymphatic drainage. The concept of AC as an immune privileged site that was physically isolated from the immune system survived until 1977 when Kaplan and Streilein [38] demonstrated that immunogenic tissue within the AC induced elevated levels of specific serum antibodies.

Extensive studies of the immune response by several laboratories indicated that the systemic immune response to antigens within the AC of the eye was altered or 'deviant' [39–41]. The characteristic deviant immune response to antigens within the AC has been termed ACAID (anterior chamber-associated immune deviation). In general, the induction of ACAID to antigens within the AC results in the clonal expansion of precusor T lymphocytes, but down-regulation of effector functions mediated by terminally differentiated T cells. Thus in mice with ACAID specific precursor cytotoxic T cells and IL-2-secreting CD4 Th cells are clonally expanded [42–44], but delayed-type hypersensitivity (DH), is suppressed [45], and terminally differentiated cytolytic T cells and memory T helper cells fail to mature [46, 47]. Mice with ACAID also possess elevated serum antibody levels of all IgG isotypes with the exception of complement binding IgG2a [48, 49]. It is believed that immune privilege within the AC is maintained by the induction of ACAID in order to protect vision through down-regulation of intraocular immunogenic inflammatory responses.

ACAID has typically been confirmed by the adoptive transfer with lymphoid cells of suppressed DH and using this criteria a wide variety of antigens injected into the AC have been found to induce ACAID: (i) cell-associated antigens (tumor-specific antigens [50], hapten modified splenocytes [51, 52], minor and/or major MHC alloantigens presented by either skin grafts [53], tumor cells [45], fetal retinal tissue [54], or corneal grafts [55]), and (ii) soluble antigens (BSA, S antigen [56], IRBP [57, 58], and HSV viral proteins [59, 60]). As in other immune privileged sites, the extended survival of allogeneic tissues within the AC can be indefinite or only prolonged slightly. The extent to which immune privilege is maintained depends upon the immunogenicity of the antigens placed within the AC; tumor allografts bearing weak transplantation antigens survive indefinitely, while survival of MHC disparate tumor allografts is prolonged only slightly. Evidence that ACAID is essential in maintaining immune privilege is provided by the direct correlation between the termination of tumor graft survival in the AC and the termination of ACAID.

Induction of ACAID

Antigenic signals can escape from the AC by passing through the trabecular meshwork into the canal of Schlem and entering the blood vasculature. Although the AC has no direct lymphatic drainage, an indirect pathway through the ciliary body has been identified in monkeys, but it is uncertain if this pathway exists in mice. Recent progress has been made in identification of the signal that induces ACAID after BSA is injected into the AC. It appears that F4/80+ cells present in the iris escape with specific antigens into the blood vasculature where they preferentially migrate to the spleen. In experiments where mice were splenectomized, F4/80 cells appear to accumulate within the peripheral blood. Moreover, if labeled F4/80 cells isolated from the iris were injected intravenously, they accumulated preferentially in the spleen [61]. These results explain earlier experiments that indicated a functional spleen was required for the induction of ACAID and immune privilege in the AC. If minor incompatible tumor cells were injected into the AC of splenectomized mice, ACAID was not induced and immune privilege was not extended to the tumor cells [62]. These results coincide with the fact that CD4 and CD8 T cells that can suppress DH are found within the spleen. Thus, it appears that F4/80 cells may function in the induction of specific suppressor cells in the spleens of mice with ACAID. In contrast with these results, Ferguson et al. [52], reported that the ACAID inducing signal that leaves the eye after AC injection of TNP coated spleen cells resides within the serum. The differences between these results may reside in the different forms of antigen injected into the AC.

The F4/80 cells present within the iris and ciliary body appear to be unique in their ability to induce ACAID [63, 64]. If F4/80 cells are isolated from the iris/ciliary body, pulsed with antigen, and injected intravenously, recipient mice fail to develop specific DH responses when immunized subcutaneously. By contrast, F4/80 cells found among peritoneal exudate cells (PEC) cannot induce ACAID using an identical experimental protocol. However, if antigen pulsed F4/80 cells obtained from the PEC are *injected into the AC*, they were now capable of inducing ACAID. These results suggested that the local microenvironment within the AC was able to convert F4/80 cells into ACAID-inducing cells. This was confirmed in experiments where F4/80 PEC were treated with aqueous humor, pulsed with antigen, and injected intravenously. When mice that had received these cells were immunized subcutaneously, they failed to display specific DH. Thus, PEC F4/80 cells pretreated with aqueous humor can induce ACAID. These series of experiments indicate that the local microenvironment within the AC is important in establishing and maintaining immune privilege ACAID.

Role of the Local Microenvironment in the Induction of ACAID

It has been known for some time that the aqueous humor that fills the AC contains immunosuppressive factors. Orginal studies indicated that aqueous humor suppressed proliferation of lymphocytes [65]. More recent experiments indicated that aqueous humor suppressed specific T cell proliferative responses to alloantigens and soluble antigens. Several laboratories are studying the composition of aqueous humor in an attempt to identify the immunosuppressive factors. Several different factors have been identified: $TGF\beta_2$, α-melanocyte-stimulating hormone (α-MSH), vasoactive intestinal polypetide (VIP), and calcitonin gene-related peptide (CGRP) [66–70]. In addition, it appears that other small molecular weight factors (less than 5 kD) may also be present. The identification of this factor(s) awaits further study. Experiments by Wilbanks et al. [71] confirmed that immunosuppressive factors within aqueous humor contribute to the induction of ACAID. As described above, if F4/80 cells isolated from PEC are pulsed with antigen and injected intravenously, they fail to induce ACAID. By contrast, if these same cells are first pretreated with either aqueous humor, $TGF\beta_2$, or α-MSH, they now are capable of inducing ACAID. Thus, the induction of ACAID by F4/80 cells is dependent upon the immunosuppressive factors present within the aqueous humor. At least some of the immunosuppressive factors found within the aqueous humor are secreted by parenchymal cells within the iris/ciliary body. F4/80 cells obtained from PEC and co-cultured with iris and ciliary body cells acquired the ability to induce ACAID when they were pulsed with antigen and injected intravenously. Moreover, supernatants from iris/ciliary body cells contain $TGF\beta_2$.

   The data from these experiments support the following mechanism for the induction of ACAID. Parenchymal cells within the iris/ciliary body cells secrete immunosuppressive factors that are contained within the aqueous humor that bathes the tissues surrounding the AC. F4/80+ cells that migrate from the blood into the iris are exposed to the suppressive factors in aqueous humor. Contact with these factors alters the ability of F4/80+ cells to process and present antigens in a conventional manner. Thus when antigens are injected into the AC, F4/80+ cells process these antigens uniquely, migrate from the eye through the blood vasculature to the spleen. Within the spleen the eye-derived F4/80+ cells induce cells that suppress the induction of DH.

   Since there are no physical barriers that prevent factors within the aqueous humor from diffusing into the posterior of the eye, it is possible that the immunosuppressive factors present with the anterior chamber are also present in the posterior of the eye. This may account for the recent results of Jiang and Streilein [72] that indicate the vitreous cavity and the subretinal space are also immune privileged sites.

   It is interesting that the ability to convert F4/80+ PEC into ACAID-inducing cells is also a property of fluids obtained from several different immune privileged sites. F4/80+ PEC pretreated with either amniotic fluid or cerebrospinal fluid (CSF) induced ACAID when they were pulsed with antigen and injected intravenously [73]. $TGF\beta_2$ is present in amniotic fluid and CSF, and treatment of these fluids with anti-$TGF\beta_2$ antibodies neutralized the ACAID-inducing of these fluids. Thus the fluids present within three different immune privileged sites: anterior chamber, fetus, and brain all appear to share a common property that can result in the down-regulation of DH responses.

Role of the Local Microenvironment in the Efferent Phase of the Immune Response

   Experiments conducted by Cousins et al. [74] indicated that the anterior segment of the eye resists development of local DH. However, this was reversed and DH developed in the eye if IFN-γ was first injected into the AC. Experiments by Streilein et al. [75] indicated that IFN-γ pretreatment altered the production of immunosuppressive factors by iris and ciliary body cells. Eyes treated with IFN-γ still contained immunosuppressive factors, however the composition changed so that aqueous humor contained less $TGF\beta_2$, and now contained PGE_2.

   It is believed that IFN-γ secreted by Tdh cells is an important mediator of DH. Taylor et al. [68, 69] examined the effect of aqueous humor immunosuppressive factors on the production of IFN-γ in vitro to determine if the resistance of the unaltered anterior chamber to DH was the result of suppression of IFN-γ production by Tdh cells. Their results indicated that aqueous humor can suppress the secretion of IFN-γ by specific T cells and this is mediated by α-MSH, VIP, and/or TGFβ_2. In this manner the immunosuppressive factors within the anterior chamber can resist the development of local DH by suppression of IFN-γ secretion by specific T cells that infiltrate into the eye.

By contrast, aqueous humor is unable to prevent lysis of tumor target cells by directly cytolytic T lymphocytes. Kaiser et al. [76] demonstrated that in the presence of aqueous humor, directly cytolytic T cells were still able to bind and lyse target cells. These results are consistent with other studies that indicate that it is very difficult to suppress cytotoxic T cells when they are terminally differentiated and poised to bind and lyse target cells. In spite of the failure of aqueous humor to prevent lysis of tumor cells, mice primed for a cytotoxic T cell response are unable to eliminate minor incompatible P815 tumor cells from the AC. Our results indicate that the failure of cytotoxic T cells to eliminate tumor cells from the AC is due to a failure of infiltrating precursor cytotoxic T cells (pTc) to terminally differentiate into cytotoxic T cells (Tc) [46]. When minor incompatible tumor cells are successfully rejected from the subconjunctival space (a nonprivileged site), pTc infiltrate the eye and terminally differentiate into Tc that lyse tumor cells [77]. Terminal differentiation of pTc coincides with the infiltration into the eye of T helper cells that secrete IL-2 + IL-4 [78]. Thus, in nonprivileged sites the final terminal differentiation step of pTc occurs within the tumor-containing site and is mediated by lymphokines secreted by T helper cells. By contrast, within the immunologically privileged AC pTc infiltrate, but fail to differentiate further and therefore are unable to lyse tumor cells. These results demonstrate that immunogenic tumors within the privileged AC escape elimination because precursor T cells fail to terminally differentiate once they enter the tumor-containing eye.

Ksander et al. [77, 78] have demonstrated that pTc differentiation is blocked within the AC (i) because of incomplete activation of T helper cells, and (ii) because aqueous humor suppresses pTc differentiation. Bando et al. [44] demonstrated that following inoculation of P815 tumor cells into the AC, specific CD4 T helper cells that secrete IL-4 failed to develop in the spleen and draining lymph nodes. The lack of specific T helper cell activity was not complete however, since specific IL-2-secreting Th cells were detected. It was interesting that specific CD4+ Th cells secreted IL-2 at levels equal to that found in lymph nodes draining SCon sites, a nonprivileged site, where tumor cells were rejected. It appeared that in mice with ACAID, the failure of IL-2-secreting cells to initiate pTc differentiation in vivo was due to suppression of Th cells that infiltrate into the tumor-containing eye. The in vitro experiments of Ksander et al. [79]

indicated that aqueous humor can suppress the differentiation of pTc into Tc in the presence of Th cells that secreted IL-2 + IL-4. Thus, immunosuppressive factors contained within aqueous humor may control the development of specific cyctotoxic T cells by preventing the secretion of lymphokines that are necessary to initiate the terminal differentiation of pTc into Tc.

Together these results indicate that the microenvironment within the AC of the eye is suppressive for the developement of both inflammatory and cytolytic effector T cells. Since we suspect that the Tdh cells that infiltrate the eye and mediate DH are the same subpopulation of Th cells that initiate differentiation of pTc, it appears that the suppressive factors within aqueous humor are directed primarily at a population of Th/dh cells. By suppressing these cells the development of both DH and CTL is blocked within the immunologically privileged anterior chamber of the eye.

In summary, the anterior chamber, vitreous cavity, and subretinal space of the mammalian eye represent immune privileged sites in which the three features expected at such sites are found: a blood:ocular barrier, absent or atypical lymphatic drainage, and a local immunosuppressive microenvironment. Studies on the anterior chamber have demonstrated that antigenic signals that escape the eye are carried to the spleen which serves as the primary lymphoid organ. The deviant immune response (ACAID) elicited by intraocular antigens serves as a prototype for immune responses to antigens placed in, or arising from, other immune privileged sites. The evidence is particularly strong that privilege in the anterior chamber of the eye includes impaired *expression* of immunity. Thus, at least at this site, both afferent and efferent limbs of the systemic immune response are modified to prevent and control intraocular immunogenic inflammation.

The Brain as an Immune Privileged Site

Historically the concept of the brain as an immune privileged site was based upon: (i) the absence of lymphatic drainage; (ii) the presence of blood-brain barrier, and (ii) the lack of class II+ cells within the brain. It was believed that the absence of normal antigen-presenting cells and the lack of lymphatic drainage prevented alloantigens within the cortex from escaping and inducing a systemic immune response. The blood-brain barrier formed by tight junctions between cerebral endothelial cells restricted the infiltration of lymphocytes and prevented initiation and development of a local immune response. Thus, the physical isolation of the brain was believed to be responsible for immune privilege. In 1914, Ebeling et al.

[80] reported that carcinoma tumor allografts injected into the brain of mice survived longer than similar tumor allografts injected subcutaneously. These experiments were followed by a wide range of experiments that indicated allogeneic and xenogeneic tumor grafts survived for prolonged periods of time within the brain. However, the results from other experiments gave highly variable results and the concept of the brain as a privileged site has been in dispute [81].

Experiments by Head and Griffen [82] provided convincing evidence that allogeneic tissue grafted onto the cerebral cortex survived for a prolonged period of time. Fisher rats grafted with immunogenic DA rat parathyroid allografts survived indefinitely (> 96 days) within the cerebral cortex, but similar grafts were rejected within 13 days when grafted into subdermal pockets. In another series of experiments, MHC-incompatible skin grafted into the cortex survived 40–50 days as assessed by the continued formation of keratin, mitotic figures, and the presence of differentiated sebaceous glands. Either parathyroid or skin allografts were rapidly rejected from the brain if rats were previously immunized by orthotopic DA skin grafts. Together these results demonstrated that, on the one hand, the extent of privilege in the brain was dependent upon the immunogenicity of the graft. Weakly immunogenic parathyroid allografts survived longer than highly immunogenic skin grafts. On the other hand, allografts were susceptible to immune-mediated rejection if the host was presensitized systemically. Thus, these results suggested that privilege in the brain existed due to a failure to induce an immune response to alloantigens within the brain. However, it was believed that there was no barrier that prevented *sensitized* lymphocytes from migrating into the brain and rejecting allografts within the cortex.

More recent results indicate that neuronal allografts such as embryonic mouse retinal grafts [83], or embryonic cerebral neocortex from rats [84], experienced prolonged survival within the brain. In the later study grafts reportedly survived even within recipients previously sensitized to the alloantigens. However, despite many reports on the prolonged survival of allogeneic grafts placed within the cortex of the brain, there are many reports of incompatible grafts being readily rejected from the brain. This has led some to conclude that the brain is not an immunologically privileged site in the 'original sense' and that only under certain conditions can incompatible grafts survive for extended periods of time [85].

The problem in examining the brain as an immune privileged site may be based upon the requirement of an intact blood-brain barrier. All previous experiments examined the survival of grafts placed in the cortex. This required surgical dissection and inevitable disruption of the blood-brain barrier which may alter the immune privileged status of the brain. The anterior chamber of the eye may provide a good example. Allogeneic tissue can be placed into the anterior chamber by injection through the avascular cornea. This maintains the normal ocular microanatomical barriers. However, if injection into the AC is attempted through a vascularized cornea, the immune privileged status of the AC is lost. Therefore, the most rigorous test of immune privilege in the brain requires placing the allogenic tissue within the cortex without disrupting the blood-brain barrier.

A model recently developed by Cserr and Knopf [86, 87] has attempted to address these issues. They have developed a rat model to determine the immune response induced to antigens placed within the brain parenchyma *without* disrupting blood-brain barriers. This is accomplished by implanting a cannula into the cortex that is maintained for at least 1 week in order to reestablish the normal microvascular barriers. Antigens are then injected directly into the brain in small volumes in order not to disrupt the local architecture. Using this protocol human serum albumin was injected into the brain and the antibody response measured within the serum. Elevated levels of specific IgG antibodies were detected and specific antibody-secreting cells were detected in the spleen and lymph nodes that drain the cranial nerves. These results indicate that antigens escape from the brain and gain access to the draining cervical lymph nodes and spleen. The likely routes of escape are (i) from the subarachnoid space across the arachnoid villi into the blood, and (ii) by intersitial fluid flow through the nasal cribriform plate into the mucous membranes lining the nasopharynx – which drains into the cervical nodes.

These investigators also used this protocol to examine the effect of antigens injected into brain on the cell-mediated immune response [88]. Rats receiving a subcutaneous immunizing dose of myelin basic protein (MBP) in adjuvant develop EAE (experimental autoimmune encephalomyelitis), a model for multiple sclerosis in which hind leg paralysis develops. The evidence suggests that EAE is mediated by an autoimmune response by CD4+ Tdh cells against central nervous system (CNS) myelin. However, if MBP is injected into the brain prior to an immunizing dose of MBP, rats fail to develop hind leg paralysis. These results suggest that injection of MBP into the brain suppressed the development of a specific DH response. It was interesting that injection of MBP into the brain had no effect on the serum antibody levels to MBP. This pattern of response – suppressed DH, but normal serum antibody levels – is very similar to the response observed to antigens injected into the anterior chamber of the eye, another immune privileged site. As described earlier, local immunosuppressive factors within the eye such as $TGF\beta_2$ are important in suppressing DH. The evidence suggests that the ability of APC to present antigen is altered after APC are exposed to $TGF\beta_2$ within the eye. These altered APC are thought to induce specific suppression of DH. A similar mechanism may exist in the brain for suppression of DH since the CSF also contains TGFβ_2.

In conclusion the current evidence suggests that soluble antigens injected into the brain with an intact blood-brain barrier are able to escape via indirect routes and gain access to region lymph nodes and the spleen. The immune response induced to these antigens is altered and consists of normal or elevated antibody responses associated with suppressed DH. This type of altered immune response is consistent with the immune response induced to soluble antigens injected into the immunologically privileged anterior chamber of the eye. Thus these results confirm that the brain is an immune privileged site. This system should allow a detailed study of the survival of allogeneic tissue and immunogenic tumors within the brain. These future experiments may resolve some of the apparent controversies on the immune privileged status of the brain.

The Testis as an Immune Privileged Site

The rationale for the immune privileged status of the testis is based on the idea that the host must be prevented from generating an immune response to autoantigens expressed on germ cells and their progeny. The study of the autoimmune response to autoantigens found in the testis has been accomplished using a model of EAO (experimental autoimmune orchitis) developed by immunizing mice with a testis homogenate in complete Freund's adjuvant (CFA) [89–91]. Active orchitis, vasitis, and aspermatogenesis occurs approximately 2 weeks after immunization and is mediated by CD4+ Tdh cells. The autoimmune response can be adoptively transferred by CD4+ T cells. The most interesting aspect of these studies was the identification of the location of the cells that express these autoantigens in the testis. At least some of the cells that express the autoantigens were germ cells found *outside* of the blood-testis barrier formed by Sertoli cells. Since autoantigenic cells exist outside the blood-testis barrier, it provides a rationale for the existence of immune privilege within the interstitial tissues of the testis.

The testis is an unusual immune privileged site in that (i) it has excellent lymphatic drainage; (ii) contains numerous class II positive APC, and (iii) does not appear to be resistant to a local inflammatory response. There is a blood-testis barrier formed by the Sertoli cells lining the seminiferous tubules. However, as indicated above, cells that express autoantigens are found in the interstitial tissue outside this barrier.

Allogeneic tissue grafted into the interstitial tissue of the testis survives for a prolonged period of time and indicates the immune privileged status of the testis. A wide variety of different types of tissue grafts have been placed in the testis with an equally wide range of results [92]. Judith Head et al. [93] provided good experimental evidence of the immune privileged status of the testis. MHC incompatible skin grafts placed into the testicular parenchyma of rats survived twice as long as similar grafts placed ortho-topically. Weakly immunogenic parathyroid grafts survived even longer in the testis, with a mean survival time of 41 days, while control parathyroid grafts were rejected within 15 days. As with other immune privileged sites, immunization of the recipient initiated rejection of the testicular grafts. These results are representative of the experiments that describe the im-mune privileged status of the testis.

Originally, the testis was thought to have poor lymphatic drainage. However, further studies indicated excellent access of interstitial fluid to the regional lymph nodes [94]. This observation originally seemed contrary to maintaining immune privilege, since lack of lymphatic drainage is a feature of other privileged sites. More recent evidence renders the finding less surprising, since the brain and the anterior chamber have been found to have indirect access to lymphatic drainage and the regional lymph nodes. It is interesting that for the anterior chamber the access to the lymphatics is indirect and access to the spleen is direct. A functional spleen is required to sustain immune privilege in the eye, and is the source of DH suppressor cells that are activated by spleen-homing cells derived from the eye. By contrast, immune privilege in the testis is actually augmented in splenec-tomized animals. This may indicate that for the testis, the source of immune regulatory cells may be the draining lymph nodes. If this were true it would coincide with the excellent lymphatic drainage of this organ.

It is believed that immune privilege is maintained in the testis by immunosuppressive factors secreted by Leydig and Sertoli cells. Consider-ing the extensive lymphatic drainage found in the testis, this may be the best example of the powerful nature of immunoregulatory factors produced locally. Originally, it was believed that local steroid production by Leydig cells suppressed the intratestis immune response and maintained immune privilege of allografts [95]. However, experiments by Selawry and Whitting-ton [96] suggest that survival of intratesticular pancreatic islet allografts occurred even in recipients in which Leydig cells and spermatogenesis was impaired. These results indicated that immune privilege does not appear to depend upon local steroid production.

Both Leydig and Sertoli cells can suppress lymphocyte proliferation in vitro [97]. Moreover they have been shown to secrete several immunosup-pressive factors. Sertoli cells secrete interleukin 1 (IL-1), $TGF\beta_2$, and other large molecular weight factors (> 400 kD). Leydig cells secrete α-MSH and ACTH. It is interesting that numerous F4/80+ cells are found in the intersitial spaces between seminiferous tubules [98]. These cells may be the equivalent of the F4/80+ cells found in the iris lining the anterior chamber of the eye. In this immune privileged site it has been shown that antigens presented by F4/80+ cells treated previously with either TGFβ_2 or α-MSH, down-regulate specific DH responses. It may be that Leydig and Sertoli cells maintain immune privilege in the testis by a similar mechanism in which F4/80 cells in the testicular parenchyma are exposed to immuno-suppressive factors secreted by Leydig and Sertoli cells. These F4/80 cells upon migration to an organized lymphoid organ can then induce specific suppression of DH responses to antigens present in the testis.

Tumors as Immune Privileged Sites

There is now conclusive evidence that at least some human sponta-neous tumors are immunogenic and induce a tumor-specific T cell response within patients. This finding presents an immunological paradox – im-munogenic tumor cells grow progressively in an immunocompetent patient that is capable of generating a T-cell response to specific antigens on the tumor. One possible explanation of this paradox, that was proposed in 1977 by Spitalny and North [99], is that tumors escape immune-mediated elimination by establishing a local microenvironment that conveys immune privelege to the developing tumor. Spontaneous tumors develop in a variety of different anatomical sites that are for the most part considered nonprivileged in that allogenic grafts placed in these sites are rapidly rejected. We propose that as spontaneous tumors form they convert the tumor site into an immune privileged site by the secretion of local immuno-suppressive factors. The preceding review of physiologically established immune privileged sites indicates that the local microenvironment plays an important role in establishing and maintaining privilege; however, it still remains to be proven whether suppressive factors alone can 'create' an immune privileged site. Two types of experimental evidence suggest that tumors establish immune privileged locally. First, alloantigenic cells in-jected into the tumor microenvironment survive and are not eliminated. Second, if immunogenic tumor cells are injected into already established privileged sites, tumors form and induce a T cell response similar to that induced by spontaneous human tumors.

There is now conclusive evidence at the cellular and molecular level that tumor specific antigens are present on at least some human sponta-neous tumors. Boon et al. [100] have demonstrated that human metastatic melanomas present tumor-specific peptides via HLA-A1 class I molecules and that these peptides are recognized by specific CD8+ T cells present among the patient's peripheral blood lymphocytes. Patients that generate an immune response to these antigens possess a higher frequency of specific precursor cytotoxic T cells among their peripheral T cells than non-tumor-bearing individuals [101]. The genes that encode these tumor-specific peptides have been cloned and sequenced, and are expressed only in tumor cells. Though tumor-specific genes are present in normal nonmalignant cells, they are silent and not transcribed [102, 103]. These experimental results clearly demonstrate that tumors can be immunogenic. However, it is important to remember that the immune response induced by these tumor antigens is unsuccessful in patients that develop metastatic disease. Understanding the mechanisms involved in how tumors escape immune-mediated destruction is likely to be very important in developing strategies that successfully manipulate the immune response to eliminate these tumors. We believe the study of immune privileged sites may provide an important model for these types of studies.

To determine whether tumor sites establish a local immune privileged environment, it is necessary to demonstrate that immunogenic tissue or cells survive within the tumor. This type of experiment was originally performed by Spitalny and North [99] in 1977. Their experiments demonstrated that mice were incapable of eliminating an inoculation of *Listeria* that was injected into a progressively growing tumor in the right hind footpad. By contrast, mice easily eliminated the same inoculation of *Listeria* injected into the tumor-free left hind footpad. The authors speculated that the failure to eliminate *Listeria* from the tumor site resulted from the secretion of factors by the tumor that suppress the function of Listeria specific T cells and macrophages.

A more elaborate series of experiments was performed recently by Perdrizet et al. [104] in 1990. UV-induced tumor cells grow progressively when inoculated into C3H/HeN mice. However, if these same tumor cells are transfected with immunogenic MHC K^{216} class I alloantigens and injected into mice with progressively growing tumors, the second tumor also grows progressively. This occurred even though it is known that C3H/HeN mice can generate an effective allospecific immune response to the MHC K^{216} class I alloantigen. Even more surprising results were obtained in a second series of experiments that utilized C3H/HeN transgenic mice that expressed the MHC K^{216} class I alloantigens in addition to endogenous normal K^k and D^k MHC class I molecules. K^{216} class I was expressed in transgenic mice on all cell types tested including: dendritic cells, fibroblasts, hepatocytes, kidney cells, spleen cells, and lymphocytes. To determine if mice that failed to eliminate tumor cells that expressed K^{216}

were capable of eliminating non-malignant tissue that expressed the same alloantigens, C3H/HeN mice with progressively growing K^{216+} tumors received grafts of skin from K^{216+} C3H/HeN transgenic mice. The surprising results indicated that mice with progressively growing K^{216+} tumors *reject* nonmalignant K^{216+} skin grafts placed on these same tumor-bearing mice. These experiments demonstrate that immunogenic cells can readily survive within the tumor microenvironment. Thus, presentation of immunogenic cells within the tumor microenvironment fails to induce an effective immune response that can eliminate the tumor cells. By contrast, if the same antigen is presented by cells on a normal skin graft, an effective alloimmune response eliminates the graft. These investigators provided further evidence that failure of the immune response to eliminate K^{216+} tumor cells was not due to (a) the lack of a strong rejection antigen; (b) release of tumor antigens that caused the host to become unresponsive, or (c) rapid proliferation of tumor cells that outstrips the capacity of the immune system to respond. They postulated that factors secreted by the tumor cells prevented the induction of an effective local immune response. These experiments demonstrate that immunogenic cells within the tumor microenvironment induce systemic immunity, but one that fails to successfully eliminate tumor cells. By definition the extended survival of the immunogenic K^{216+} tumor cells within the tumor microenvironment indicates that the tumor environment is acting as an immune privileged site.

Additional circumstantial evidence that tumors can create immune privileged sites is obtained from experiments in our laboratory that examined the immune response induced to immunogenic tumor cells injected into an already established immune privileged site. The immune response induced to these tumors is remarkably similar to the response that is observed among patients with spontaneous human tumors. When minor histoincompatible P815 tumor cells are inoculated into the anterior chamber of the eye they grow progressively and escape immune-mediated elimination despite the fact that they induce the clonal expansion of tumor-specific precursor cytotoxic T cells in these mice. Precursor cells migrate systemically and infiltrate into the tumor-containing eye, but fail to terminally differentiate into cytolytic T cells that can eliminate tumor cells [46]. Thus, immunogenic tumor cells escape elimination by cytotoxic T cells within the immunologically privileged anterior chamber because precursor T cells fail to terminally differentiate into cytolytic T cells. The failure of precursor cells to differentiate locally coincides with (i) the lack of memory T helper cells that secrete lymphokines required to initiate precursor differentiation [47], and (ii) the suppression of precursor differentiation by the local microenvironment within the anterior chamber [79]. In general this type of immune response is very similar to the immune response observed among patients with metastastic skin melanoma. These patients possess high frequencies of tumor-specific precursor cytotoxic T cells in their peripheral blood. These precursor cells infiltrate into the tumor site, but there is no evidence that these cells terminally differentiate into cytolytic T cell. Moreover, it is difficult to demonstrate that these patients possess tumor-specific CD4 T helper cells. Together these results indicate that spontaneous immunogenic tumors induce a specific albeit unique spectrum of T cell responses that are similar in many ways to immune responses induced within immune privileged sites.

If our prediction is correct and tumors establish an immune privilege-like microenvironment in order to escape from immune elimination, it is likely that tumors accomplish this by the local secretion of immunosuppressive factors. The study of immune privileged sites may provide some insights into the factors secreted by these tumors. One factor present within all immune privileged sites discussed in this review is TGFβ_2. It is interesting to note that recent reports indicate that some metastatic human melanomas secrete TGFβ_2.

Conclusions

In conclusion, there is now sufficient data to dispel some of the old concepts of immune privilege. Immune privilege is *not maintained* by sequestering antigens from the host's immune response. Antigens within privileged sites can escape and induce an immune response in the host. Moreover, immune privilege is *not maintained* by preventing the infiltration of effector lymphocytes into privileged sites. All privileged sites appear to be accessible to infiltrating primed antigen-specific lymphocytes. Since antigens at privileged sites can reach and make an impact on the systemic immune apparatus, and since immune effector modalities (cells and molecules) can gain access to privileged sites, how is immune privilege created and sustained? The evidence reviewed here indicates that the following factors and features make variable contributions to the creation and maintenance of privilege at the discussed organs, tissues and sites: (i) a blood:tissue barrier that restricts blood-borne cells and molecules, including sensitized lymphocytes and antibodies, from unregulated access to the site; (ii) an atypical, or even deficient, lymphatic drainage route that permits and promotes the escape of antigens from the site directly into the blood stream; (iii) reduced (or even completely deficient) expression of MHC-encoded molecules on certain cells at the site, rendering them invulnerable to attack by immune T cells; and (iv) a unique local microenvironment that contains locally produced factors that suppress and regulate both the induction and the expression of immunity, especially that mediated by CD4$\pm$ T cells that initiate immunogenic inflammation.

One rationale that has been advanced for the creation of immune privileged sites, such as the eye and brain, is based on the knowledge that the specialized, neural functions of these organs are highly vulnerable to the deleterious effects of immunogenic inflammation. Consequently, it has been reasoned that immune privilege has been created physiologically at these sites in order to prevent immunogenic inflammation from destroying ocular and cerebral function; despite these immune deficits, other types of immune effector modalities (cytotoxic T cells, non-complement-fixing antibodies) are generated in order to provide these organs at least partial immune protection from invading pathogens. However, each privileged site described – eye, brain, fetus and placenta, testis – contains cells that express autoantigens or alloantigens that can evoke intense and destructive immune responses. Thus, another rationale for the creation of immune privilege as a physiologic feature of these disparate sites is that privilege is necessary to prevent the development of autoimmune diseases or to mitigate against abortions.

Whichever rationale eventually serves best to explain the existence of immune privilege under physiologic circumstances, it is provocative to consider the possibility that immune privilege of immunogenic malignant cells becomes a destructive force that preserves and protects the tumor from immune surveillance. It is our belief that the study of immune privilege, as physiologically created at specialized sites and organs, may serve to illuminate pathogenic mechanisms that are important in the process by which primary and metastatic cancers emerge and cause disease. Moreover, by understanding the molecular and cellular bases of immune privilege, we anticipate that new approaches to immunotherapy may help in the treatment of cancer, autoimmune diseases, and diseases of immunopathogenic origin.

References

1 Woodruff, M.F.A.: Transplatation immunity and the immunological problem of pregnancy. Proc. R. Soc. *148:* 68 (1958).
2 Redman, C.W.G.: HLA-DR antigen on human trophoblast. A review. Am. J. Reprod. Immunol. *3:* 175 (1983).
3 Hunt, J.S.; Andrews, G.K.; Wood, G.W.: Normal trophoblasts resist induction of class I HLA. J. Immunol. *138:* 2481 (1987).
4 Feinman, M.A.; Kliman, H.J.; Main, E.K.: HLA antigen expression and induction by gamma-interferon in cultured human trophoblasts. Am. J. Obstet. Gynecol. *157:* 1429 (1987).

5 Faulk, W.P.; Temple, A.: Distribution of beta-2-microglobulin and HLA in chorionic villi of human placentae. Nature *262:* 799 (1976).

6 Goodfellow, P.N.; Barnstable, C.J.; Bodmer, W.F.; Snary, D.E.; Crumpton, M.J.: Expression of HLA system antigens on platentae. Transplatation *22:* 555 (1976).

7 Hunt, J.S.; Fishback, J.L.; Andrews, G.K.; Wood, G.W.: Expression of class I HLA genes by trophoblast cells: analysis by in situ hybridization. J. Immunol. *140:* 1293 (1988).

8 Hunt, J.S.; Fishback, J.L.; Chumbley, G.; Loke, Y.W.: Identification of class I MHC mRNA in human first trimester trophoblast cells by in situ hybridization. J. Immunol. *144:* 4420 (1990).

9 Kovats, S.; Main, E.L.; Librach, C.; Stubblebine, M.; Fisher, S.J.; DeMars, R.: A class I antigen, HLA-G, expressed in human trophoblasts. Science *248:* 220 (1990).

10 Ellis, S.A.; Palmer, M.S.; McMichael, A.J.: Human trophoblast and the choriocarcinoma cell line BeWo express a truncated HLA class I molecule. J. Immunol. *144:* 731 (1990).

11 Ellis, S.A.; Strachan, T.; Palmer, M.S.; McMichael, A.J.: Complete nucleotide sequence of a unique HLA class I C locus product expressed on the human choriocarcinoma cell line BeWo. J. Immunol. *142:* 3281 (1990).

12 Grabowska, A.; Chumbley, G.; Carter, N.; Loke, Y.W: Interferon-gamma enhances mRNA and surface expression of class I antigen on human extravillous trophoblast. Placenta *11:* 301 (1990)

13 Loke, Y.W.; King, A.: Recent developments in the human maternal-fetal immune interaction. Curr. Opin. Immunol. *3:* 762 (1991).

14 Zuckermann, F.A.; Head, J.R.: Expression of MHC antigens on murine trophoblast and their modulation by interferon. J. Immunol. *137:* 846 (1986).

15 Zuckermann, F.A.; Head, J.R.: Murine trophoblast resists cell-mediated lysis. I. Resistance to allospecific cytotoxic T lymphocytes. J. Immunol. *139:* 2856 (1987).

16 Zuckermann, F.A.; Head, J.R.: Murine trophoblast resist cell-mediated lysis. II. Resistance to natural cell-mediated cytotoxicity. Cell. Immunol. *166:* 274 (1988).

17 Drake, B.L.; Head, J.R.: Murine trophoblast cells are not killed by tumor necrosis factor-alpha. J. Reprod. Immunol. *17:* 93 (1990).

18 Mitchison, N.A.: The effect on the offspring of maternal immunization in mice. J. Genet. Hum. *51:* 406 (1953).

19 Clarke, A.G : The effect of maternal preimmunization on pregnancy in the mouse. J. Reprod. Fertil. *24:* 369 (1971).

20 Gambel, P.; Croy, B.A.; Moore, W.D.; Hunziker, R.D.; Wegmann, T.G.; Rossant, J.: Characterization of immune effector cells present in early murine decidua. Cell. Immunol. *93:* 303 (1985).

21 Wegmann, T.G.: The cytokine basis for cross-talk between the maternal immune and reproductive system. Curr. Opin. Immunol. *2:* 765 (1990).

22 Wegmann, T.G.; Lin, H.; Guilbert, L.; Mosmann, T.R.: Bidirectional cytokine interactions in the maternal-fetal relationship: is successful pregnancy a Th2 phenomenon? Immunol. Today *14:* 353 (1993).

23 Drake, B.L.; Head, J.R.: Murine trophoblast can be killed by lymphokine-activated killer cells. J. Immunol. *143:* 9 (1989).

24 Croy, B.A.; Gambel, P.; Rossant, J.; Wegmann, T.G.: Characterization of murine decidual natural killer cells and their relevance to the success of pregnancy. Cell. Immunol. *93:* 315 (1985).

25 Croy, B.A.; Waterfield, A.; Wood, W.; King, G.J.: Normal murine and procine embryos recruit NK cells to the uterous. Cell. Immunol. *115:* 471 (1988).

26 King, A.; Wellings, V.; Gardner, L.; Loke, Y.W.: Immunocytochemical characterization of the unusual large granular lymphocytes in human endometrium throughout the menstrual cycle. Hum. Immunol. *24:* 195 (1989).

27 King, A.; Balendran, N.; Wooding, P.; Carter, N.P.; Loke, Y.W.: Phenotypic and morphologic charterization of novel CD3−, CD56 bright+ lymphocytes in the pregnant human uterous. Dev. Immunol. *1:* 169 (1991).

28 King, A.; Loke, Y.W.: Human trophoblast and JEG Choriocarcinoma cells are sensitive to lysis by IL-2 stimulated decidual NK cells. Cell. Immunol. *129:* 435 (1990).

29 Croy, B.A.: Granulated metrial gland cells – Interesting cells found in the pregnant uterus. Am. J. Reprod. Immunol. *23:* 19 (1990).

30 Stewart, I.J.: Granulated metrial gland cells in the mouse placenta. Placenta *11:* 263 (1990).

31 Parr, E.L.; Young, L.H.Y.; Parr, M.B.; Young, J.D.E.: Granulated metrial gland cells of pregnant mouse uterus are natural killer-like cells that contain perforin and serine esterases. J. Immunol. *14:* 2365 (1990).

32 Gendron, R.L.; Baines, M.G.: Immunohistochemical analysis of decidual natural killer cells during spontaneous abortion in mice. Cell. Immunol. *11:* 147 (1988).

33 Gendron, R.L.; Farookhi, R.; Baines, M.G.: Resorption of CBA/J × DBA/2 mouse conceptuses in CBA/J uteri correlates with failure of the feto-placental unit to suppress natural killer cell activity. J. Reprod. Fertil. *89:* 277 (1990).

34 Clark, D.A.; Flanders, K.C.; Banwatt, D.; Millar-Brook, W.; Manuel, J.: Active suppression of host-versus-graft reaction in pregnant mice. IX. Soluble suppressor activity obtained from allopregnant mice decidua that blocks the cytolytic effector response to IL-2 is related to transforming growth factor beta. J. Immunol. *141:* 3833 (1988).

35 Lea, R.G.; Flanders, K.C.; Harley, C.B.; Manuel, J.; Banwatt, D.; Clark, D.A.: Release of a transforming growth factor (TGF)-b2-related suppressor factor from postimplantation murine decidual tissue can be correlated with the detection of a subpopulation of cells containing RNA for TGF-b2. J. Immunol. *148:* 778 (1991).

36 Redline, R.W.; McKay, D.B.; Vazquez, M.A.; Papaioannou, V.E.: Lu, C.Y.: Macrophage functions are regulated by the substratum of murine decidual stromal cells. J. Clin. Invest. *85:* 1951 (1990).

37 Greene, H.S.N.; Lund, P.K.: The heterologous transplantation of human cancers. Cancer Res. *4:* 352 (1944).

38 Kaplan, H.J.; Streilein, J.W.: Immune response to immunization via the anterior chamber of the eye. I. Fl lymphocyte-induced immune deviation. J. Immunol. *118:* 809 (1977).

39 Streilein, J.W.: Immune privilege as the result of local tissue barriers and immunosuppressive microenvironments. Curr. Opin. Immunol. *5:* 428 (1983).

40 Streilein, J.W.: Immune regulation and the eye: a dangerous compromise. FASEB J. *1:* 199 (1987).

41 Niederkorn, J.Y.: Immune privilege and immune regulation in the eye. Adv. Immunol. *48:* 191 (1990).

42 Ksander, B.R.; Streilein, J.W.: Analysis of cytotoxic T cell responses to intracameral allogenic tumors. Invest. Ophthalmol. Vis. Sci. *30:* 323 (1989).

43 Niederkorn, J.Y.; Streilein, J.W.: Alloantigens placed into the anterior chamber of the eye induce specific suppression of delayed-type hypersensitivity but normal cytotoxic T lymphocyte and helper T lymphocyte responses. J. Immunol. *131:* 2670 (1983).

44 Bandon, Y.; Ksander, B.R.; Streilein, J.W.: Characterization of specific T helper cell activity in mice bearing alloantigenic tumors in the anterior chamber of the eye. Eur. J. Immunol. *21:* 1923 (1991).

45   Streilein, J.W.; Niederkorn, J.Y.; Shadduck, J.A.: Systemic immune unresponsiveness
     induced in adult mice by anterior chamber presentation of minor histocompatibility
     antigens. J. Exp. Med. *152:* 1121 (1980).
46   Ksander, B.R.; Streilein, J.W.: Failure of infiltrating precursor cytotoxic T cells to
     acquire direct cytotoxic function in immunologically privileged sites. J. Immunol. *145:*
     2057 (1990).
47   Bando, Y,; Ksander, B.R.; Streilein, J.W.: Incomplete activation of lymphokine-produc-
     ing T cells by alloantigenic intraocular tumours in anterior chamber-associated immune
     deviation. Immunology *78:* 266 (1993).
48   Niederkorn, J,Y.; Streilein, J.W.: Analysis of antibody production induced by allogeneic
     tumor cells inoculated into the anterior chamber of the eye. Transplantation *33:* 573
     (1982).
49   Wilbanks G.A.; Streilein, J.W.: Distinctive humoral immune responses following ante-
     rior chamber and intravenous administration of soluble antigen. Evidence for active
     suppression of IgG2-secreting B lymphocytes. Immunology *71:* 566 (1990).
50   Niederkorn, J.Y.; Streilein, J.W.; Kripke, M.L.: Promotion of syngeneic intraocular
     tumor growth in mice by anterior chamber-associated immune deviation. JNCI *71:* 193
     (1983).
51   Ferguson, T.A.; Waldrep, J.C.; Kaplan, J.J.: The immune response and the eye II. The
     nature of T suppressor cell induction of anterior chamber associated immune deviation
     (ACAID). J. Immunol. *139:* 352 (1987).
52   Ferguson, T.A.; Hayashi, J.D.; Kaplan, H.J.: The immune response and the eye III.
     Anterior chamber associated immune deviation can be adoptively transferred by serum.
     J. Immunol. *143:* 821 (1989).
53   Medawar, P.: Immunity to homologous grafted skin III. The fate of skin homografts
     transplanted to the brain, to subcutaneous tissue, and to the anterior chamber of the
     eye. Br. J. Exp. Pathol. *29:* 58 (1948).
54   Jiang, L.Q.; Streilein, J.W.: Immunologic privilege evoked by histoincompatible intra-
     cameral retinal transplants. Regional Immunology *3:* 121 (1991).
55   Sonoda, Y.; Streilein, J.W.: Impaired cell-mediated immunity in mice bearing healthy
     orthotopic corneal allografts. J. Immunol. *150:* 1727 (1993).
56   Mizuno, K.; Clark, A.F.; Streilein, J.W.: Anterior chamber-associated immune deviation
     induced by soluble antigens. Invest. Ophthalmol. Vis. Sci. *30:* 1112 (1989).
57   Hara, Y.; Caspi R.R.; Wiggert B.; Chan C.-C.; Streilein J.W.: Use of ACAID to
     suppress interphotoreceptor retinoid binding protein-induced experimental autoimmune
     uveitis. Curr. Eye Res. II: suppl., p. 97 (1992).
58   Hara, Y.; Caspi, R.R.; Wiggert, B.; Chan, C.-C.; Wilbanks, G.A.; Streilein, J.W.:
     Suppresion of experimental autoimmune uveitis in mice by induction of anterior
     chamber-associated immune deviation with interphotorecepton retinoid-binding protein.
     J. Immunol. *148:* 1685 (1992).
59   Whittum, J.A.; McCulley, J.P.; Niederkorn, J.Y.; Streilein, JW.: Ocular disease induced
     in mice by anterior chamber inoculation of herpes simplex virus. Invest. Ophthalmol.
     Vis. Sci. *25:* 1065 (1984).
60   Atherton, S.; Kanter, M.Y.; Streilein, J.W.: ACAID requires realy replication of HSV-1
     in the injected eye. Curr. Eye Res. *10:* suppl., p. 75 (1991).
61   Wilbanks, G.A.; Streilein, J.W.: Macrophages capable of inducing anterior chamber
     associated immune deviation demonstrate spleen-seeking migratory properties. Regional
     Immunol. *4:* 130 (1992).
62   Streilein, J.W.; Niederkorn, J.Y.: Induction of anterior chamber-associated immune
     deviation requires an intact, functional spleen. J. Exp. Med. *153:* 1058 (1981).

63 Wilbanks, G.A.; Mammolenti, M.; Streilein, J.W.: Studies on the induction of anterior chamber-associated immune deviation (ACAID). II. Eye-derived cells participate in generating blood-borne signals that induce ACAID. J. Immunol. *146:* 3018 (1991).

64 Wilbanks, G.A.; Streilein, J.W.: Studies on the induction of anterior chamber-associated immune deviation (ACAID). I. Evidence that an antigen-specific, ACAID-inducing, cell-associated signal exists in the peripheral blood. J. Immunol. *146:* 2610 (1991).

65 BenEzra, D.; Sachs, U.: Growth factors in aqueous humor of normal and inflamed eyes of rabbits. Invest. Ophthalmol. Vis. Sci. *13:* 868 (1973).

66 Granstein, R.D.; Staszewski, R.; Knisely, T.L.; Zwira, E.; Nazareno, R.; Latina, M.; Albert, D.M.: Aqueous humor contains transforming growth factor-beta and a small (< 3,500 daltons) inhibitor of thymocyte proliferation. J. Immunol. *144:* 3021 (1990).

67 Cousins, S.W.; McCabe, M.M.; Danielpour, D.; Streilein, J.W.; Identfication of trans-forming growth factor-beta as an immunosuppressive factor in aqueous humor. Invest. Ophthalmol. Vis. Sci. *32:* 2201 (1991).

68 Taylor, A.W.; Streilein, J.W.; Cousins, S.W.: Identification of alpha-melanocyte stimu-lating hormone as a potential immunosuppressive factor in aqueous humor. Curr. Eye Res. *11:* 1199 (1992).

69 Taylor, A.W.; Streilein, J.W.; Cousins, S.W.: Neuropeptides contribute to the immuno-suppressive activity of aqueous humor. Invest. Ophthalmol. Vis. Sci. *34:* 903 (1993).

70 Granstein, R.D.; Knisely, T.L.; Hosoi, J.: Hydrocortisone is a probable mediator of the inhibitory activity of aqueous humor on langerhans cell antigen presenting function. Invest. Ophthalmol. Vis. Sci. *34:* 903 (1993). (Abstract)

71 Streilein, J.W.; Bradley, D.: Analysis of immunosuppressive properties of iris and ciliary body cells and their secretory products. Invest. Ophthalmol. Vis. Sci. *32:* 2700 (1991).

72 Jiang, L.Q.; Streilein, J.W.: Immune privilege extended to allogeneic tumor cells in the vitreous cavity. Invest. Ophthalmol. Vis. Sci. *32:* 224 (1991).

73 Wilbanks, G.A.; Streilein, J.W.: Fluids from immune privileged sites endow macro-phages with the capacity to induce antigen-specific immune deviation via a mechanism involving transforming growth factor-b. Eur. J. Immunol. *22:* 1031 (1992).

74 Cousins, S.W.; Trattler, W.B.; Streilein, J.W.: Immune privilege and suppression of immunogenic inflammation in the anterior chamber of the eye. Curr. Eye Res. *10:* 287 (1991).

75 Streilein, J.W.; Cousins, S.W.; Bradley, D: Effect of intraocular gamma-interferon on immunoregulatory properties of iris and ciliary body cells. Invest. Ophthalmol. Vis. Sci. *33:* 2304 (1992).

76 Kaiser, C.; Ksander, B.R.; Streilein, J.W.: Inhibition of lymphocyte proliferation by aqueous humor. Regional Immunol. *4:* 130 (1992).

77 Ksander, B.R.; Streilein, J.W.: Recovery of activated cytotoxic T cells from minor H incompatible tumor graft rejection sites. J. Immunol. *143:* 426 (1989).

78 Ksander, B.R.; Acevedo, J.; Streilein, J.W.: Local T helper cell signals by lymphocytes infiltrating intraocular tumors. J. Immunol. *148:* 1955 (1992).

79 Ksander, B.R.; Miki, S.; Streilein, J.W.: Normal iris and ciliary body cells suppress the terminal differentiation of tumor-specific precursor cytotoxic T cells. Invest. Ophthal-mol. Vis. Sci. *33:* 1283 (1992).

80 Barker, C.F.; Billingham, R.E.: Immunologically privileged sites. Adv. Immunol. *25:* 1 (1977).

81 Head, J.R.; Billingham, R.E.: Immunologically privileged sites in transplantation im-munology and oncology. Perspect. Biol. Med. *29:* 115 (1985).

82 Head, J.R.; Griffen, S.T.: Functional capacity of solid tissue transplants in the brain: Evidence for immunological privilege. Proc. R. Soc. *224:* 375 (1985).

83   Rao, K.; Lund, R.D.; Kunz, H.W.; Gill, T.J.: The role of MHC and non-MHC antigens in the rejection of intracerebral allogeneic neural grafts. Transplantation *48:* 1018 (1989).

84   Poltorak, M.; Freed, W.J.: BN rats do not reject F344 brain allografts even after systemic sensitization. Ann. Neurol. *29:* 377 (1991).

85   Gill, T.J.; Lund, R.D.: Implantation of tissue into the brain – an immunologic perspective. JAMA *261:* 2674 (1989).

86   Cserr, H.F.; Knopf, P.M.: Cervical lymphatics, the blood-brain barrier and the immunoreactivity of the brain: A new view. Immunol. Today *13:* 507 (1992).

87   Cserr, H.F.; Harling-Berg, C.J.; Knopf, P.M.: Drainage of brain extracellular fluid into blood and deep cervical lymph and its immunological significance. Brain Pathol. *2:* 269 (1992).

88   Harling-Berg, C.J.; Knopf, P.M.; Cserr, H.F.: Myelin basic protein infused into cerebrospinal fluid suppresses experimental autoimmune encephalomyelitis. J. Neuroimmunol. *35:* 45 (1991).

89   Kohno, S.; Munoz, J.A.; Williams, T.M,; Teuscher, C.; Bernard, C.C.A.; Tung, K.S.K.: Immunopathology of murine experimental allergic orchitis. J. Immunol. *130:* 2675 (1983).

90   Mahi-Brown, C.A.; Yule, T.D.; Tung, K.S.K.: Adoptive transfer of murine autoimmune orchitis to naive recipients with immune lymphocytes. Cell. Immunol. *106:* 408 (1987).

91   Yule, T.D.; Montoya, G.D.; Russell, L.D.; Williams, T.M.; Tung, K.S.K.: Autoantigenic germ cells exist outside the blood-testis barrier. J. Immunol. *141:* 1161 (1988).

92   Maddocks, S.; Setchell, B.P.: Recent evidence for immune privilege in the testis. J. Reprod. Immunol. *18:* 9 (1990).

93   Head, J.R.; Neaves, E.B.; Billingham, R.E.: Immune privilege in the testis. I. Basic parameters of allograft survival. Transplantation *36:* 423 (1983).

94   Head, J.R.; Neaves, W.B.; Billingham, R.E.: Reconsideration of the lymphatic drainage of the rat testis. Transplantation *35:* 91 (1983),

95   Head, J.R.; Billingham, R.E.: Immune privilege in the testis, II. Evaluation of potential local factors. Transplantation *40:* 269 (1985).

96   Selawry, H.P.; Whittington, K.B.: Prolonged intratesticular islet allograft survival is not dependent on local steroidogenesis. Horm. Metab. Res. *20:* 562 (1988).

97   Pollanen, P.; Euler, M.; Soder, O.: Testicular immunoregulatory factors. J. Reprod. Immunol. *18:* 51 (1990).

98   Yule, T.D.; Mahi-Brown, C.A.; Tung, K.S.K.: Role of testicular autoantigens and influence of lymphokines in testicular antoimmune disease. J. Reprod. Immunol. *18:* 89 (1990).

99   Spitalny, G.L.; North, R.J.: Subversion of host defense mechanisms by malignant tumors: An established tumor as a privileged site for bacterial growth. J. Exp. Med. *145:* 1264 (1977).

100 Boon, T.: Toward a genetic analysis of tumor rejection antigens. Adv. Cancer Res. *58:* 177 (1992).

101 Coulie, P.G.; Somville, M.; Lehmann, F.; Hainaut, P.; Brasseur, F.; Devos, R.; Boon, T.: Precursor frequency analysis of human cytolytic T lymphocytes directed against autologous melanoma cells. Int. J. Cancer *50:* 289 (1992).

102 Traversari, C.; Van der Bruggen. P.; Luescher, I.F.; Lurquin, C.; Chomez, P.; Van Pel, A.; De Plaen, E.; Amar-Costesec,A.; Boon, T.: A nonapeptide encoded by human gene MAGE-1 is recognized on HLA-A1 by cytolytic T lymphocytes directed against tumor antigen MZ2-E. J. Exp. Med. *176:* 1453 (1992).

103 Brasseur, F.; Marchand, M.; Vanwijck, R.; Hérin, M.; Lethé, B.; Chomez, P.; Boon, T.: Human gene MAGE-1, which codes for a tumor-rejection antigen, is expressed by some breast tumors. Int. J. Cancer 52: 839 (1992).

104 Perdrizet, G.A.; Ross, S.R.; Strauss, H.J.; Singh, S.; Koeppen, H.; Schreiber, H.: Animals bearing malignant grafts reject normal grafts that express through gene transfer the same antigen. J. Exp. Med. 171: 1205 (1990).

Dr. Bruce R. Ksander, Schepens Eye Research Institute, 20 Staniford Street, Boston, MA 02114 (USA)

Granstein RD (ed): Mechanisms of Immune Regulation.
Chem Immunol. Basel, Karger, 1994, vol 58, pp 146–192

Suppressor Cells and Immunity[1]

David R. Webb[a], *Ellen Kraig*[b], *Bruce H. Devens*[a]

[a]Institute of Immunology and Biological Sciences, Syntex Discovery Research,
Palo Alto, Calif., and [b]Department of Cell and Structural Biology,
University of Texas Health Science Center, San Antonio, Tex., USA

Historical Background

The last 20 years have been extremely exciting for immunologists. We
have solved the structure of the antibody molecule and the T-cell antigen
receptor (TcR); elucidated the nature of the genes that code for immunoglobulins; defined functional subsets of lymphocytes into B and T
cells; and elucidated the structure and function of major histocompatibility
antigens. We have identified and purified a large number of the cytokines
that regulate the growth and function of immunocompetent cells. We have
also begun to understand the nature of the interactions that take place
between immunocompetent cells that lead to the various manifestations of
immunity. Nevertheless, there still remain many aspects of immune responses that are not well understood. Among these is the problem of how
the immune response is turned off, turned down or tolerized. The general
problem of tolerance is a very old issue having been raised in the early
years of this century by Ehrlich [1]. In recent years, it has become clear that
the term tolerance actually covers many different immunological phenomena. Central tolerance to self antigens has clearly been established to take
place via clonal deletion in the thymus [2]. However, it is equally well
established that self-reactive clones of T cells and B cells do exist in the
periphery but are kept under tight control and do not normally mount an
anti-self response [3–5]. Thus, questions arise as to how these self-reactive
lymphocytes are kept in check. Possibly related to this is the generation of
diversity that takes place in the V-region of immunoglobulin genes in B-cell
populations; what is the mechanism that prevents the immune system from
recognizing new antibody V-regions as 'foreign' and deleting or killing

[1] Supported in part by NIH Grant AI 22181 (E.K.) and by a grant from the Robert
Welch Foundation (E.K.).

them? Additionally, one might ask whether the mechanisms responsible for turning off immune responses following antigen stimulation are in any way related to the processes involved in peripheral tolerance. A great many thoughtful workers have addressed these and related questions over the years; this review will focus its attention on one aspect of immune regulation that appears to play a key role in regulating both peripheral tolerance and developing immune responses, namely antigen-specific and nonspecific suppressor cells.

It is important at the outset to note that work on the cellular basis of immune suppression has proven to be a fertile area. There are now well over 4,000 references in the Medline data base [D.R.W. and B.H.D, personal data base search] that address some aspect of cell-based suppression. These date from the early 1970s to the present day. Despite the lack of attention to suppression in immunology courses and at most national or international meetings (not to mention the lack of funding), publication of studies involving suppressor cells continues apace with remarkably little diminution. As that is the case, we wish to make clear that the present review cannot possibly be exhaustive in the sense of covering all of the reports published. Rather, it is our intent to give an overview of how the field has developed, what are the major issues that have divided immunologists into 'believers' and 'nonbelievers', and what progress has been made in establishing the veracity of the suppressor cell concept as an important component of immunocompetent cell regulation.

The first golden age of cellular immunology, as it has been termed, [6] occurred in the 1960s. In the course of studies on adoptive transfer of immune competent cells, several workers noted that T cells from tolerant animals could transfer tolerance that was antigen-specific [6]. This work gave rise to the concept of infectious tolerance and set the stage for the next series of studies that led to the concept that infectious tolerance was due to a specific type of cell, termed the suppressor cell.

The observations referred to above, showed that certain lymphoid cells were incapable of transferring responsiveness to an antigen; on the contrary, they seemed capable of blocking the development of an immune response in the recipient animals. A few immunologists, notably Benacerraf [7], Allison [8] and, in particular, Gershon [6], hypothesized that these 'regulator cells' might perform a vital function in controlling the immune response and in regulating peripheral (induced) tolerance. In a seminal review paper published in 1974, Gershon [6], outlined the case for these regulator cells which he termed suppressor cells. Studies in his own laboratory using a hamster tumor model had shown the rapid development of tumor-specific anergy following resection. This rapid onset of unresponsiveness reminded Gershon and co-workers of the immune paralysis observed by others using large doses of antigen. This, coupled with the observations of Davies [9] that T cells were involved in the immune response to SRBC as helper cells, led Gershon to hypothesize that T cells might be involved in the development of suppression. Through a series of animal experiments in which irradiated mice were reconstituted with or without T cells, he and his co-workers demonstrated that only when T cells were present could high doses of antigen (SRBC) lead to immune paralysis. Subsequent studies showed that these paralyzed mice had B cells capable of making anti-SRBC antibodies but only when fresh T cells from nonparalyzed donors were present. These experiments were interpreted to suggest that the presence of T cells during tolerance induction led to the development of a population of suppressor T cells that, via the release of soluble mediators, specifically prevented B cells from producing antibody to antigenic challenge. Subsequent to those observations, many other workers repeated and extended these observations, further solidfying the concept of the antigen-specific suppressor T cell [10–16]. The postulation by Gershon and co-workers that the suppressor activity resided in a soluble factor is a topic to which we will return later. Finally, in the same volume, Herzenberg and Herzenberg [16] and Allison [8] further reviewed a variety of immune response models and concluded that the evidence supporting the existence of antigen-specific suppressor cells was compelling. The Herzenberg and Herzenberg paper is of interest in that it was one of the first papers to suggest that one potential mechanism of suppression is T cell-mediated killing of self-reactive T cells. As will be pointed out later, the distinction, or lack thereof, between suppression and cytolysis is important since the effector cells in both cases can be CD8+.

At the same time as Gershon [6] and others [10–16] were studying immune responses to classical antigens such as SRBC, *Ascaris* or natural protein antigens (e.g., BSA, OVA, KLH, etc.), another group of investigators had taken a very different approach. Studies initiated by Sela and co-workers [17] at the Weizmann Institute using synthetic polypeptides of defined amino acid composition were quickly recognized as providing a powerful new tool for the dissection of the genetic influences in immunity by McDevitt and Chintz [18] and Benacerraf et al. [19]. These studies, which begun as an effort to understand better the nature of immunogenicity, broadened considerably when it was discovered that certain strains of mice and guinea pigs seemed unable to respond to certain synthetic polymers. The lack of responsiveness was traced to genes that lay within the major histocompatibility complex (MHC) [18, 19] thus providing for the first time a connection between immunogenicity and a subset of genes that came to be called immune response (Ir) genes. These genes were eventually shown to encode the class II molecules, I-A and I-E and first characterized at the molecular level, in the mouse, by Steinmetz et al. [20]. There were other genes that were tentatively identified at this same time as being part of the Ir gene locus, these were the I-B, I-C and the I-J loci. Most of these genes were defined on the basis of antisera that could be raised only across selected H-2 strains and represented apparent special cases although they mapped by classical recombination studies in the I-A/I-E region. Since the story of I-J is central to the history of suppression, it will be covered in more depth further on in this discussion; it has also been the subject of recent reviews [21, 22].

Returning to the studies on synthetic polypeptide antigens, the role of such antigens in understanding and dissecting suppressor cells in immune regulation was explored to a large extent in only a few groups. Perhaps the group that carried out the most extensive studies over the years derived from the laboratory of Benacerraf and co-workers, Pierce, Kapp, Dorf, Greene, Germain and Sy among others, at Harvard Medical School [23–32]. This early work established that animals could be divided into two groups, responders and nonresponders based on their capacity to make antibodies to synthetic polypeptide antigens or carriers. For nonresponders, to some antigens, there appeared to be a hole in the repertoire, that is they lacked the innate capacity to make antibodies complimentary to the synthetic antigen determinants [29]. In other cases, there appeared to be present in the nonresponders, antigen-reactive B cells and T cells as well as competent macrophages, yet antibodies were not made [23–26, 30–32]. In one of the best studied examples, using the random terpolymer L-glutamic acid^{60}-L-alanine^{30}-L-tyrosine^{10} (GAT), it was shown that nonresponder mice (H-2^q and H-2^s, for example) possessed GAT-specific B cells and T cells and had macrophages that could process GAT [23–26, 30–32]. Furthermore, nonresponders to GAT could be converted to responders simply by immunizing the mice with GAT-mBSA. However, nonresponder mice primed with GAT and then challenged with GAT-mBSA failed to make GAT-specific antibody [25]. This inhibition of the nonresponder antibody response by priming with GAT could be adoptively transferred with thymocytes, lymph nodes or extracts of the thymus [29]. These results indicated that the nonresponder status was actively maintained by a population of cells later proved to be suppressor T cells, and that extracts of these cells could also transfer nonresponsiveness. Extensive studies carried out by Benacerraf and co-workers [23–26, 28] and by Kapp, Pierce and co-workers [29–32] clearly established that for GAT and other, related polymers (e.g. GT or GA) nonresponders selectively activated at least two populations of T-suppressor cells that could be distinguished by cell surface phenotype and time of appearance. These suppressor cells blocked GAT-specific T-helper cells and B cells from cooperating in the production of anti-GAT antibodies. Remarkably, the suppression appeared to be antigen-specific. That is, GAT-specific suppressor cells had no effect on the response to SRBC for example. In other laboratories, workers such as Dorf and Benacerraf [14] and Tada and co-workers, notably Taniguchi [10, 33–40], and Gershon et al. [6, 13], were detecting the existence of similar populations of antigen-specific suppressor cells in studies of delayed-type hypersensitivity (DTH) and IgE-mediated immune responses. Suppressor cells were also being identified in mice injected with natural proteins such as KLH [41] or lysozyme [42]. In some cases, the suppressor cells were not only antigen-specific but also haplotype-restricted in that they would only suppress immune responses in strains of mice having the same MHC class II haplotype [29, 35, 38]. Allotype restrictions were also reported [38, 39]. These studies all pointed clearly to the existence of populations of T cells that seemed to function at very specific, negative regulators of immunity. Moreover, the use of cell or tissue extracts to mimic the effects of suppressor cells implied that there existed soluble factors that served to communicate between suppressor cell populations and their target T cells, B cells or macrophages.

The existence of soluble factors that could transfer antigen-specific suppression presented additional difficulties. By the early 1980s it had been clearly shown that the regulation of immune responses was dependent on soluble cytokines that controlled both cellular differentiation and growth [43–47]. However, none of these cytokines could be shown to be antigen-specific [43–47]. Moreover, many laboratories had shown that nonspecific suppressor factors were also produced in response to antigenic or mitogenic challenge [10, 44]. This forced those studying antigen-specific suppression to argue that it represented a special case (antigen-specific help had been described in the early 1970s but had proved very elusive [48]). While considerable discussion took place concerning the existence and function of antigen-specific suppressor factors in vivo [49], a technical development took place that seemingly made the argument moot for a while; that was the development of a cell line, AKR thymoma BW5147 (HGPRT−) that could be used as a hybridizing partner to produce T-cell hybridomas [36]. Within a relatively brief period of time, several different groups reported T-suppressor hybridomas that constitutively produced antigen-specific suppressor factors that reflected the antigen specificity of the parental T-suppressor cell lineage.

Prior to this time very little had been done to characterize suppressor factors biochemically; however, it now became possible to attempt to purify and study these curious molecules. In addition, it lent considerable veracity to complex circuitry schemes such as those published in a seminal review by Germain and Benacerraf [11] in 1981. This review represented a major attempt to synthesize the then disparate results from a wide variety of laboratories into a cohesive hypothesis about the nature of antigen-specific suppressor cells and their products. It presented both the differentiation of suppressor cells in response to antigen challenge and attempted to define discreet subpopulations of antigen-specific suppressor cells on the basis of cell surface phenotype and sensitivity to radiation or cyclophosphamide. It also attempted to link antigen-specific suppressor cells to nonspecific suppressor cells and factors. Although this hypothesis is now over 10 years old, it is still the clearest summation of the state of suppressor cells studied through 1981. The paradigm of a pathway of suppressor cell interactions flowing from specific to nonspecific suppressor cells and factors must still be viewed as a possibility. And, the overall hypothesis that it takes at least two types of cells (e.g. inducer and effector) in order to generate antigen-specific or nonspecific suppression is probably correct. Despite all these data, however, the underlying reason why some strains of mice are responders and some are nonresponders remains unresolved [22] although clonal deletion and/or anergy may very well account for some of these observations.

With the availability of T-suppressor hybridomas, two major issues that had not been well developed became approachable. One had to do with the biochemical nature of the antigen-specific suppressor factors; the other focused on the nature of the gene product of the I-J locus. These two problems were related because many laboratories [10–14] reported that antigen-specific suppressor factors (TSFs) could be bound by anti-I-J antisera. These issues will be discussed in more detail below.

Demonstration of Suppressor Cells in in vivo Immune Responses

Suppression has been historically described as an in vivo phenomenon as discussed above. Ever since the initial recognition of the role of Ts in the regulation of immune responses, investigators have continued to report myriad examples of Ts regulation in in vivo immune responses. The following section details some of the more recent findings in infectious diseases, autoimmunity, cancer, normal immune responses, and various other systems to make the point that suppression of immunity due to cellular regulation is a normal as well as a pathological part of the immune system.

Role of Suppressor Cells in Infectious Disease

It has been well documented for many years, as noted above, that suppressor cells play an important role in the response of the body to a variety of infectious diseases, including viral, fungal, and parasitic.

For example, Junin virus infection of adult BALB/c mice has been shown to lead to the development of splenic T cells (Thy-1+, Ly-1+2−) that transferred nonspecific suppression of DTH responses. Lethal viral infection was associated with strong DTH response. These Ts additionally triggered an antigen-specific Ts upon antigen challenge [50, 51]. Also, in herpes simplex virus type 2 infection, UV-B irradiation induced Ts which decreased the proliferative response of immune lymph node cells. Adoptive transfer of L3T4+ splenocytes suppressed proliferative responses in vivo. The suppressive cells in this protocol were Lyt-2+, suggesting a multiple cell pathway [52]. In rabbits, infection with the oncogenic rabbit fibroma virus led to severe immunosuppression. Splenic T cells from these animals made a nonspecific TSF [53, 54]. In a murine model of cytomegalovirus infection which is highly immunosuppressive, there occurred a reactivation of dormant *Toxoplasma gondii* resulting in pneumonia. An influx of CD8+ suppressive lymphocytes into the lung was described. These lymphocytes suppressed the ability of splenocytes to proliferate to *T. gondii* antigens and mitogens in vitro [55].

Mycobacterial infection with *Mycobacterinum lepraemurium* resulted in splenic CD8+ lymphocytes which inhibit proliferation of splenocytes from uninfected mice to both specific antigen and mitogen stimulation. Culture of spleen cells from infected mice produced supernatants with IFN-γ, as well as TNF-α and TNF-β. Suppression was reduced with antibody to IFN-γ [56, 57]. *Mycobacterium tuberculosis* extracts were shown to activate CD8+ Ts in vitro to produce soluble mediators of suppression [58]. A series of recent studies have characterized Ts activity in clones from leprosy patients. These studies have been recently reviewed by Bloom et al. [59]. They demonstrate that clones of CD8+ Ts, which are antigen-specific, can block the response of CD4+ antigen-specific clones. These Ts were shown to produce IL-4 and antibody to IL-4 abrogated the suppressive effect of these clones.

Suppressor cells have recently been described in a number of fungal infections. CBA/J mice injected with an extract of mannan from a fungus suppressed the antigen-specific DTH response. The Ts population was demonstrated to be Thy-1+ Lyt-2+ by adoptive transfer [60]. In the rat a Ts cell was induced by *Cryptococcus neoformans* infection. This Ts cell produced a soluble TSF that led to a decrease in cell surface I-A expression of peritoneal macrophages [61]. Populations of Ts inducers and Ts effectors have been described in this infection that inhibit cell-mediated responses to human serum albumin [62]. Similarly, mice injected with antigen from *Paracoccidioides brasiliensis* developed Ts that down-regulated the DTH response to *P. brasiliensis*. These cells were L3T4+, Lyt-1+2−, I-J+ T cells that produced a TSF that induced a Ts effector population. The Ts effectors were L3T4−, Lyt-2+, I-J+ [63, 64].

Numerous reports have appeared indicating a role for suppressor cells in parasitic diseases. Infection with the metazoan parasite *Echinococcus multilocularis* resulted in poor proliferative responses and lymphokine production by splenic lymphocytes. A population of CD8+ (dull) cells from these spleens decreased the in vitro response to concanavalin A (conA) in normal splenocytes [65]. A population of CD8+ lymphocytes taken from the lungs of mice infected with *T. gondii* blocked antigen- and mitogen-induced proliferative responses of splenocytes from infected mice. Coincident with the appearance of this CD8+ population in the lung, disease reactivation is seen in the mice [66]. There are many reports describing Ts activity in schistosomiasis. Ts are readily demonstrated from infected mice. Recent characterizations of the suppressive system in mice infected with *Schistosomula mansoni* described Ts involved in granulomatous hypersensitivity producing a two chain TSF recognized by mAb 14-12 and anti-I-J [67]. Ts cells were induced in vitro in an *S. mansoni* model by soluble egg antigen. These cells blocked antigen-induced proliferative responses but were not cytotoxic [68]. Ts were described in studies of patients with *Schistosomula japonicum* and a line of Ts was established that produced a TSF [69, 70]. Ts have also been frequently described in *Trypanosoma cruzi* infection. These Thy-1+, Lyt-2+ Ts decreased IL-2 production in response to mitogen stimulation. Culture of the Ts resulted in loss of the Ts activity unless antigen was added to the culture media [71, 72]. Culture supernatants from splenic T cells of infected mice contained a factor which was inhibitory to DTH responses [73, 74].

Role of Suppressor Cells in Autoimmune Disease

A key question in autoimmune disease research is the cellular and molecular basis by which responses to self antigens develop in the face of powerful mechanisms that exist specifically to prevent anti-self responses. One of the means demonstrated to prevent anti-self responses is by active suppression. Recent studies have demonstrated a role for suppressor cells in a number of animal models of autoimmune disease as well as in patients with these diseases. For example, in a murine collagen-induced arthritis model it was shown that adoptively transferable antigen-specific Ts arise. At least two populations of Ts were demonstrated in this model [75]. Similarly, in a rat model of collagen arthritis, Ts were described [76].

In murine experimental autoimmune thryoiditis (EAT) a population of CD4+ Ts were described following induction of disease with deaggregated mouse thyroglobulin. Resistance to EAT was obtained when mice were vaccinated with irradiated primed spleen cells via both CD4+ and CD8+ Ts populations acting cooperatively [77, 78]. Immunization of mice with various retinal antigens results in the development of experimental autoimmune uveitis [EAU]. Injection of retinal antigen intracamerally led to development of a population of Ts which transferred protection from EAU induction adoptively to naive recipients and reversed the inflammation in mice with ongoing EAU [79]. In experiments where EAU was suppressed by anti-MHC class II antibody, this suppression was transferable adoptively with splenocytes [80]. A Ts line was generated from rats primed with retinal soluble antigen. This line inhibited the proliferation of uveitogenic Th lymphocytes in response to antigen. On adoptive transfer this same line was able to reduce the severity of ongoing disease. The Ts line was not cytotoxic and produced a TSF [81].

Rats immunized with myelin basic protein (MBP) develop experimental autoimmune encephalomyelitis (EAE). Spleen or lymph node cells from rats which have recovered from acute EAE transfer protection against EAE to naive rats. These CD4+ Ts block IFN-γ production of EAE effector cells. Primed B cells are also required to transfer protection [82, 83]. A Ts line of CD8+ phenotype was isolated from SJL/J mice that had recovered from EAE. This line inhibited proliferation of MBP-sensitized T cells in vitro and adoptive transfer of the line led to a diminution of EAE induction [84]. In another variation of the EAE model it was shown that oral administration of MBP suppresses disease via CD8+ Ts that adoptively transfer protection. These Ts produce TGF-β and anti-TGF-β antiserum abrogated oral tolerance [85].

An autoimmune disease occurs in the NOD strain of mice that resembles type 1, insulin-dependent diabetes mellitus. Suppressor cell activation was shown to be defective in these mice. Cytokine release was markedly diminished in the NOD mouse. A clone derived from the islets of disease resistant male NOD mice was protective in adoptive transfer experiments. The Ts clone also produced a TSF [86].

Role of Suppressor Cells in the Normal Immune Response
The immune system regulates itself in many ways when stimulated by antigen. Studies done over many years have demonstrated that among the complex cellular interactions that occur, Ts are activated that can downregulate the response to the immunizing antigen. Recent studies continue to demonstrate a role of suppressor cells in the regulation of normal immune responses.

For example, the humoral immune response to noninfectious antigens has been shown to be regulated by Ts activity. Studies using SRBC as antigen demonstrated a population of adoptively transferable, noncytolytic thymic Ts cells in the rabbit capable of inhibiting plaque-forming cell responses. These Ts secrete an antigen-specific TSF that acts on antigen-forming B cells [87, 88]. In mice immunized with a subimmunogenic dose of dextran, a population of CD8+ T cells suppressed the antibody response following adoptive transfer to immunized recipients [89]. T cells from mice immunized with pneumococcal polysaccharide produced a transferable factor that suppressed the antibody response to the antigen in an antigen-specific fashion. This factor was absorbed by B cells specific for the antigen but not with antigen [90]. In another approach to the role of Ts in vivo, mice were treated with anti-CD8 monoclonal antibody. This led to an increase in the production of IgG1, IgG2a and IgG2b, suggesting a regulatory role for CD8+ cells in the regulation of immunoglobulin production [91]. An effort was made to estimate the frequency of Ts in the normal unimmunized mouse by adding T cells into a primary plaque-forming cell response. This study demonstrated that such cells were detectable albeit at a low frequency [92].

The role of Ts in the maintenance of tolerance to self was examined in non-autoimmune mice in an assay of responsiveness to mouse red blood cells (MRBC). Depletion of Ly-2+ T cells with antibody and complement led to an increase of IgM and IgG anti-MRBC antibody-forming cells. An anti-MRBC response was seen in splenocytes from 12-month-old, autoimmune NZB mice as well as splenocytes from young NZB mice treated with Ly-2+ plus complement. Addition of isolated Ly-2+ cells restored the normal regulation of the autoimmune response [93]. In a similar study, depletion of Lyt-2+ cells by in vivo antibody treatment enhanced endogenous lymph node and splenocyte T-cell proliferation and increased the response to self Ia antigens in vitro. Depletion of the Lyt-2+ cells in vitro increased the responsiveness in a syngeneic mixed lymphocyte response while addition of irradiated Lyt-2+ cells inhibited the response of L3T4 cells to self Ia antigens [94].

The cellular immune response has also been shown to be regulated by the development of Ts cells. In studies in which mice were immunized with high doses of minor histocompatibility antigens, it was shown that a transferable population of cells developed that suppressed the cellular responses to antigen. Transfer of these cells was shown to reduce the ability of immune cells to produce IL-2 and IL-3. In vitro coculture experiments with the Ts resulted in suppression of IL-2 production [95]. In a rat transplant model, liver grafts between DA into PVG strain rats are not rejected. Glass-adherent populations from spleen cells taken early (5–28 days) from graft recipient animals suppress rat mixed lymphocyte reactions in vitro in a nonspecific fashion. At a later time (>20 weeks) a T-cell population of antigen-specific Ts was demonstrated [96]. Heart grafts in the same strain combination were not rejected when the recipients were given anti-PVG serum on the day of grafting. Cell transfer studies into irradiated recipients demonstrated that CD4+ cells from such animals restored graft rejection up to 50 days posttransplant. After 50 days they were incapable of restoring rejection of PVG hearts but did restore rejection of third-party (Wistar-Furth) grafts. These CD4+ Ts inhibited the capacity of naive cells to restore rejection [97]. In studies in our laboratory [B.H.D. and D.R.W., submitted] two populations of Ts were shown to arise following stimulation of allogeneic responses in vivo in C57Bl/6 mice, a population of radiation-sensitive, nonspecific Ts producing the lymphokine soluble immune response suppressor (SIRS) (described below) as well as a population of radiation-resistant antigen-specific Ts recognized by novel marker mAb 984D4.6. These populations of Ts developed following a wave of cytolytic T-cell activity. In an allogeneic mouse skin transplant model a population of Ts developed following the rejection of the graft which adoptively transferred significant extension of graft survival [B.H.D. and D.R.W., in preparation].

The role of genetic factors in the regulation of the immune response was described above for the immune response gene phenomenon of high and low responders. A similar genetic control of immune responsiveness had been described in the human. In a family and population study it was shown that there exist high and low responders to natural antigens. CD8+ Ts were shown to arise in low responders to streptococcal cell wall antigen and depletion of the CD8+ cells restored the response to this antigen. The gene controlling this response was in strong linkage disequilibrium with alleles of the HLA-DQ locus. Similar results were seen using schistosomal antigen, *M. leprae* antigen, tetanus toxoid, *Cryptomeria* pollen antigen, and hepatitis B. Anti-HLA-DQ monoclonal antibody restored the responsiveness in vitro in low responders, while anti-HLA-DR antibody inhibits the response of high responders. Furthermore, *Streptococcus* cell wall-specific HLA-DQ-restricted CD4+ cell lines activated CD8+ Ts that down-regulated CD4+ T cells [98].

Role of Suppressor Cells in Cancer

Tumors have numerous ways to avoid destruction by the immune system. Data developed over many years have suggested that one such mechanism is by activation of suppressor cells that block immune responses to the tumor. There is a large number of such reports and we will not attempt to describe all of them, but a few have been selected to show the wide variation shown both in the type of suppressor cells and the responses regulated by these cells.

Suppressor cells have been described in response to tumors that block various antitumor responses. BALB/c mice in response to immunization with mitomycin c-treated MOPC 104E plasmacytoma generated very weak CTL responses. In vitro sensitization to this tumor led to good activity. A Ts population was described in the in vivo exposed mice that blocks CTL generation in vitro [99]. Clones of CD4+ cells from melanoma-involved lymph node selectively down-regulate the induction of cytolytic immune responses against autologous tumor [57, 100]. Clones of CD4−/CD8− antigen-specific Ts were described in a murine UV-induced skin tumor system [101, 102]. Also, NK activity was decreased in BALB/c mice infected with Moloney murine leukemia virus. Splenocytes from infected mice were shown to block NK activity in vitro. T-cell lines from infected mice also were shown to block NK activity [103]. CD8+ cells from lymph nodes of tumor-bearing SJL/J mice suppressed the proliferation of syngeneic CD4+ cells in response to mitogen or antigen. Anti-CD8 treatment of mice with limited numbers of CD4+ cells showed increased growth of the B-cell lymphomas that requires CD4 cell proliferation. These data suggest a nonspecific Ts population [104]. Various reports have detailed the function of soluble factors that suppress immune responses in tumor-bearing animals or human cancer. Suppressor factor activity was described in the serum and in tumor-bearing spleen cell cultures in a murine system [105]. One such factor has been characterized from a Ts clone from a patient with acute lymphoblastic leukemia [106]. These reports illustrate the wide range of Ts activities effecting a variety of tumor responses.

A genetic difference in Ts activity has also been described between BALB/c mouse strains. The BALB/cAnPt mice develop plasmacytomas readily upon mineral oil injection while the BALB/cJ mice are relatively resistant. In BALB/cAnPt mice a population of Ts developed in response to the plasmacytoma ADJ-PC-5 but not in BALB/cJ mice [107].

Characterization of Suppressor Cells in vitro

Suppressor Cell Clones

One of the difficulties in the study of suppressor cells over the years has been that while methods have emerged for cloning of T-helper and cytotoxic cell lines, it has been much more difficult to obtain clones of suppressor cells. Much of the earlier work was done using T-T hybridomas with the attendant problems of stability and complex regulation as discussed below. More recently, a number of laboratories have succeeded in producing clones of both human and murine suppressor cells and studies have been done with these clones. In an elegant series of studies in the mouse by Kolsch and co-workers [108, 109], clones of Ts have been established from tolerized mice which are MHC class II-restricted and CD4+. These clones block helper function of Th clones for immunoglobulin production. These clones have been shown to function in vivo by injecting the clone at the time of immunization. The clone blocks proliferation to antigen in vitro as well as production of antibody to the immunizing antigen in vivo. No effect was seen on antibody production to an unrelated antigen [108, 109]. Asano, Tada and co-workers [110–112] have generated murine clones of both CD4+ and CD8+ phenotype which block the proliferative response of CD4+ clones cultured with APC and antigen. Ts inhibition blocked an early event in cell triggering and an increase in intracellular calcium. Th clones did not effect Ts clones nor could Ts clones suppress other Ts clones. When Ts clones were triggered with anti-CD3 antibody they produced a TSF that blocked the proliferation and calcium response of CD4+ clones. A series of CD8+ Ts clones were studied and shown to have no cytolytic or helper activity and to produce no IL-2 or IL-4. These Ts clones blocked the production of IL-2 by CD4+ Th clones [110–113]. Sehon's group [113–115] has generated clones from mice tolerized with polyethylene glycol-conjugated antigens. These clones block production of immunoglobulin in a MHC class I-restricted and antigen-specific fashion. All clones from this study were Thy-1.2+, CD4−, CD5−, CD8+, CD3+, TcR-$\alpha\beta$+, and were not cytolytic [113–115]. Clones of other phenotypes have also been described. Ts clones specific for minor histocompatibility antigens were MHC class II-restricted, CD4+, CD8+, TcR-$\alpha\beta$+, and noncytolytic [116]. Ts clones down-regulating the response to self MHC were established from adult bone marrow and bear the CD4−, CD8−, Thy-1+, MHC class I+, TcR-γ+. These clones block generation of CTL activity toward self MHC but not other MHC types. In a skin allograft model these clones along with IL-3 significantly prolonged allograft survival in an antigen-specific fashion [117]. Clones of nonspecific Ts have also been produced by Strober and co-workers [118, 119]. Natural suppressor cell clones blocked IL-2 production in mixed lymphocyte cultures, but not mitogen-stimulated proliferation. Characterization of these clones demonstrated a CD3+, CD4−, CD8−, TcR-$\alpha\beta$+ phenotype. These clones showed both suppressive activity as well as the ability to induce thymocyte proliferation when IL-2 and PHA or IL-4 were added to the cultures [118, 119].

Clones and lines of Ts have also been described in the human. An antigen-specific clone RLB9-7, from a patient with allergic contact dermatitis, was described that was CD3+, CD4+, CD8+, and TCR-$\alpha\beta$+. The clone proliferated in response to antigen, and was MHC class I-restricted. The proliferation was blocked by antibody to CD8 but not CD4 suggesting that only CD8 was functionally associated with the TcR. The clone was NK−, a poor producer of IL-2 and IFN-γ. This clone plus antigen blocked the mitogen response of autologous but not allogeneic PBL. The immunoglobulin response of autologous lymphocytes to tetanus toxoid was also blocked. Thus, the clone was nonspecific in effector function but triggering was antigen-dependent [120].

In another case, both Ts and Th clones were made from a lepromatous leprosy patient. The antigen-specific Ts clones proliferated in response to antigen and were CD4+ and class II-restricted. Proliferation to antigen was blocked by mAb to TcR-$\alpha\beta$+, IL-2R, CD4, and MHC class II. These clones produced IFN-γ. Although the clones had lytic machinery, the suppression was not via a lytic mechanism. The major distinction between the Th and Ts clones in this system was that the Ts clones lacked CD28 on their surface [121, 122]. A study by Bloom's laboratory [123–125] analyzed the CD8+ Ts clones derived from leprosy patients demonstrating the clones to be Tcr-$\alpha\beta$+, HLA-DQ-restricted, and to produce primarily IL-4.

In other studies, antigen nonspecific lines of CD8+ Ts were obtained from mitogen-stimulated PBL. These lines blocked proliferation of CD4+ cells stimulated with mitogen, OKT3, or tetanus toxoid, but failed to block mixed lymphocyte responses. Stimulated CD4+ cells produced normal amounts of IL-2, but failed to express the high affinity IL-2R on their surface [126]. Clones were derived from PBL which were CD3+ CD4− CD8− and CD3+ CD4+ CD8+. All clones were CD29+, CD11b+, and NKH1+. These clones were MHC-unrestricted and produced low amounts of IL-2 and IFN-γ, but normal TNF in response to stimulation with anti-CD3. When stimulated with mAb to CD3 some of the clones were able to support, weakly, B-cell differentiation. When stimulated with PWM, all clones suppressed B-cell differentiation supported by mitomycin c-treated fresh CD4+ lymphocytes [127].

Another group reported lines of MBP-specific Ts that were specific for Th clones to the same antigen. These were CD4+ (class II-restricted) or CD8+ (class I-restricted) and blocked antigen-driven proliferation of the Th clones [128]. In another study, a human CD8+ Ts line was generated by incubation of a CD4+ clone specific for streptococcal cell wall antigen with CD8+ cells. This activation was mediated by a factor from the CD4+, MHC class II-retricted, clone. The activated cells were nonspecific suppressors, class I-restricted, and showed no cytolytic activity [129].

Several human clones have been described that have multiple functions. In one series of clones it was shown that as cells were cultured over a period of time they became anergic. These aged clones still proliferated in response to antigen and they continued to produce IFN-γ and GM-CSF, but they lost the ability to produce IL-2 or up-regulate IL-2 receptors. Further, they induced suppressive activity in fresh lymphocytes. Induction of Ts activity required cell adhesion as mAB to CD11a blocked induction. Anti-CD4 also blocked induction, as did mAB to the IL-2 receptor. Induced Ts were allounrestricted, CD4+, and CD45R+ [130, 131]. In another study a CD4+ clone was described with bifunctional properties. The clone provided help for differentiation of autologous B cells into Ig-secreting cells in the absence of PWM, but blocked such differentiation when incubated with B cells activated with PWM. This Ts function was effective even when added late into B cell cultures [132]. Bifunctional clones were also described in another series of reports. A series of autologous autoreactive clones were generated to a CD4+ clone. The antigen-specific proliferative response of the clones is blocked by antisera to MHC class II. Under some conditions suppression of the proliferative response of the target clone was observed while other growth conditions led to stimulation [133–135].

The studies described here are by no means an exhaustive listing of all Ts clones described in the literature, but are illustrative of the diversity in Ts phenotypes and activities recently discovered. Although it is still more difficult to establish Ts clones than Th clones, the number of such clones is rapidly expanding and should lead to a clearer understanding of the various mechanisms of suppression.

Cytokines and Suppression

The current status of the biochemistry and molecular biology of specific and nonspecific suppressor factors will be described later. Nevertheless, suppressor cells must also be viewed in terms of their relationship to other lymphocytes and the interactions between cell populations producing various lymphokines. A paradigm explaining many of the regulatory interactions of immune responses was proposed recently by Mossmann et al. [136]. Briefly, this model describes CD4+ T-helper cells as falling into two categories termed Th1 and Th2, which produce different sets of lymphokines. The Th1 subset produces primarily IL-2 and IFN-γ, while the Th2 cells produce IL-4, IL-5, IL-6, and IL-10. Th1 cells are involved in DTH and inflammatory responses and support IgM and IgG2a, while Th2 cells are involved in allergic responses and support IgE and IgG1. In addition, the Th1 and Th2 type cells mutually regulate each other. IFN-γ produced by Th1 cells down-regulates production of lymphokines by Th2 cells, while IL-4 and IL-10 produced by Th2 cells down-regulate lymphokine production by Th1 cells. Although these phenomena are more difficult to dissect in the human immune system, recent data from a number of labs have shown that the basic pattern exists in man as well as mice. These data have been recently reviewed for the murine and human systems [43, 136, 137]. The basic model has recently been extended by Bloom et al. [59, 138] to explain the phenomenon of suppression and includes CD8+ T cells. In this model CD8+ T cells are divided into two populations, type 1 which suppress B cells and have CTL function and type 2 which provide B-cell help and suppress DTH. These CD8+ T cells produce lymphokines as do CD4+ T cells in order to carry out these functions. Bloom et al.

[59, 138] use this model in an attempt to explain the Ts phenomenon and place it within the context of current thinking on lymphokine regulation. Recent reviews describe this proposal in more detail.

A number of studies have demonstrated a role for lymphokines in the induction of Ts activity. These studies suggest a direct role for cytokines in culture models by leading to enhanced activation of Ts activity or the inhibition of induction of Ts activity when neutralizing or receptor-blocking monoclonal antibodies are added. For example, the role of IL-2 in the activation of Ts was studied in a variety of systems. In one study, human CD4+ T cells were stimulated with immobilized monoclonal anti-CD3 antibody (64.1). While low concentrations of this antibody stimulated help for B-cell responses, high concentrations led to CD4+ Ts activity. IL-2 was required for the generation of Ts activity. These Ts directly blocked *Staphylococcus aureus* and IL-2 stimulation of B cells [139]. Another study demonstrated a positive effect of IL-2 on the secretion of a TSF from a suppressor T-cell hybridoma which did not utilize IL-2 for growth [140]. Several studies of nonspecific suppression have demonstrated that IL-2 can activate Ts. Naive murine lymph node cells incubated with IL-2 generated increased inhibitory activity in a population of CD8+ cells, and IL-2 administration in vivo was potentiated Ts cell activity [141]. In another study, naive murine splenocytes incubated with IL-2 developed nonspecific Ts activity. Antibody to the IL-2 recepor blocked the IL-2-induced Ts development [142, 143]. Antibody to the human IL-2R (anti-TAC) blocked the induction of conA-induced CD8+ Ts and of Epstein-Barr virus-induced Ts and of antigen-specific anti-SRBC Ts induced by high concentration of SRBC [144, 145]. In clinical trials of IL-2, the ability to generate Ts activity in mixed lymphocyte cultures was increased in patients treated with IL-2 [146–147]. In contrast to these studies, other studies have shown that IL-2 is not required for the generation of Ts. Anti-IL-2R antibodies reacting with different epitopes on the p55 β chain were used in a heterotopic cardiac allograft study in rats. These antibodies extended graft survival while Ts activity was selectively spared. Interestingly, ART-18 treatment led to a CD4+8– adoptively transferable Ts while ART-65 led to a CD4–8+ transferable Ts [148]. In a human model, antigen-specific Ts were generated in a mixed lymphocyte culture with PBL in the presence of antibody to the IL-2R. These Ts were predominantly CD4+CD45RA+, but required CD8+ cells for activity [149].

Other cytokines besides IL-2 have been demonstrated to have a role in activation of Ts activity. Both rhIFN-α and rhIFN-β increased the ability to generate nonspecific Ts activity with lymphocytes from patients with multiple sclerosis [150]. Nonspecific Ts were stimulated by IFN-α and IFN-γ in a murine system [151]. Murine natural suppressors are also stimulated by IFN-γ and blocked by antibody to IFN-γ. The stimulation seen with IL-2 was also blocked by antibody to IFN-γ, suggesting that IL-2 acted as an inducer of IFN-γ [152]. However, in another protocol, injection of anti-IFN-γ antibody resulted in increased activation of CD8+ Ts which suppress Th activity [153].

A second group of studies have examined the role of cytokines in the effector activity of Ts. As described above, workers have investigated both the cytokines produced by Ts clones and lines as well as the effect of neutralizing antibodies on suppression. For example, one study demonstrated that normal murine bone marrow, when cultured with IL-3 and GM-CSF, generated Ts that blocked splenic mitogen blastogenesis. Supernatant from these Ts were also suppressive blocking splenic blastogenesis and NK activity. These supernatants contained TGF-β and anti-TGF-β antibodies reduced the suppression [154]. A similar study done in a model of EAE also demonstrated a role for TGF-β in suppression. In this study, tolerance was induced in mice by oral MBP and an adoptively transferable population of CD8+ Ts demonstrated. These cells produced a TSF in vitro which was inhibited by anti-TGF-β antibodies. This antiserum abrogated the protective effect of oral tolerization and led to an increased severity of disease when administered to nontolerized mice [85].

In another system, a CD8+ Ts clone was shown to produce a soluble TSF which suppressed the proliferation of a Th1 clone. This TSF was purified and determined to be IL-10. An anti-IL-10 monoclonal antibody completely blocked the suppression [155]. Another CD8+ Ts population, generated by culture with components of *M. tuberculosis*, produced supernatants containing IL-4 and IL-6. Production of IL-1, TNF-α, IL-2 and IFN-γ were suppressed. Addition of IL-6 to cultures also decreased cytokine production while addition of mAb to IL-6 restored the production of IL-1β and TNF-α [156].

In studies of nonspecific Ts activity, we [D.R.W., B.D.H., et al.] demonstrated that antiserum to the cytokine SIRS was completely able to block the CD8+ Ts-mediated suppression of a mixed lymphocyte response induced in vitro by incubation of normal murine splenocytes with conA or IL-2. SIRS added to cultures without Ts was suppressive of the response and this suppression was reversed by the anti-SIRS antiserum [142–143].

The examples described of effector activites of Ts mediated by cytokines all concern suppressive activity mediated by CD8+ T cells. There are numerous examples as well of regulatory functions by CD4+ T cells as described in the reviews on the Th1 and Th2 phenotypes of lymphocyte [43, 136]. Now that clones of Ts are available for study, the pattern of lymphokines produced by Ts clones of both the CD4+ and CD8+ phenotypes is under investigation in much the same fashion as studies done on panels of CD4+ clones. For example, one study investigated two Ts clones and showed production of high levels of IFN-γ and TNF, low levels of IL-3 and IL-6, and no IL-2, IL-4, IL-5 or IL-9 [109]. A study of stimulated natural suppressor clones showed that they produced IL-3, TGF-β, IFN-γ, GM-CSF, and TNF-α, but not IL-1, IL-2, IL-4, IL-5, IL-6, IL-7, or IL-10 [118]. The lymphokine profile from a series of CD8+ Ts clones from leprosy patients were analyzed by Bloom and co-workers [123–125] in comparison with alloreactive CD8+ CTL clones. The CTL clones produced IFN-γ but not IL-4. All the Ts clones were HLA-DQ-restricted, TcR-$\alpha\beta$+, and produced IL-4 and little or no IFN-γ. Anti-IL-4 antibody blocked Ts suppression of Th1 clones. Thus, these CD8+ Ts clones acted similarly to Th2 clones. Studies such as these combined with functional studies should enable the dissection of the mechanisms by which Ts are regulated by cytokines and function through the production of cytokines.

Other studies have focused on the identification of new cytokines as products of Ts cells. Several investigators, as discussed below, have demonstrated a role for released TcR molecules, most common TcR-α chains, as effectors in Ts function. Other studies have focused on the characterization of new factors.

Antigen-Specific Suppressor Factors (TSF)

TSF were described almost simultaneously with suppressor cells [6, 10, 12]. The initial observations were made using extracts of tissues such as the thymus from mice subjected to tolerogenic immunizations and from short-term T-cell cultures [6, 10, 12]. The early studies established that TSF derived from extracts could bind to antigen-linked Sepharose or to immunoabsorbent columns using anti-I-J antisera or anti-idiotype antisera where available. Moreover, Tada, Taniguchi and co-workers [35–40] showed that TSF consisted of a disulfide-linked heterodimer, one chain which bound to antigen and a second chain that reacted with anti-I-J and contained the haplotype restriction element. A major problem that concerned many observers of this work was that although TSF could be recovered from cell or tissue extracts, it was difficult to obtain from culture fluids of cells that were clearly suppressive. This led some immunologists to wonder whether such molecules played a significant biological role in immune regulation. Pierres and Germain [28], among others, were able to demonstrate that by removing adherent cells from cell cultures containing GAT-specific T-suppressor cells, GAT-TSF could be recovered from the culture fluid. This implied that the failure to detect soluble TSF was due to its rapid adsorption or uptake by adherent cells. Further biochemical characterization would have proven difficult had it not been for the development of T-cell hybridomas, as mentioned earlier. As will be shown, the development of T-suppressor hybridomas was both a blessing and a curse for immunologists interested in studying suppression.

Several groups in the early 1980s were able to produce T-suppressor hybridomas that constitutively released antigen-specific TSF in seemingly large amounts as assayed by dilution analysis [12–15, 30, 36, 40]. This allowed the use of standard protein purification and analytical methods to be applied to the characterization of TSF [157–166]. One of the more concerted efforts in this regard was carried out in the laboratory of one of us (D.R.W.) in collaboration with the laboratory of Pierce and Kapp [157–166]. The problems and issues that arose from those studies may be used to illustrate the general nature of the difficulties encountered by many.

The purification of antigen-specific TSF to chemical homogeneity required that there be a consistent, reproducible source of TSF. T-cell hybridomas that constitutively produced antigen-specific TSF appeared to be the ideal source since these cells could be grown in normal cell culture media without the need for added growth factors, cell feeder layers, or antigen stimulation [30, 36, 40, 158, 163, 164]. In studies with GAT-specific suppressor cells, Kapp, Pierce and co-workers [163, 164] generated large panels of T-suppressor cell hybridomas that constitutively produced a variety of GAT-TSF molecules and released them into the culture fluids. Dilution analysis of these cell culture fluids suggested that the hybridoma cultures contained many times the amount of activity recovered from primary cell cultures or from tissues obtained from non-responder or responder mice immunized with GAT [162]. However, it was impossible to analyze what percent of the hybridoma cells were actually producing GAT-TSF. Moreover, the hybridomas showed a disturbing tendency to lose cell surface markers usually associated with the various types of T-suppressor cells [163], thus raising some questions about hybridoma stability – a problem encountered by many other workers studying T-helper or cytolytic T-cell hybridomas.

Despite these concerns, the tissue culture bioassays seemed to indicate that the T-suppressor hybridomas were reasonably stable for the production of GAT-TSF. To further control the stability of GAT-TSF producing hybridomas, cultures were grown for a limited number of generations (generally no more than 4 weeks or approximately 30 generations) and then discarded. In order to insure objectivity in analyzing samples of GAT-TSF during purification, the samples were coded and assayed for biological activity in a separate laboratory from that carrying out the purification [157]. Purification of the GAT-TSF-1 (suppressor-inducer factor) from culture fluids of the Ts hybridoma 258C4.4 was carried out using ion-exchange chromatography and reverse-phase HPLC [157]. Since the protein was purified from culture fluids containing a 10% fetal calf serum, the final purification was 5.2×10^7 fold from 6 liters of culture fluid (1.8×10^3 ng). The molecular weight of the purified GAT-TSF-1 was estimated by SDS-PAGE to be 29 kD. In order to obtain the approximately 1-nM amounts needed for protein sequencing would require approximately 20–30X the amount of culture fluid used for the original purification. During the course of these studies, a change in the purification scheme resulted in the introduction of a serum protein contaminant, apolipoprotein A-1, into the final stages of purification. Final resolution was accomplished by flat-bed isoelectric focusing in a granulate gel [166]. Scale-up created problems of cell line stability such that our large preparations (150 liters) yielded barely sufficient protein for N-terminal sequencing. N-terminal sequence analysis suggested the presence of a blocked amino-terminus. Repeated attempts to obtain sequence data for the GAT-TSF-1 over a 6-year period resulted in one five-residue sequence from a trypsin digest [Turck and Webb, unpubl. data]. The difficulties encountered led us to abandom this approach to obtain structural information.

Similar studies were performed using a hybridoma 762B3.7 that produced a two-chain, disulfide-linked heterodimeric GAT-TSF [165]. We were able to confirm the earlier work reported by Taniguchi et al. [40], that effector TSF (TSF-2) is also antigen-specific, with one chain (α) binding antigen, and the second chain [X] possessing the MHC-restricting element. These factors bore a striking resemblance to the TcR (about which more later), except that they were found in the culture fluid and would bind to free antigen. Once again, lack of hybridoma stability made it very difficult to obtain sufficient protein for full characterization [164, 165]. Other researchers reported variations on the theme described above. For example, Green et al. [13] at Yale found that in response to SRBC challenge, a two-chain suppressor factor was generated that was the product of two different cells. Nevertheless, they also established that one chain was antigen-specific while the other chain carried the restriction element. These investigators and their collaborators had some success in producing hybridoma ascites and/or serum that contained at least the antigen-binding chain of SRBC-TSF [13]. However, no protein sequence information was obtained by any of the groups studying TSF. Thus, the link between the soluble antigen-specific TSF at the protein level and TCR genes remains to be established. What was produced were a series of monoclonal antibodies specific for TSF that have, in some cases, proven to be useful reagents for exploring the physiological role of T-suppressor cells (see below).

Use of TCR Immunoglobulin Genes by Antigen-Specific T Cells
Like B cells, Ts cells had been shown to bind soluble antigen and they frequently expressed cross-reactive idiotypic determinants present on B-cell

immunoglobulin (Ig) directed to the same antigen, so it was initially presumed that Ts cells would utilize Ig or related genes to encode their antigen receptors. However, it was demonstrated directly that Ts cells do not rearrange or express Ig genes [167]. Moreover, in systems where cross-reactive idiotype has been demonstrated, the Ts cell does not express the VH gene segment used by the B cell [168, 169].

Since Ts cells express Thy-1 and markers of the T-cell lineage, it seemed more likely that they would utilize TcR genes to encode their receptors for antigen. Yet, the use of TcR-α and -β genes by Ts cells has been very controversial. In two early studies, there was no evidence of Ts-derived TcR-β chain gene rearrangement in suppressor T-cell hybridomas [170, 171]. However, in certain hybridomas as well as in some murine and human Ts clones, the cell responsible for suppression did not appear to be CD3+ and express the $\alpha\beta$-TcR [102, 124, 172–179].

The failure to detect TcR in the earlier studies may have been due in part to the inherent instability of the hybridomas [180]. For example, one of us (E.K.) directly assessed chromosome loss in C4.4, a GAT-specific hybridoma, kindly provided by Kapp. Although C4.4 continued to express suppressor activity at a high level, there was no unique TcR-β chain present. Additionally, using restriction fragment length polymorphisms to analyze 6 different genetic loci, we were only able to detect the BW5147-derived chromosomes; all Ts-derived homologues had been lost [Kraig, unpubl.]. Thus, it is possible that fewer than 10% of the cells in a hybridoma culture may actually contain the functional chromosomes; such a contribution would likely be missed by Southern blot analysis.

Given these considerations, Dorf and co-workers [181, 182] and Weiner et al. [177] attempted to enrich for suppressive activity by selecting for that subset of cells that remained CD3+. In this way, they were able to characterize the TcR genes being expressed in the Ts hybridomas. Fairchild, Moorhead and co-workers [174, 175] provided convincing evidence in a class I-restricted Ts system, that both the α chain and the β chain were involved; the β chain imparted MHC restriction. Of the many other systems that have been studied the most convincing data derived from Collins, Dorf and co-workers [182, 183] and Green and co-workers [184]. Using either chromosome loss variants [182, 183] or antisense oligonucleotide inhibition [184] to eliminate synthesis of individual chains, they were able to demonstrate that the TcR-β chain was dispensable, but that the TcR-α chain was required for Ts function. In a complementary set of experiments, transfer of the TcR-α chain back into cells was sufficient to restore the suppressive phenotype [182, 185]. The importance of the α chain had been strongly suggested by Taniguchi's group [186], since his laboratory had shown a bias in the TcR-α chain expressed by KLH-specific Ts cells. However, in spite of the apparent consensus that the α chain plays an important role, it is not clear whether this polypeptide is acting alone, whether it combines with a second chain (which may or may not be β) or whether it is a modified form of α. In a study from Ishizaka's lab [187], evidence is provided of a Ts factor that reacts with anti-TcR reagents as well as with monoclonals to TSF (14–12 or 14–30); so the TcR that is suppressive may be an altered form. The molecular tools to address this possibility are now available.

Novel Approaches to Cloning the X Chain

The consensus view would appear to be the suppressor T cells may be similar to helper and cytotoxic T cells in expressing the TCR-αβ heterodimer. However, as discussed above, an α chain in the absence of a β chain is sufficient for antigen-specific function. Is the α chain associated with another polypeptide? perhaps an I-J-bearing chain? To address this possibility directly, one of us (E.K.) took a nonbiased approach in an attempt to clone other relevant polypeptides. Briefly, a large cDNA expression library was constructed in λgt11 using RNA from a GAT-specific hybridoma generated by Sorensen et al. [163], termed 372D6.5. The library was screened with a pool of monoclonal antibodies, including anti-I-J and antifactor reagents. Ten cDNA clones gave unambiguous positive reactions in the serological screens and were analyzed further. One of these cDNAs, designated pS35, was shown by Southern blots to rearrange in some Ts hybridomas. Thus, we consider pS35 an ideal candidate for a novel Ts antigen receptor gene, it was selected with Ts-specific serological reagents, and its gene undergoes rearrangement as do the other receptor genes. The short cDNA had only a single open reading frame, that was unique when compared to protein sequences in the national data bases; we conclude that pS35 is not Ig, TcR or a known oncogene. The pS35 sequence is found by PCR analysis using RNA from the CD3+ Ts hybridomas characterized and kindly provided by Drs. Taniguchi and Dorf, suggesting a more general function [Kraig, unpubl.]. Recently, genomic clones homologous to the pS35 cDNA have been isolated and studies are now underway to determine whether this represents a novel rearranging gene family.

Phenotypic Studies of Suppression
The I-J Enigma

The discovery of th I-J locus has recently been discussed by Murphy and co-workers [21, 22] among others [49] and will not be discussed in depth here. Suffice it to say, by reciprocal immunization using splenocytes of either B10.A(3R) origin or B10.A(5R), one could obtain antisera that could distinguish between these two seemingly identical H-2 mice. This difference was detected on only a subset ($< 10\%$) of T cells (and not on B cells); thus I-J was not thought to be another class II molecule. Although I-J was displayed primarily on T cells with suppressor or inducer function [10–15, 21] non-T and non-B cells reactive with anti-J antisera have been reported [22]. I-J+ T cells were detected solely by cell killing experiments using the anti-I-J antisera plus guinea pig complement or by blocking with anti-I-J antisera, and observing a loss of cell function. A number of laboratories tried at the time to radiolabel cell proteins and immunoprecipitate an I-J molecule but none were successful [22]. By contrast, it was possible to construct a Sepharose-anti-I-J column and remove antigen-specific TSF from a T-suppressor cell extract or from a T-suppressor hybridoma cell culture fluid. The material could then be eluted and could be shown to be active [10–12, 21, 36, 38, 40, 166]. This would only work with cell extracts or culture fluids obtained from animals or cells of the correct H-2 haplotype [11]. Thus, the purification and characterization of TSF offered a way to possibly understand the I-J puzzle and at the same time understand how T-suppressor cells could function to block an immune response in an antigen-specific fashion.

Although the genetic studies strongly argued that the I-J gene was located within the I-A/I-E region of the murine MHC, molecular biological approaches produced data conflicting with this conclusion. Experiments using cosmid clones that covered the region between I-A and I-E had shown that there was insufficient DNA to encode an additional molecule (I-J) and that no RNA homologous to this region was detectable in T-cell hybrids [188]. In addition, Kobori et al. [189] sequenced the DNA in this region from the B10.A(5R) and B10.A(3R) recombinant strains that had been used initially to map I-J and showed that the recombination breakpoints that defined the region were no more than 2 kb apart and were located within the intron of the $E\beta$ gene.

These experiments were used to support the hypothesis originally put forward by Schrader [190], and expanded by others [188, 191] that the I-J serological determinant on Ts cells is most likely not encoded by the MHC, but is instead a receptor on Ts cells that is involved in MHC recognition. This hypothesis is supported by two recent experiments demonstrating that the I-J phenotype can be altered using either a chimeric [192] or a transgenic [193] animal. The latter experiment is perhaps the most convincing and elegant approach to I-J in the literature. Flood et al. [193] showed that the presence of an $E_\alpha{}^k$ transgene was sufficient to confer the I-J^k phenotype on a mouse of non H-2^k background, suggesting that not only does the MHC play a role in shaping I-J, but that the gene(s) involved must lie within the 8.2 kb fragment used in construction of the $E_\alpha{}^k$ mouse.

Perhaps I-J is controlled by two or more gene products: one, within the MHC, expressed on the antigen-presenting cells, and one, mapping elsewhere, that is expressed by the Ts cell itself. Such a scenario was suggested by a study undertaken by Hayes et al. [194]; they demonstrated that the expression of I-J^k could be affected by a second genetic locus, designated Jt which mapped to chromosome 4, near the Fv-1 locus. These data however remain somewhat controversial as the strains used did not test equivalently for all investigators [191, 193]. Nonetheless, it is interesting to note that a locus Ssm-1 has recently been mapped on chromosome 4, near Fv-1, which controls the patterns of DNA methylation which affect transcription of genes mapping elsewhere [195]. No relationship was suggested between Jt and Ssm-1, but the possibility of a modifier locus that acts at the DNA level on the MHC could now be pursued as well.

To add to the enigma surrounding I-J was the discovery by Ishizaka's group [196] in 1986 than an I-J+ factor that induces suppression in the IgE system, glycosylation inhibitory factor (GIF), is bound by monoclonal antibodies to lipomodulin, a general inhibitor of phospholipase. It has been similarly demonstrated that the I-J chain in other suppressor systems can be identified by antilipomodulin serological reagents [197]. It remains unclear whether this is due to fortuitous cross-reactivity, modification of the I-J chain, or a functional association between an MHC-restricted polypeptide and lipomodulin. The molecular nature of I-J can be addressed only by direct cloning or purification efforts like those that have been attempted in our labs and were described in detail earlier. Finally, it should be noted that Tada and co-workers [198] have recently characterized a disulfide-linked, heterodimeric, cell surface molecule on helper T-cell clones that reacts with monoclonal anti-I-J antibodies. This molecule is not the antigen-receptor on these cells and its function remains undefined.

It was originally assumed that characterization of I-J$^+$ TSF would ultimately resolve at least part of the puzzle concerning I-J. That has so far not proven to be the case. In those cases where TSF have been extensively purified and studied (see above) there still has not been a specific gene product identified with I-J. What can be said with certainty is that proteins with molecular weight of 29–33 kD that associate with antigen-binding chains or which recognize antigen themselves [166] can react with both polyvalent or monoclonal anti-I-J antibodies. Such polypeptide chains can be shown to have a variety of biological effects, e.g. controlling I-J-based restriction and/or IgVh restriction, but their precise molecular nature and the identification of the genes that code for such molecules has yet to be accomplished. Fortunately, the problem encountered in determining the molecular nature of the I-J$^+$ chain of TSF have been overcome recently for the antigen-binding chain of TSF [174–177, 182–185, 187, 199, 200]. However, this did not occur without considerable confusion and misunderstanding as to what was the nature of antigen-specific TSF as discussed earlier.

New Markers of Ts

A key question in the study of Ts is whether it is possible to make a phenotypic identification of suppressor cells as distinct from other T cells. The antisera and monoclonal antibodies to I-J as described above were early markers for Ts cells, and despite the difficulties in identification of the molecular target for the antibody, I-J continues to be used in many studies to identify Ts. A number of other antibodies have also been developed which can distinguish Ts from other cells. Ts have been described in human cell populations by Schlossman and co-workers [201–203] using monoclonal antibodies 2H4 and anti-MOL. The 2H4 mAb recognizes the CD45R 220 kD glycoprotein and has been described on a CD4+ population of cells which blocked T-cell help for Ig production. CD8+ cells were shown to be required for this Ts activity. CD8+ Ts cells were described as two populations based on the CD11 marker recognized by the anti-MOL antibody. One population was CD11− and required CD4+ CD45R+ cells to elicit Ts activity and the second was CD11+, did not require CD45R+ cells, and was activated by IL-2 [201, 202]. CD45R+ CD8+ Ts were identified in the autologous mixed lymphocyte reaction [203].

Another mAb, 4B4, recognizing the CD29 marker was also used in combination with CD45R to define Ts subsets. Suppressor inducing CD4+ cells were shown to bear CD45RA+, CD29− phenotype while inducers of help were CD45RA−, CD29+. CD8+ Ts effector cells were CD11b−, CD45RA+, and CD29− [204, 205]. A mAb to a rat CD45 isoform of 220 kD, RTS-1, was described which identified a population of CD8+ Ts which inhibited mitogen-induced Ig production of B cells [206]. Another monoclonal antibody used to detect Ts activity in the mouse is the 14–12 mAb. This mAb was shown to block the in vivo induction of Ts by several antigens including bovine serum albumin and trinitrobenzene sulfonic acid. This mAb also converted nonresponder strains of mice for the GAT antigen into responders [207, 208].

An additional set of monoclonal antibodies was described that recognized discreet populations of murine Ts cells. The mAb 984D4.6 was shown to ablate, via complement-mediated lysis, antigen-specific Ts activity generated to alloantigen in vivo or in vitro. A second mAb, 2441, was also shown to lyse, via complement-mediated killing, Ts activity to alloantigen. When separate populations of cells treated with mAb 984D4.6 and mAb 2441 were recombined, suppression was re-established suggesting that two separate, collaborating populations of Ts were recognized by these monoclonal antibodies [209, 210].

A monoclonal antibody to the natural suppressor cell has been developed, mAB 1E5.B5, which recognized some, but not all, murine NS cells [211]. A series of human Ts clones from a lepromatous leprosy patient were isolated and shown to be CD4+ CD28− in phenotype. CD28− has been proposed as the necessary costimulatory signal in Th activation and the lack of CD28 could prevent Th function and permit Ts activity [121]. Similarly, a population of CD11b+ CD28− cells were shown to be suppressive when induced with mitogen [212].

Antigen Recognition and Activation of Ts

Another area of investigation has been in the mechanisms by which the immune system selects the Ts pathway rather than the Th pathway. The role of Ir gene regulation of the Ts pathway has been discussed above. A number of other studies have addressed this question as well. Sercarz and Kyzych [42] have investigated the antigenic determinants on β-galactosidase and hen egg lysozyme which induce Ts vs. Th activity. These studies have demonstrated that the determinants on an antigen which induce Ts are not the same as those which induce Th. In some cases, removal of the suppressogenic determinant was able to render it nonfunctional and permit the immunization of strains of mice which were nonresponders to the native protein [213]. The ability of Ts to suppress the function of Th was also different for Ts recognizing different epitopes on an antigen, some Ts suppressed a wide range of Th while some only Th for a selective epitope [214]. The reasons why a particular epitope is selected and Ts develop in some cases while Th prevails in others is not understood but could relate to the particular APC population and MHC determinants.

The role of the antigen-presenting cell (APC) in the development of Ts has also been investigated. Many studies have shown that antigen-specific Ts require APC and antigen in order to be activated. The exact requirements for these APCs show considerable variation from study to study. One group used human monocyte hybridomas to investigate the triggering of Th vs. Ts cells. Both HLA-DR+ and HLA-DQ+DP+ hybridomas stimulated T-cell proliferation. HLA-DR+ cells stimulated primarily CD4+ proliferation while HLA-DR−DP+DQ+ cells stimulated CD8+ cells. Functional studies showed the CD8+ cells to be Ts [215]. A mouse study also demonstrated that different MHC restriction elements served to trigger different functional responses. An I-E restricted cell line was shown to trigger both Th and Ts responses while a variant of this line triggered H-2D-restricted Th and H2K-restricted Ts activity [216]. Both B and adherent cells stimulated Ts activity, and in some cases these cells were functional even when they were glutaraldehyde-fixed or irradiated to block metabolic activity [217, 218]. The physiological conditions of APC were important for selection of Ts vs. Th in a study demonstrating that UV-irradiated APC generated a TSF from a Th clone while nonirradiated APC led to helper function [218].

In a series of studies, Dorf and co-workers [219–222] investigated macrophage hybridomas coupled with antigen for the ability to provide help for a DTH response or to generate Ts activity. Most clones generated Th activity but one clone, 63, failed to induce DTH but instead induced Ts. Clone 63 could induce immunity when mice were first given anti-I-J or when the clone was given to I-J-incompatible recipients. A second macrophage hybridoma, clone 59, normally induced immunity. When this clone was incubated with Th1 supernatant the clone became capable of inducing Ts, while incubation with Th2 supernatant had no effect in this regard. IFN-γ was found to be sufficient to cause clone 59 to become competent to induce Ts activity. PMA could substitute for IFN-γ and inhibitors of PKC activation blocked the induction of Ts generation. IFN-γ activation of thymic stromal cells was also shown to lead to induction of Ts. Other cell types shown to stimulate Ts activity were human thyroid cells activated with IFN-γ which were poor inducers of cytotoxic and helper activity, and human decidual tissue APCs exposed to fetal tissue [223, 224].

Nonspecific Suppression

As mentioned earlier, during the time that antigen-specific suppression was being investigated, many other investigators were noting the existence of nonspecific suppression [225–228]. Reports extending back to the early 1970s had shown for example, that mitogen-activated spleen cells would suppress in vitro antibody-forming cell responses [225–228]. In some cases, antigen-stimulated cells were demonstrated to be nonspecifically suppressive. In addition, it was noted that other, nonlymphoid cells, particularly tumor cells, could produce substances that inhibited immune responses [229]. The initial identification of the nature of these nonspecific suppressor cells and factors was retarded by the lack of appropriate reagents, e.g., antibodies and lack of interest on the part of most biochemists and molecular biologists.

In the mid-1970s, prostaglandins were found to be produced during immune responses by macrophages and were determined to be responsible for some aspects of immune suppression [230, 231]. This was due in large part to the existence of specific cyclo-oxygenase inhibitors that blocked prostaglandin production specifically and thus made excellent biochemical probes. Later, other products of arachidonic acid metabolism, such as the leukotrienes, were also found to be capable of regulating immunocompetent cell function [231, 232]. Recently, it has been shown that PGE_2 acts on T cells to block the expression of the IL-2 receptor, and the production of IL-2 which helps to explain its effects on lymphocyte proliferation. The effects of PGE_2 on B-cell function are more complex in the sense that virgin B cells are also sensitive to inhibition of proliferation and antibody production [233–236]. However, memory B cells respond to PGE_2 by increasing the production of IgG [237]. It has also been demonstrated that PGE_2 will stimulate the activation of other nonspecific suppressor factors [238].

As T-cell hybridomas and T-cell clones became more universally available, the analysis of nonspecific regulatory factors improved both at the cellular and biochemical levels [239]. Early in the 1970s for example it had become clear that murine spleen cells stimulated with the lectin conA contained a population of highly active nonspecific suppressor cells. Extensive study using the reagents available at that time established that the suppressor cells were Ly-2+ T cells [10]. These cells released a factor termed soluble immune response suppressor (SIRS) that could inhibit in vitro antibody-forming cell responses and the generation of CTL [225, 226, 228, 240]. The early observations suggested that SIRS acted on a splenic adherent cell, probably a macrophage, which then released a second factor that suppressed directly the cells involved in generating an antibody response or those involved in generating CTL. By the late 1970s it became clear that the Ly-2+ T cells generated SIRS and the macrophage processed it via an oxidation-dependent step (SIRSox) to its bioactive form [241, 242]. The formation of SIRSox could be accomplished by treating SIRS with H_2O_2 directly; inhibition of SIRSox formation was observed in the presence of catalase or a variety of electron acceptors; its bioactivity was also inhibited by sulfhydryl reagents such as β-mercaptoethanol [239, 241–243]. In the early 1980s a hybridoma that produced SIRS constitutively was made and was then used for further biochemical studies of the nature of SIRS [239, 243–246]. SIRS, as produced by the hybridoma 393D2.6, was found to be a small (MW = 10 kD) polypeptide that required Fe for activity and that exhibited multiple isoelectric points, possibly related to its oxidation state [245]. A 20-amino acid N-terminal sequence of SIRS was obtained in the late 1980s and a polyclonal rabbit antipeptide antiserum was made that recognized authentic SIRS and could also neutralize SIRS activity. During this same period, human SIRS was tentatively identified based on production by CD8+ T cells in response to conA and its sensitivity to β-mercaptoethanol [247]. It was shown that human SIRs could be found in the serum and urine of patients with steroid-sensitive postnephrotic syndrome [248], and thus clearly established this factor as one of the cirulating cytokines. Further studies showed that the rabbit antipeptide antibody made against the murine SIRS sequence also reacted with human SIRS, suggesting a close structural relationship [51]. To date, attempts to obtain a cDNA for either human or mouse SIRS have not been successful, for reasons that are not entirely clear. Nevertheless, using the SIRS-specific antipeptide antisera it has been possible to show that the production of SIRS by conA-stimulated T cells is dependent on the production of IL-2; indeed, IL-2 alone can stimulate the production of SIRS by naive murine spleen cells [142, 143]. Further studies have established that SIRS is produced in splenic mixed lymphocyte cultures as well as in cultures stimulated by conA, in both cases anti-IL-2 receptor antibody blocks SIRS production [142, 143]. Thus, a new loop of regulatory cell interactions has been proposed to exist between CD4+, Th1 cells that produce IL-2 and CD8+ noncytolytic T cells that produce SIRS coupled with adherent cells that are required, presumably to modify the SIRS to SIRSox [143].

In addition to the nonspecific Ts network involving SIRS, a number of other nonspecific Ts have been described. One such suppressor called natural suppressor cells (NS) were characterized as a population of Ts in murine bone marrow that is wheat germ agglutinin-positive (WGA+) and Thy-1−, CD3−, CD4−/8−, B220−, sIg−, MHC class II−, and asialo-GM1−. A novel phenotypic marker, mAb 1E5.B5 for some NS cells, has recently been described [211]. These Ts nonspecifically inhibit mixed lymphocyte responses, plaque-forming cell responses to SRBC, and mitogen-induced proliferative responses, but are noncytolytic. The activity of NS cells was enhanced by IL-3 and GM-CSF but blocked by INF-γ [249]. NS were described in the spleens of mice following bone marrow transplant [250–252]. LPS injection to mice enhanced bone marrow NS activity [253, 254]. Splenic populations of mice have NS activity upon exposure to radiation, chronic graft-versus-host disease, or cancer. Culture of normal adult mouse spleen cells for 2–3 weeks resulted in NS cell induction [255]. NS cells produce a TSF that blocks murine Ab responses in vitro. This TSF leads to a decrease in IL-2-produced decreasing proliferative responses [249, 256]. Supernatants from cloned murine NS cells activated with PMA and calcium ionophore blocked proliferation and IL-2 production in mixed lymphocyte cultures. This activity has recently been purified, a partial sequence obtained, indicating it to be a unique cytokine [119, 257].

The regulatory loop identified with the partially characterized cytokine SIRS, must also be viewed in the context of other nonspecific inhibitory cytokines including TGF-β, IFN-γ, IL-4 and, most recently, IL-10 (see above). In the late 1980s, although the concept of a specific population of suppressor cells had become unpopular, increasingly, researchers were finding evidence, using T-cell clones, that these clones could act as suppressor cells under certain circumstances. This has led, ironically, to the reinvigoration of the concept of suppressor cell-based regulation with an interesting twist. As recently presented by Bloom et al. [59, 138], this hypothetical model does away with a unique, separate population of Ts and instead uses current concepts of Th1/Th2 as well as noncytolytic CD8+ T cells as the principal mediators of suppression (see above).

Several different mechanisms for limiting the reactivity of self-reactive B and T lymphocytes have been proposed. These may be categorized as models involving (a) deletion of self-reactive clones; (b) anergy or functional down-regulation, and (c) thymus during ontogeny. Other T cells may be anergized (rendered nonresponsive) by antigen presented in a certain context. Anergy appears important in tolerance to peripheral antigens not seen in the thymus; the mechanisms involved have not yet been elucidated. The third possibility, that of a set of cells whose function is to down-regulate immune responsiveness (suppressor T cells for example) has not been rigorously excluded. In fact, antigen-specific Ts cells have been implicated in peripheral tolerance, transplantation, and susceptibility to certain tumors. Thus far, there is no other explanation for the ability to transfer the tolerant state from one animal to another.

However, it is possible that these mechanisms overlap. For example, a Ts cell could represent a Th cell that received the antigenic signal under a particular set of circumstances leading to the 'suppressive' mode. If, like Th cells they use TCR, then they probably derive from a common lineage. On the other hand, if Ts cells use a unique polypeptide receptor, then they are independent in origin and may have a specific function in immune regulation.

Veto Cells?

In the midst of the studies outlined above on antigen-specific and nonspecific suppressor cells, it was proposed that both immune and some nonimmune cells possessed the capacity to block or 'veto' the response of autoreactive lymphocytes [259–260]. In this system, the antigen recognition capacity of the veto cell is not engaged, unlike conventional suppressor T cells that clearly use the TCR. Instead, the veto cell is recognized by the autoreactive lymphocyte; this recognition is unidirectional (unlike suppressor cells) [261]. The act of recognition by the autoreactive T cell leads the veto cell to engage a tolerogenic pathway that inactivates the recognizing cell. As described by Fink et al. [262], this mechanism is very attractive in that it does not involve the invocation of a specialized lineage of suppressor cells with yet another set of antigen receptors. Instead, tolerance can be maintained by mechanisms such as cytolysis. In the original experiments carried out by Miller and Derry [259] the veto cells were detected among the spleen cells of nude mice that were mixed with lymph node responder cells and irradiated allogeneic target cells. Thymocytes and bone marrow cells, but not spleen cells, from normal mice also possessed veto activity. In a later study, Muraoka et al. [263] in Miller's laboratory found that a substantial veto activity could be derived from bone marrow colonies of T cells. These cells were Thy-1— and derived from radiosensitive precursors. Thus, in these original publications, Miller and co-workers rather narrowly and specifically defined a population of suppressor or veto cells. Since then, however, others have sought to broaden this definiton so that the distinction between suppressor cells and veto cells has blurred. This may be due, in part, to experimental manipulations in which defining whether a given cell engages its antigen receptor is difficult or moot. In these experiments [reviewed in 262], the common feature is the activation of CTL clones to act as veto cells via klling upon the activation of responder cells (that are destined to become CTL) via irradiated stimulator cells. The suppression is highly specific and not easily reversible. It is not clear however, exactly how veto cells suppress responses and do so in an antigen-specific fashion.

A possible mechanism may be related to the studies of Fast [141] and is supported by the studies of Nakamura dn Gress [264] as well as work carried out by two of us (B.H.D. and D.R.W.). These data show that IL-2 is capable of directly stimulating CD8+ T cells to release a suppressor factor. In our laboratory, we have shown that this factor is a small molecular weight cytokine, SIRS (see above). In in vitro models we have described above, such a cell was shown to be stimulated in cultures containing mitogens, antigens or anti-CD3 but was still activated via an IL-2-dependent process [142, 143]. This cell was termed a CD8+ nonspecific suppressor cell. Hence, we have placed a question mark around the heading for this section. The need for additional terminology to describe a phenomenon that is unquestionably suppressive seems unnecessarily confusiong. The distinction that these cells are activated to suppress via a mechanism that does not necessarily engage their antigen-receptor is hairsplitting. The fact remains that this is an interesting population of cells that can suppress responses in a potent and long-lasting fashion and undoubtedly they play a role, as do other types of suppressor cells, in maintaining control over potentially anti-self reactive T cells and, possibly, B cells.

Conclusions

It should be clear from the above discussion that suppression represents a normal physiological response of immunocompetent cells to antigenic stimulation. It should be equally clear that suppression may take many forms and involve a variety of immunocompetent cell types and cell subsets. The clearest examples of cell-based suppression are to be found in the production of cytokines such as IL-4, IFN-γ and IL-10 by Th-1 and Th-2 lymphocytes. That having been said, we wish to emphazise that there remains considerable evidence that additional mechanisms of cell-based immune suppression exist and play a role in immune regulation. There remain many issues still to be addressed in this regard. First, are there phenotypic markers that can define more discreetly, T-suppressor cells? Do such cells act via an antigen-specific mechanism and not via the release of nonspecific cytokines? Can the soluble TSF that are detected in hybridoma culture fluids and in culture fluids from Ts clones, be shown to play a normal physiological role in the whole animal? There remains an abundant amount of work to determine how all these various regulatory cells interact to control immunity. Even more daunting is the task to determine what role suppression plays in the pathogenesis of disease.

References

1   Ehrlich, P.: On immunity with special reference to cell life. Proc. R. Soc. B *66:* 424 (1990).
2   Kappler, J.; Roehm, N.; Marrack, P.: T cell tolerance by clonal elimination in thymus. Cell *49:* 273–285 (1987).
3   Lord, E.M.; Dutton, R.W.: The properties of plaque-forming cells from autoimmune and normal strains of mice with specificity for autologous erythrocyte antigens. J. Immunol. *115:* 1199–1205 (1975).
4   Lord, E.M.; Dutton, R.W.: Antigen suppression of the in vitro development of plaque-forming cells to autologous erythrocyte antigens. J. Immunol. *115:* 1631–1635 (1975).
5   Zimecki, M.; Webb, D.R.: The role of prostaglandins in the control of the immune response to an autologous red blood cell antigen (Hb). Clin. Immunol. Immunopathol. *8:* 420–429 (1977).
6   Gershon, R.K.: T cell control of antibody production. Contemp. Top. Immunobiol. *3:* 1–40 (1974).
7   Benacerraf, B.; Dorf, M.E.: Genetic control of specific immune responses and immune suppression by I-region genes. Cold Spring Harbor Symp. Quant. Biol. *41:* 465–475 (1976).
8   Allison, A.C.: The roles of T and B lymphocytes in self-tolerance and autoimmunity. Contemp. Top. Immunobiol. *3:* 227–242 (1974).
9   Davies, A.J.S.: The thymus and the cellular basis of immunity. Transplant Rev. *1:* 43–91 (1969).
10 Tada, T.; Okumura, K.: The role of antigen-specific T cell factors in the immune response. Adv. Immunol. *28:* 1–87 (1979).
11 Germain, R.N.; Benacerraf, B.: A single major pathway of T-lymphocyte interactions in antigen-specific immune suppression. Scand. J. Immunol. *13:* 1–10 (1981).
12 Webb, D.R.; Kapp, J.A.; Pierce, C.W.: The biochemistry of T-cell factors. Annu. Rev. Immunol. *1:* 423–438 (1983).

13 Green, D.R.; Flood, P.M.; Gershon, R.K.: Immunoregulatory T-cell pathways. Annu. Rev. Immunol. *1:* 439–463 (1983).

14 Dorf, M.E.; Benacerraf, B.: Suppressor cells and immunoregulation. Annu. Rev. Immunol. *2:* 127–158 (1984).

15 Asherson, G.L.: Colizzi, V.; Zembala, M.: An overview of T-suppressor cell circuits. Annu. Rev. Immunol. *4:* 37–68 (1986).

16 Herzenberg, L.A.; Herzenberg, L.A.: Short-term and chronic allotype suppression in mice. Contemp. Top. Immunobiol. *3:* 41–76 (1974).

17 Sela, M.: Immunological studies with synthetic polypeptides. Adv. Immunol. *5:* 29–129 (1966).

18 McDevitt, H.O.; Chintz, A.: Genetic control of the antibody response: Relationship between immune response and histocompatibility (H-2) type. Science *163:* 1207–1217 (1969).

19 Benacerraf, B.; Waltenbaugh, C.; Theze, J.; Kapp, J.; Dorf, M.E.: The I region genes in genetic regulation; in Sercarz, E.; Herzenberg, L.A.; Fox, C.F. (eds): The Immune System. II. Regulatory Genetics. New York, Academic Press, 1977, pp. 1–363.

20 Steinmetz, M.; Minard, K.; Horvath, S.; McNicholas, J.; Frelinger, J.; Wake, C.; Long, E.; Mach, B.; Hood, L.: Molecular map of the immune response region from the major histocompatibility complex of the mouse. Nature *300:* 35–42 (1982).

21 Murphy, D.B.; Horowitz, M.C.; Homer, R.J.; Flood, P.M.: Genetic, serological, and functional analysis of I-J molecules. Immunol. Rev. *83:* 79–103 (1985).

22 Murphy, D.B.: The I-J puzzle. Annu. Rev. Immunol. *5:* 405–427 (1987).

23 Kapp, J.A.; Pierce, C.W.; Benacerraf, B.: Genetic control of immune responses in vitro. I. Development of primary and secondary plaque-forming cell responses to the random terpolymer *L*-glutamic acid[60]-*L*-alanine[30]-*L*-tyrosine[10] (GAT) by mouse spleen cells in vitro. J. Exp. Med. *138:* 1107–1120 (1973).

24 Kapp, J.A.; Pierce, C.W.; Benacerraf, B.: Genetic control of immune responses in vitro. II. Cellular requirements for the development of primary plaque-forming cell responses to the random terpolymer *L*-glutamic acid[60]-*L*-alanine[30]-*L*-tyrosine[10] (GAT) by mouse spleen cells in vitro. J. Exp. Med. *138:* 1121–1136 (1973).

25 Kapp, J.A.; Pierce, C.W.; Benacerraf, B.: Genetic control of immune responses in vitro. VI. Experimental conditions for the development of helper T cell activity specific for the terpolymer *L*-glutamic acid[60]-*L*-alanine[30]-*L*-tyrosine[10] (GAT) in nonresponder mice. J. Exp. Med. *142:* 1282–1296 (1978).

26 Kapp, J.A.; Pierce, C.W.; Schlossman, S.; Benacerraf, B.: Genetic control of immune responses in vitro. V. Stimulation of suppressor T cells in nonresponder mice by the terpolymer *L*-glutamic acid[60]-*L*-alanine[30]-*L*-tyrosine[10] (GAT). J. Exp. Med. *140:* 648–662 (1974).

27 Greene, M.I.; Nelles, M.J.; Sy, M.-S.; Nisonoff, A.: Regulation of immunity to the azobenzenarsonate hapten. Adv. Immunol. *32:* 253–300 (1982).

28 Pierres, M.; Germain, R.N.: Antigen-specific T cell-mediated suppression. IV. Role of macrophages in generation of *L*-glutamic acid[60]-*L*-alanine[30]-*L*-tyrosine[10] (GAT)-specific suppressor T cells in responder mouse strains. J. Immunol. *121:* 1306–1314 (1978).

29 Kapp, J.A.; Araneo, B.A.: Antigen-specific suppressor T cell interactions. I. Induction of an MHC-restricted suppressor factor specific for *L*-glutamic acid[50]-*L*-tyrosine[50]. J. Immunol. *128:* 2447–2452 (1982).

30 Sorensen, C.M.; Pierce, C.W.: Antigen-specific suppression in genetic responder mice to *L*-glutamic acid[60]-*L*-alanine[30]-*L*-tyrosine[10] (GAT). Characterization of conventional and hybridoma-derived factors produced by suppressor T cells from mice injected as neonates with syngeneic GAT-pulsed macrophages. J. Exp. Med. *156:* 1691–1710 (1982).

31 Pierce, C.W.: Sorensen, C.M.; Kapp, J.A.: Identification of suppressor T cells in virgin non-responder spleen cells responsible for primary unresponsiveness to *L*-glutamic acid^{60}-*L*-alanine^{30}-*L*-tyrosine^{10} (GAT). J. Immunol. *151:* 64–70 (1988).

32 Pierce, C.W.; Sorensen, C.M.; Kapp, J.A.: T cell subsets regulating antibody responses to *L*-glutamic acid^{60}-*L*-alanine^{30}-*L*-tyrosine^{10} (GAT) in virgin and immunized nonresponder mice. J. Immunol. *143:* 29–35 (1985).

33 Taniguchi, M.; Miller, J.F.A.P.: Specific suppression of the immune response by a factor obtained from spleen cells of mice tolerant to human gamma globulin. J. Immunol. *120:* 21–27 (1978).

34 Sy, M.S.; Dietz, M.H.; Germain, R.N.; Benacerraf, B.; Greene, M.I.: Antigen- and receptor-driven regulatory mechanisms. IV. Idiotype-bearing I-J+ suppressor T cell factors induce second-order suppressor T cells which express anti-idiotypic reeptors. J. Exp. Med. *151:* 1183–1195 (1980).

35 Taniguchi, M.; Hayakawa, K.; Tada, T.: Properties of antigen-specific suppressive T-cell factor in the regulation of antibody response of the mouse. II. In vitro activity and evidence for the I region gene product. J. Immunol. *116:* 542–548 (1976).

36 Taniguchi, M.; Saito, T.; Tada, T.: Antigen-specific suppressive factor produced by a transplantable I-J bearing T-cell hybridoma. Nature *278:* 555–556 (1979).

37 Okumura, K.; Takemori, T.; Tokuhisa, T.; Tada, T.: Specific enrichment of the suppressor T cell bearing I-J determinants: Parallel, functional and serological characterizations J. Exp. Med. *146:* 1234–1245 (1977).

38 Tada, T.; Hayakawa, K.; Okumura, K.; Taniguchi, M.: Coexistence of variable region of immunoglobulin heavy chain and I-region gene products on antigen-specific suppressor T cells and the suppressor T cell factor. A minimal model of functional antigen receptor of T cells. Mol. Immunol. *17:* 867–877 (1980).

39 Tokuhisa, T.; Taniguchi, M.: Constant region determinants on the antigen-binding chain of the suppressor T cell factor. Nature *298:* 174–175 (1982).

40 Taniguchi, M.; Takei, I.; Tada, T.: Functional and molecular organization of an antigen-specific suppressor factor derived from a T cell hybridoma. Nature *283:* 227–228 (1980).

41 Takemori, T.; Tada, T.: Properties of antigen-specific suppressor T cell factor in the regulation of antibody response in the mouse. I. In vivo activity and immunochemical characterizations. J. Exp. Med. *142:* 1241–1253 (1975).

42 Sercarz, E.; Kyzych, U.: The distinctive specificity of antigen-specific suppressor T cells. Immunol. Today *12:* 111–118 (1991).

43 Mossmann, T.R.; Coffman, R.L.: Th1 and Th2 cells: Different patterns of lymphokine secretion lead to different functional properties. Annu. Rev. Immunol. *7:* 145–173 (1989).

44 Kehrl, J.H.; Taylor, A.; Kim, S.J.; Fauci, A.S.: Transforming growth factor-β is a potent negative regulator of human lymphocytes. Ann. N.Y. Acad. Sci. *628:* 345–353 (1991).

45 Mossmann, T.R.; Moore, K.W.: The role of IL-10 in cross-regulation of Th1 and Th2 responses; in Ash, C.; Callagher, R.B. (eds): Immunoparasitology Today. Cambridge, Elsevier Trade Journals, 1991, pp. A49–53.

46 Paul, W.E.: Interleukin-4: A prototypic immunoregulatory lymphokine. Blood *77:* 1859–1870 (1991).

47 Swain, S.L.; Bradley, L.M.; Croft, M.; Tonkonogy, S.; Atkins, G.; Weinberg, A.; Duncan, D.D.; Hedrick, S.M.; Dutton, R.W.; Huston, G.: Helper T-cell subsets: Phenotype, function and the role of lymphokines in regulating their development. Immunol. Rev. *123:* 115–144 (1991).

48   Taussig, M.: Antigen-specific T-cell factors. Immunology *41:* 759–785 (1980).

49   Dorf, M.E.: Kuchroo, V.K.; Collins, M.: Suppressor T cells: Some answers but more questions. Immunol. Today *13:* 241–243 (1992).

50   Campetella, O.E.; Barrios, H.A.; Galassi, N.V.: Suppressor T-cell population induced by Junin virus in adult mice. Immunology *64:* 407–412 (1988).

51   Campetella, O.E.; Galassi, N.V.; Barrios, H.A.: Junin virus-induced non-specific suppressor cells interact with unrelated antigen-specific suppressor cells. Immunology *74:* 14–19 (1991).

52   Aurelian, L.; Yasumoto, S.; Smith, C.C.: Antigen-specific immune-suppressor factor in herpes simplex virus type 2 infections of UV-B-irradiated mice. J. Virol. *62:* 2520–2524 (1988).

53   Strayer, D.S.; Dombrowski, J.: Immunosuppression during viral oncogenesis. V. Resistance to virus-induced immunosuppressive factor. J. Immunol. *141:* 347–351 (1988).

54   Strayer, D.S.; Korber, K.; Dombrowski, J.: Immunosuppression during viral oncogenesis. IV. Generation of soluble virus-induced immunologic suppressor molecules. J. Immunol. *140:* 2051–2059 (1988).

55   Pomeroy, C.; Filice, G.A.; Hitt, J.A.; Jordan, M.C.: Cytomegalovirus-induced reactivation of *Toxoplasma gondii* pneumonia in mice: Lung lymphocyte phenotypes and suppressor function. J. Infect. Dis. *166:* 677–681 (1992).

56   Lemieux, S.; Gosselin, D.; Lusignan, Y.; Turcotte, R.: Early accumulation of suppressor cell precursors in the spleen of *Mycobacterium lepraemurium*-infected mice and analysis of their in vitro-induced maturation. Clin. Exp. Immunol. *81:* 116–122 (1990).

57   Chakraborty, N.G.; Twardzik, D.R.; Sivanandham, M.; Ergin, M.T.; Hellstrom, K.E.; Mukherji, B.: Autologous melanoma-induced activation of regulatory T cells that suppress cytotoxic response. J. Immunol. *145:* 2359–2364 (1990).

58   Sussman, G.; Wadee, A.A.: Production of a suppressor factor by CD8+ lymphocytes activated by mycobacterial components. Infect. Immun. *59:* 2828–2835 (1991).

59   Bloom, B.R.; Modlin, R.L.; Salgame, P.: Stigma variations: Observations on suppressor T cells and leprosy. Annu. Rev. Immunol. *10:* 453–488 (1992).

60   Gerner, R.E.; Childress, A.M.; Human, L.G.; Domer, J.E.: Characterization of *Candida albicans* mannan-induced, mannan-specific delayed hypersensitivity suppressor cells. Infect. Immun. *58:* 2613–2620 (1990).

61   Masih, D.T.; Sotomayor, C.E.; Cervi, L.A.; Riera, C.M.; Rubinstein, H.R.: Inhibition of I-A expression in rat peritoneal macrophages due to T-suppressor cells induced by *Cryptococcus neoformans*. J. Med. Vet. Mycol. *29:* 125–128 (1991).

62   Masih, D.T.; Sotomayor, C.E.; Rubinstein, H.R.; Riera, C.M.: Immunosuppression in experimental cryptococcosis in rats. Induction of efferent T suppressor cells to a non-related antigen. Mycopathologia *114:* 179–186 (1991).

63   Jimenez-Finkel, B.E.; Murphy, J.W.: Characterization of efferent T suppressor cells induced by *Paracoccidioides brasiliensis*-specific afferent T suppressor cells. Infect. Immun. *56:* 744–750 (1988).

64   Jimenez-Finkel, B.E.; Murphy, J.W.: Induction of antigen-specific T suppressor cells by soluble *Paracoccidioides brasiliensis* antigen. Infect. Immun. *56:* 734–743 (1988).

65   Iizaki, T.; Kobayashi, S.; Ogasawara, K.; Day, N.K.; Good, R.A.; Onoe, K.: Immune suppression induced by protoscoleces of *Echinococcus multilocularis* in mice. Evidence for the presence of CD8 dull suppressor cells in spleens of mice intraperitoneally infected with *E. multilocularis* J. Immunol. *147:* 1659–1666 (1991).

66   Pomeroy, C.; Miller, L.; McFarling, L.; Kennedy, C.; Filice, G.A.: Phenotypes, proliferative responses, and suppressor function of lung lymphocytes during *Toxoplasma gondii* pneumonia in mice. J. Infect. Dis. *164:* 1227–1232 (1991).

67 Perrin, P.J.; Phillips, S.M.: The molecular basis of granuloma formation in schistosomiasis. I. A T cell-derived suppressor effector factor. J. Immunol. *141:* 1714–1719 (1988).
68 Fidel, P.L., Jr.; Boros, D.L.: Regulation of granulomatous inflammation in murine
   schistosomiasis. IV. Antigen-induced suppressor T cells down-regulate proliferation and
   IL-2 production. J. Immunol. *145:* 1257–1264 (1990).
69 Ohta, N.; Itagaki, T.; Minai, M.; Hirayama, K.; Hosaka, Y.: *Schistosoma japonicum*
   egg-antigen-specific T cell lines in man. Induction of helper and suppressor T cell lines
   and clones in vitro in a patient with chronic *Schistosomiasis japonica*. J. Clin Invest. *81:*
   775–781 (1988).
70 Li, Z.J.; Luo, D.D.; Dai, J.Z.; Zheng, L.L.; Wang, X.H.; Yang, Y.X.: Changes in T cell
   subsets and T suppressor cell function and their relationship in human *Schistosomiasis
   japonica*. J. Tongi Med. Univ. *11:* 135–140 (1991).
71 Tarleton, R.L.: *Trypanosoma cruzi*-induced suppression of IL-2 production. I. Evidence
   for the presence of IL-2 producing cells. J. Immunol. *140:* 2763–2768 (1988).
72 Tarleton, R.L.: *Trypanosoma cruzi*-induced suppression of IL-2 production. II. Evidence
   for a role of suppressor cells. J. Immunol. *140:* 2769–2773 (1988).
73 Gao, X.M.; Schmidt, J.A.; Liew, F.Y.: Suppressive substance produced by T cells from
   mice chronically infected with *Trypanosoma cruzi*. III. Genetic restriction and further
   characterization. J. Immunol. *141:* 989–995 (1988).
74 Liew, F.Y.; Schmidt, J.A.; Liu, D.S.; Millott, S.M.; Scott, M.T.; Dhaliwal, J.S.; Croft,
   S.L.: Suppressive substance produced by T cells from mice chronically infected with
   Trypanosoma cruzi. II. Partial biochemical characterization. J. Immunol. *140:* 969–973
   (1988).
75 Kresina, T.F.: Antigen-specific down-regulation of murine collagen-induced arthritis:
   T-suppressor cell circuits in arthritis immunotherapy. Int. Rev. Immunol. *4:* 91–106
   (1988).
76 Takagishi, K.; Hotokebuchi, T.; Arai, K.; Arita, C.; Kaibara, N.: Collagen arthritis in
   rats: The importance of humoral immunity in the initiation of the disease and perpetuation of the disease by suppressor T cells. Int. Rev. Immunol. *4:* 35–48 (1988).
77 Nabozny, G.H.; Flynn, J.C.; Kong, Y.C.: Synergism between mouse thyroglobulin- and
   vaccination-induced suppressor mechanisms in murine experimental autoimmune thyroiditis. Cell. Immunol. *136:* 340–348 (1991).
78 Flynn, J.C.; Kong, Y.C.: In vivo evidence for CD4+ and CD8+ suppressor T cells in
   vaccination-induced suppression of murine experimental autoimmune thyroiditis. Clin.
   Immunol. Immunopathol. *60:* 484–494 (1991).
79 Hara, Y.; Caspi, R.R.; Wiggert, B.; Chan, C.C.; Wilbanks, G.A.; Streilein, J.W.:
   Suppression of experimental autoimmune uveitis in mice by induction of anterior
   chamber-associated immune deviation with interphotoreceptor retinoid-binding protein.
   J. Immunol. *148:* 1685–1692 (1992).
80 Rao, N.A.; Atalla, L.; Fong, S.L.; Chen, F.; Linker-Israeli, M.; Steinman, L.: Antigen-
   specific suppressor cells in experimental autoimmune uveitis. Ophthalmic Res. *24:* 92–98
   (1992).
81 Caspi, R.R.; Kuwabara, T.; Nussenblatt, R.B.: Characterization of a suppressor cell line
   which downgrades experimental autoimmune uveoretinitis in the rat. J. Immunol. *140:*
   2579–2584 (1988).
82 McDonald, A.H.; Swanborg, R.H.: Antigen-specific inhibition of immune interferon
   production by suppressor cells of autoimmune encephalomyelitis. J. Immunol. *140:*
   1132–1138 (1988).
83 Karpus, W.J.; Swanborg, R.H.: Protection against experimental autoimmune encephalomyelitis requires both CD4+ T cell suppressor cells and myelin basic protein-
   primed B cells. J. Neuroimmunol. *33:* 173–177 (1991).

84   Ofasu-Appiah, W.; Mokhtarian, F.: Characterization of a T-suppressor cell line that downgrades experimental allergic encephalomyelitis in mice. Cell. Immunol. *135:* 143–153 (1991).

85   Miller, A.; Lider, O.; Roberts, A.B.; Sporn, M.B.; Weiner, H.L.: Suppressor T cells generated by oral tolerization to myelin basic protein suppress both in vitro and in vivo immune responses by the release of transforming growth factor-beta after antigen-specific triggering. Proc. Natl Acad. Sci. USA *89:* 421–425 (1992).

86   Pankewycz, O.G.; Guan, J.X.; Benedict, J.F.: A protective NOD islet-infiltrating CD8 + T cell clone, IS 2.15, has in vitro immunosuppressive properties. Eur. J. Immunol. *22:* 2017–2023 (1992).

87   Talor, E.; Jodouin, C.A.; Richter, M.: Cells involved in the immune response. XXXVI. The thymic antigen-specific suppressor cell in the immunized rabbit is a T cell with receptors for FcG and the antigen and its acts, via a secreted suppressor factor, directly on the immune splenic AFC B cell to inhibit antibody secretion. Immunology *64:* 253–259 (1988).

88   Talor, E.; Richter, M.: Cells involved in the immune response. XXXV. The antigen-specific antibody response in the rabbit is suppressed by thymocytes of allogeneic immunized rabbits (ITSC) and by the non-toxic suppressor factor (ITSF) secreted by these thymocytes. Clin. Immunol. Immunopathol. *48:* 150–160 (1988).

89   Haslov, K.R.; Fauntleroy, M.B.; Stashak, P.W.; Taylor, C.E.; Baker, P.J.: T cells regulate the IgM antibody response to BALB/c mice to dextran B1355. Immunobiology *182:* 100–115 (1990).

90   Taylor, C.E.; Bright, R.: Production of suppressor factor by T-cells from mice immunized with pneumococcal polysaccharide. Adv. Exp. Med. Biol. *225:* 247–252 (1987).

91   Coutelier, J.P.: Enhancement of IgG production elicited in mice by treatment with anti-CD8 antibody. Eur. J. Immunol. *21:* 2617–2620 (1991).

92   Melchers, I.: Limiting dilution analysis of T cells suppressing the primary antibody response to sheep erythrocytes. J. Mol. Cell. Immunol. *3:* 1–12 (1987).

93   Miller, R.D.; Calkins, C.E.: Suppressor T cells and self-tolerance. Active suppression required for normal regulation of anti-erythrocyte autoantibody responses in spleen cells for nonautoimmune mice. J. Immunol. *140:* 3779–3785 (1988).

94   Nagarkatti, P.S.; Nagarkatti, M.; Mann, L.W.; Jones, L.A.; Kaplan, A.M.: Characterization of an endogeneous Lyt-2 + T-suppressor-cell population regulating autoreactive T cells in vitro and in vivo. Cell. Immunol. *112:* 64–77 (1988).

95   Yin, L.; Chain, B.M.: Suppression of lymphokine production in anti-minor histocompatibility antigen responses. Cytokine *3:* 5–11 (1991).

96   Yoshimura, S.; Gotoh, S.; Kamada, N.: Immunological tolerance induced by liver grafting in the rat: Splenic macrophages and T cells mediate distinct phases of immunosuppressive activity. Clin. Exp. Immunol. *85:* 121–127 (1991).

97   Pearce, N.W.; Spinelli, A.; Gurley, K.E.; Dorsch, S.E.; Hall, B.M.: Mechanisms maintaining antibody-induced enhancement of allografts. II. Mediation of specific suppression by short-lived CD4 + T cells. J. Immunol. *143:* 499–506 (1989).

98   Sasazuki, T.: HLA-linked immune suppression genes. Jinrui Idengaku Zasshi *35:* 1–13 (1990).

99   Miura, T.; Ghanta, V.K.; Hiramoto, R.N.: Host response to myeloma. I. Induction of cytotoxic and suppressor T cells by in vivo immunization with MOPC 104E plasmacytoma. Cancer Invest. *6:* 29–37 (1988).

100   Mukherji, B.; Guha, A.; Chakraborty, N.G.; Sivanandham, M.; Nashed, A.L.; Sporn, J.R.; Ergin, M.T.: Clonal analysis of cytotoxic and regulatory T cell responses against human melanoma. J. Exp. Med. *169:* 1961–1976 (1989).

101 Kraig, E.; Kannapell, C.C.; Fischbach, K.; Zellmer, V.; Trial, J.: Two ultraviolet tumor-specific suppressor cell clones. One expresses Ig RNA and the other expressed T cell receptor-alpha and -beta RNA. J. Immunol. *145:* 2050–2056 (1990).

102 Trial, J.; McIntyre, B.W.: Suppressor cell clones specific for ultraviolet radiation-induced tumors. Function and surface proteins. J. Immunol. *145:* 2044–2049 (1990).

103 Ofir, R.; Weinstein, Y.; Rager-Zisman, B.: Down-regulation of natural killer cell activity in MoLV leukemogenesis: Evidence for tumor-cell-mediated suppression. Nat. Immun. Cell Growth Regul. *7:* 77–86 (1988).

104 Thrush, G.R.; Butch, A.W.; Lerman, S.P.: CD8 suppressor cell activity and its effect on CD4 helper cell-dependent growth of SJL/J B-cell lymphomas. Cell. Immunol. *122:* 555–562 (1989).

105 Greer, J.M.: Halliday, W.J.: Distinctive molecular markers and biological activities in two tumour-specific murine suppressor factors. Immunol. Cell. Biol. *69:* 135–143 (1991).

106 Ponzoni, M.; Montaldo, P.G.; Lanciotti, M.; Castagnola, E.; Cirillo, C.; Cornaglia-Ferraris, P.: Purification to homogeneity and biochemical characterization of two suppressor factors from human malignant T-cells. Biochem. Biophys. Res. Commun. *150:* 702–710 (1988).

107 Lugering, N.; Kolsch, E.: Differences in the induction of tumor-specific T suppressor lymphocytes among the peritoneal exudate cells in histocompatible BALB/cAnPt and BALB/cJ mouse sublines. Immunobiology *184:* 106–110 (1991).

108 Degwert, J.; Heuer, J.; Kolsch, E.: Specific in vivo suppression of lymph node cell proliferation and humoral immune responses by cloned antigen-specific T suppressor cells. J. Immunol. *140:* 1448–1453 (1988).

109 Pauels, H.G.; Austrup, F.; Becker, C.; Schmitt, E.; Rude, E.; Kolsch, E.: Lymphokine profile and activation pattern of two unrelated antigen- or idiotype-specific T suppressor cell clones. Eur. J. Immunol. *22:* 1961–1966 (1992).

110 Utsunomiya, N.; Nakanishi, M.; Arata, Y.; Kubo, M.; Asano, Y.; Tada, T.: Unidirectional inhibition of early signal transduction of helper T cells by cloned suppressor T cells. Int. Immunol. *1:* 460–463 (1989).

111 Tada, T.; Hu, F.Y.; Kishimoto, H.; Furutani-Seiki, M.; Asano, Y.: Molecular events in the T cell-mediated suppression of the immune response. Ann. N.Y. Acad. Sci. *630:* 20–27 (1991).

112 Hu, F.Y.; Asano, Y.; Sano, K.; Inoue, T.; Furutani-Seiki, M.; Tada, T.: Establishment of stable CD8+ suppressor T cell clones and the analysis of their suppressive function. J. Immunol. Methods *152:* 123–134 (1992).

113 Mokashi, S.; Holford-Strevens, V.; Sterrantino, G.; Jackson, C.J.; Sehon, A.H.: Down-regulation of secondary in vitro antibody responses by suppressor T cells of mice treated with a tolerogenic conjugate of ovalbumin and monomethoxypolyethylene glycol, OVA(mPEG)13. Immunol. Lett. *23:* 95–102 (1989).

114 Takata, M.; Maiti, P.K.; Bitoh, S.; Holford-Strevens, V.; Kierek-Jaszczuk, D.; Chen, Y.; Lang, G.M.; Sehon, A.H.: Down-regulation of helper T cells by an antigen-specific monoclonal Ts factor. Cell. Immunol. *137:* 139–149 (1991).

115 Chen, Y.; Takata, M.; Maiti, P.K.; Rector, E.S.; Sehon, A.H.: Characterization of suppressor T cell clones derived from a mouse tolerized with conjugates of ovalbumin and monomethoxypolyethylene glycol. Cell. Immunol. *142:* 16–27 (1992).

116 Nanda, N.K.; Thomson, E., Mason, I.: Murine suppressor T cell clones specific for minor histocompatibility antigens express CD4, CD8, and alpha beta T cell receptor molecules. Int. Immunol. *2:* 1063–1071 (1990).

117 Takahashi, T.; Mafune, K.; Maki, T.: Cloning of self-major histocompatibility complex antigen-specific suppresssor cells from adult bone marrow. J. Exp. Med. *172:* 901–909 (1990).

118 Van Vlasselaer, P.; Fischer, M.; Strober, S.; Zlotnik, A.: Regulation of thymocyte proliferation by alpha beta TcR+ CD3+ CD4– CD8– cloned natural suppressor cells. Cell. Immunol. *136:* 1–15 (1991).

119 Van Vlasselaer, P.; Niki, T.; Strober, S.: Identification of a factor(s) from cloned murine natural suppressor cells that inhibit IL-2 secretion during antigen-driven T cell activation. Cell. Immunol. *138:* 326–340 (1991).

120 Kalish, R.S.; Morimoto, C.: Urushiol (poison ivy)-triggered suppressor T cell clone generated from peripheral blood. J. Clin. Invest. *82:* 825–832 (1988).

121 Li, S.G.; Ottenhoff, T.H.; Van den Elsen, P.; Koning, F.; Zhang, L.; Mak, T.; De Vries, R.R.: Human suppressor T cell clones lack CD28. Eur. J. Immunol. *20:* 1281–1288 (1990).

122 Li, S.G.; Elferink, D.G.; de Vries, R.R.: Phenotypic and functional characterization of human suppressor T-cell clones. II. Activation by *Mycobacterium leprae* presented by HLA-DR molecules to alpha beta T-cell receptors. Hum. Immunol. *28:* 11–26 (1990).

123 Salgame, P.; Modlin, R.; Bloom, B.R.: On the mechanism of human T cell suppression. Int. Immunol. *1:* 121–129 (1989).

124 Salgame, P.; Convit, J.; Bloom, B.R.: Immunological suppression by human CD8+ T cells is receptor dependent and HLA-DQ restricted. Proc. Natl Acad. Sci. USA *88:* 2598–2608 (1991).

125 Salgame, P.R.; Abrams, J.S.; Clayberger, C.; Goldstein, H.; Modlin, R.L.; Convit, J.; Bloom, B.R.: Differing lymphokine profiles of functional subsets of human CD4 and CD8 T-cell clones. Science *254:* 279–282 (1991).

126 Aune, T.M.; Pogue, S.L.: Generation and characterization of continuous lines of CD8+ suppressor T lymphocytes. J. Immunol. *142:* 3731–3739 (1989).

127 Patel, S.S.; Wacholtz, M.C.; Duby, A.D.; Thiele, D.L.; Lipsky, P.E.: Analysis of the functional capabilities of CD3+CD4–CD8– and CD3+CD4+CD8+ human T cell clones. J. Immunol. *143:* 1108–1117 (1989).

128 Zhang, J.W.; Schreurs, M.; Medaer, R.; Raus, J.C.: Regulation of myelin basic protein-specific helper T cells in multiple sclerosis: Generation of suppressor T cell lines. Cell. Immunol. *139:* 118–130 (1992).

129 Fukunaga, M.; Kirayama, K.; Sasazuki, T.: Activation of human CD8+ suppressor T cells by an antigen-specific CD4+ T-cell line in vitro. Hum. Immunol. *25:* 157–168 (1989).

130 Pawelec, G.; Brocker, T.; Busch, F.W.; Buhring, H.J.; Fernandez, N.; Schneider, E.M.; Wernet, P.: 'Tolerization' of human T-helper cell clones by chronic exposure to alloantigen: Culture conditions dictate autocrine proliferative status but not acquisition of cytotoxic potential and suppressor-induction capacity. J. Mol. Cell. Immunol. *4:* 21–34 (1988).

131 Pawelec, G.: CD4+ CD45R– suppressor-inducer T-cell clones: Requirements for cellular interaction, proliferation and lymphokines for the induction of suppression in peripheral blood mononuclear cells. Immunology *69:* 536–541 (1990).

132 Kotani, H.; Mitsuya, H.; Benson, E.; James, S.P.; Strober, W.: Activation and function of an autoreactive T cell clone with dual immunoregulatory activity. J. Immunol. *140:* 4167–4172 (1988).

133 Naor, D.; Essery, G.; Tarcic, N.; Kahan, M.; Feldmann, M.: Interactions between autologous T cell clones. Cell. Immunol. *128:* 490–502 (1990).

134 Naor, D.; Essery, G.; Kahan, M.; Feldmann, M.: T-cell clone anti-clone interactions. Effects on suppressor and helper activities. J. Autoimmun. 2(suppl): 3–14 (1989).

135 Naor, D.; Essery, G.; Tarcic, N.; Kahan, M.; Feldmann, M.; Regulatory interactions among autologous T cell clones. Human bifunctional T cell clones regulate the activity of an autologous T cell clone. Ann. N.Y. Acad. Sci. 636: 135–146 (1991).

136 Mossmann, T.R.; Schumacher, J.H.; Street, N.F.; Budd, R.; O'Garra, A.; Fong, T.A.T.; Bond, M.W.; Moore, K.W.M.; Sher, A.; Fiorentino, D.F.: Diversity of cytokine synthesis and function of mouse CD4+ T cells. Immunol. Rev. 123: 209–229 (1991).

137 Peltz, G.: A role for CD4+ T-cell subsets producing a selective pattern of lymphokines in the pathogenesis of human chronic inflammatory and allergic diseases. Immunol. Rev. 123: 23–35 (1991).

138 Bloom, B.R.; Salgame, P.; Diamond, B.: Revisiting and revising suppressor T cells. Immunol. Today 13: 131–136 (1992).

139 Hirohata, S., Davis, L.S.; Lipsky, P.E.: Role of IL-2 in the generation of CD4+ suppressors of human B cell responsiveness. J. Immunol. 142: 3104–3112 (1989).

140 Freire-Moar, J.; Kapp, J.A.; Webb, D.R.: Effect of IL-2 on suppressor factor production. Lymphokine Res. 8: 9–18 (1989).

141 Fast, L.D.: Generation and characterization of IL-2-activated veto cells. J. Immunol. 149: 1510–1515 (1992).

142 Terajima, C.; Koontz, A.; Kelley, M.; Webb, D.R.; Devens, B.H.: Cytokine effects on the down-regulation of cytolytic T cell responses in vitro. Lymphokine Res. 9: 499–506 (1990).

143 Fukuse, S.; Kelley, M.; Terajima, C.; Webb, D.R.; Devens, B.H.: Interleukin-2 stimulates the development of anergy via the activation of non-specific suppressor T-cells. Int. Arch. Allergy Appl. Immunol. 99: 411–415 (1992).

144 Oh-Ishi, T.; Goldman, C.K.; Misiti, J.; Waldmann, T.A.: Blockade of the interleukin-2 receptor by anti-Tac antibody inhibits the generation of antigen-nonspecific suppressor T cells in vitro. Proc. Natl Acad. Sci. USA 85: 6478–6482 (1988).

145 Oh-Ishi, T.; Goldman, C.K.; Misiti, J.; Waldmann, T.A.: The interaction of interleukin-2 with its receptor in the generation of suppressor T cells in antigen-specific and antigen-nonspecific systems in vitro. Clin. Immunol. Immunopathol. 52: 447–459 (1989).

146 Pawelec, G.; Schwulera, U.; Lenz, H.; Owsianowski, M., Buhring, H.J.; Schlag, H.; Schneider, E.; Schaudt, K.; Ehninger, G.: Lymphokine release, suppressor cell generation, cell surface markers, and cytotoxic activity in cancer patients receiving natural interleukin-2. Mol. Biother. 2: 44–49 (1990).

147 Pawelec, G., Lenz, H.J.; Schneider, E.; Buhring, H.J.; Rehbein, A.; Baumgartner, P.; Ehninger, G.: Clinical trial of natural human lymphocyte-derived interleukin-2 in cancer patients: Effects on cytokine production and suppressor cell status. Biotherapy 3: 309–318 (1991).

148 Di Stefano, R.; Mouzaki, A.; Araneda, D.; Diamantstein, T.; Tilney, N.L.; Kupiec-Weglinski, J.W.: Anti-interleukin-2 receptor monoclonal antibodies spare phenotypically distinct T suppressor cells in vivo and exert synergistic biological effects. J. Exp. Med. 167: 1981–1986 (1988).

149 Tan, P.; Anasetti, C.; Martin, P.J.; Hansen, J.A.: Alloantigen-specific T suppressor-inducer and T suppressor-effector cells can be activated despite blocking the IL-2 receptor. J. Immunol. 145: 485–488 (1990).

150 Noronha, A.; Toscas, A.; Jensen, M.A.: Contrasting effects of alpha, beta, and gamma interferons on nonspecific suppressor function in multiple sclerosis. Ann. Neurol. 31: 103–106 (1992).

151 Devens, B.H.; Semenuk, G.; Webb, D.R.: Antipeptide antibody specific for the N-terminal of soluble immune response suppressor neutralizes concanavalin A and IFN-induced suppressor activity in an in vitro cytolytic T lymphocyte response. J. Immunol. *141:* 3148–3155 (1988).

152 Holda, J.H.; Maier, T.; Claman, H.N.: Evidence that IFN-gamma is responsible for natural suppressor activity in GVHD spleen and normal bone marrow. Transplantation *45:* 772–777 (1988).

153 Frasca, D.; Adorini, L.; Landolfo, S.; Doria, G.: Enhancement of suppressor T cell activity by injection of anti-IFN-gamma monoclonal antibody. J. Immunol. *140:* 4103–4107 (1988).

154 Young, M.R.; Wright, M.A.; Coogan, M.; Young, M.E.; Bagash, J.: Tumor-derived cytokines induce bone marrow suppressor cells that mediate immunosuppression through transforming growth factor beta. Cancer Immunol. Immunother. *35:* 14–18 (1992).

155 Hisatsune, T.; Minai, Y.; Nishisima, K.; Enomoto, A.; Moore, K.W.; Yokota, T.; Arai, K.; Kaminogawa, S.: A suppressive lymphokine derived from Ts clone 13G2 is IL-10-Lymphokine Cytokine Res. *11:* 87–93 (1992).

156 Sussman, G.; Wadee, A.A.: Supernatants derived from CD8+ lymphocytes activated by mycobacterial functions inhibit cytokine production. The role of interleukin-6. Biotherapy *4:* 87–95 (1992).

157 Krupen, K.; Araneo, B.; Kapp, J.; Stein, S.; Wieder, K.J.; Webb, D.R.: Purification and characterization of a monoclonal T cell suppressor factor specific for *L*-glutamic acid^{60}-*L*-alanine^{30}-*L*-tyrosine^{10} (GAT). Proc. Natl Acad. Sci. USA *79:* 1254–1258 (1982).

158 Webb, D.R.; Araneo, B.A.; Healy, C.; Kapp, J.A.; Krupen, K.; Nowowiejski, I.; Pierce, C.; Sorensen, C.; Stein, S.; Wieder, K.J.: Purification and biochemical analysis of antigen-specific suppressor factors isolated from T-cell hybridomas; in Current Topics in Microbiology and Immunology, vol. 100. New York, Springer, 1982, pp. 53–59.

159 Sorensen, C.M.; Pierce, C.W.; Webb, D.R.: Purification and characterization of an *L*-glutamic acid^{60}-*L*-alanine^{30}-*L*-tyrosine^{10} (GAT)-specific suppressor factor from genetic responder mice. J. Exp. Med. *158:* 1034–1047 (1983).

160 Healy, C.T.; Kapp, J.A.; Webb, D.R.: Purification and biochemical analysis of antigen-specific suppressor factors obtained from the supernatant fluid, membrane or cytosol of a T-cell hybridoma. J. Immunol. *131:* 2843 (1983).

161 Webb, D.R.; Kapp, J.A.; Pierce, C.W.: Antigen-specific factors: An overview. Methods Enzymol. *116:* 295–303 (1985).

162 Krupen, K., Turck, C.W.; Stein, S., Kapp, J.A.; Webb, D.R.: GAT antigen-specific suppressor factors. Methods Enzymol. *116:* 325–340 (1985).

163 Kapp, J.A.; Sorensen, C.M.; Pierce, C.W.; Trial, J.; Schreffler, D.C.; Webb, D.R.: Characterization of hybridomas producing single and two chain antigen-specific suppressor T cell factors; in Taussig, M. (ed.): T-Cell Hybridomas. Boca Raton, CRC Press, 1985.

164 Webb, D.R.; Kapp, J.A.; Pierce, C.W.: Biochemical and genetic analysis of antigen-specific suppressor factors obtained from T-cell hybrids; in Taussig, M. (ed.): T-Cell Hybridomas. Boca Raton, CRC Press, 1985.

165 Turck, C.W.; Kapp, J.A.; Webb, D.R.: Purification and partial characterization of a monoclonal 'second order' suppressor factor specific for *L*-glutamic acid^{60}-*L*-alanine^{30}-*L*-tyrosine^{10}. J. Immunol. *135:* 3232–3237 (1985).

166 Krupen, K.I.; Kapp, J.A.: Bellone, C.J.; Jendrisak, G.; Webb, D.R.: Direct demonstration that an antigen-specific suppressor factor purified from extracts of T-cell hybridomas binds antigen and bears I-J determinants. Lymphokine Res. *7:* 429–444 (1988).

167 Kronenberg, M.; Steinmetz, M.; Kobori, J.; Kraig, E.; Kapp, J.A.; Pierce, C.W.; Suzuki, G.; Tada, T.; Hood, L.: RNA transcripts for I-J polypeptides are apparently not encoded between the I-A and I-E subregions of the murine major histcompatibility complex. Proc. Natl Acad. Sci. USA 80: 5704–5708 (1983).

168 Nakanishi, K.; Kazuhisa, S.; Yaoita, Y.; Maeda, K.; Kashiwamura, S.-I.; Honjo, T.; Kishimoto, T.: A T15-idiotype-positive T suppressor hybridoma does not use the T15 VH gene segemnt. Proc. Natl Acad. Sci. USA 79: 6984 (1982).

169 Kraig, E.; Kronenberg, M.; Kapp, J.A.; Pierce, C.W.; Abruzzini, A.F.; Sorensen, C.M.; Samelson, L.E.; Schwartz, R.H.; Hood, L.E.: T and B cells that recognize the same antigen do not transcribe similar heavy chain variable region gene segments. J. Exp. Med. 157: 192–209 (1983).

170 Kronenberg, M.; Goverman, J.; Haars, R.; Malissen, M.; Kraig, E.; Phillips, L.; Delovitch, T.; Suciu-Foca N.; Hood, L.: Rearrangement and transcription of the β-chain genes of the T-cell antigen receptor in different types of murine lymphocytes. Nature 313: 647–653 (1985).

171 Hedrick, S.M.; Germain, R.N.; Bevan, M.J.; Dorf, M.; Engel, I.; Fink, P.; Gascoigne, N.; Heber-Katz, E.; Kapp, J.; Kaufmann, Y.; Kaye, J.; Melchers, F.; Pierce, C.; Schwartz, R.H.; Sorensen, C.; Taniguchi, M.; David, M.M.: Rearrangement and transcription of a T-cell receptor beta chain gene in different T-cell subsets. Proc. Natl Acad. Sci. USA 82: 531–535 (1985).

172 O'Hara, R.J., Jr., Sherr, D.H.; Dorf, M.E.: In vitro generation of suppressor T cells. J. Immunol. 141: 2935–2942 (1988).

173 Toyonaga, B.; Yanagi, Y.; Suciu-Foca, N.; Minden, M.; Mak, T.: Rearrangements of T-cell receptor gene YT35 in human DNA from thymic leukaemia T-cell lines and functional T-cell clones. Nature 311: 385–387 (1984).

174 Fairchild, R.L.; Kubo, R.T.; Moorhead, J.W.: Soluble factors in tolerance and contact sensitivity to 2,4-dinitrofluorobenzene in mice. IX. A monoclonal T cell suppressor molecule is structurally and serologically related to the α/β cell receptor. J. Immunol. 141: 3342–3348 (1988).

175 Fairchild, R.L.; Kubo, R.T.; Moorhead, J.W.: DNP-specific/class I MHC-restricted suppressor molecules bear determinants of the T cell receptor α and β chains. J. Immunol. 145: 2001–2009 (1990).

176 Imai, K.; Kanno, M.; Kimoto, H.; Shigemoto, K.; Yamamoto, S.; Taniguchi, M.: Sequence and expression of trancripts of the T-cell antigen receptor α-chain gene in a functional, antigen-specific suppressor-T-cell hybridoma. Proc. Natl Acad. Sci. USA 83: 8708–8712 (1986).

177 Weiner, D.B.; Liu, J.; Hanna, N.; Bluestone, J.A.; Coligan, J.E.; Williams, W.V.; Greene, M.I.: CD3-associated heterodimeric polypeptides on suppressor hybridomas define biologically active inhibitory cells. Proc. Natl Acad. Sci. USA 85: 6077–6081 (1988).

178 Modlin, R.L.; Brenner, M.B.; Krangel, M.S.; Duby, A.D.; Bloom, B.D.: T cell receptors of human suppressor cells. Nature 329: 541–545 (1987).

179 De Santis, R.; Givol, D.; Hsu, P.-L.; Adorini, L.; Doria, G.; Appella, E.: Rearrangement and expression of the α- and β-chain genes of the T-cell antigen receptor in functional murine suppressor T-cell clones. Proc. Natl Acad. Sci. USA 82: 8638 (1985).

180 Lonai, P.; Rechavi, G.; Arman, E.; Givol, D.: Retention and loss of immunoglobulin heavy chain alleles in helper T cell hybridoma clones. EMBO J. 2: 781–786 (1983).

181 Kuchroo, V.K.; Steele, J.K.; Billings, P.R.; Selvaraj, P.; Dorf, M.E.: Expression of CD3-associated antigen-binding receptors on suppressor T cells. Proc. Natl Acad. Sci. USA 85: 9209 (1988).

182   Kuchroo, V.K.; Byrne, M.C.; Atsumi, Y.; Greenfield, E., Connolly, J.B.; Whitters, M.J.;
      O'Hara, R.M., Jr.; Collina, M.; Dorf, M.E.: T-cell receptor α chain plays a critical role
      in antigen-specific suppressor cell function. Proc. Natl Acad. Sci. USA 88: 8700–8704
      (1991).
183   Collins, M.; Kuchroo, V.K.; Whitters, M.J.; O'Hara, R.M., Jr.; Kelleher, K., Kubo,
      R.T.; Dorf, M.E.: Expression of functional α/β T cell receptor gene rearrangements in
      suppressor T cell hybridomas correlates with antigen binding but not with suppressor
      function. J. Immunol. 145: 2809–2819 (1990).
184   Zheng, H.; Sahai, B.M.; Kilgannon, P.; Fotedar, A.; Green, D.R.: Specific inhibition of
      cell surface T cell receptor expression by antisense oligodeoxynucleotides and its effect
      on the production of an antigen-specific regulatory T cell factor. Proc. Natl Acad. Sci.
      USA 86: 3785–3762 (1989).
185   Green, D.R.; Bissonnette, R.; Zheng, H.; Onda, T.; Echeverri, F.; Mogil, R.J.; Steele,
      J.K.; Voralia, M.; Fotedar, A.: Immunoregulatory activity of the T cell receptor α chain
      demonstrated by retroviral gene transfer. Proc. Natl Acad. Sci. USA 88: 8475–8479
      (1991).
186   Koseki, H.; Imai, K.; Ichikawa, T.; Hayata, I.; Taniguchi, M.: Predominant use of a
      particular alpha chain in suppressor T cell hybridomas specific for keyhole limpet
      hemocyanin. Int. Immunol. 1: 557–564 (1989).
187   Iwata, M.; Katamura, K.; Kubo, R.T.; Grey, H.M.; Ishizaka, K.: Relationship between
      T cell receptors and antigen binding factors. II. Common antigenic determinants and
      epitope recognition shared by T cell receptors and antigen binding factors. J. Immunol.
      143: 3917–3924 (1989).
188   Kronenberg, M.; Kraig, E.; Siu, G.; Kapp, J.A.; Kappler, J.A.; Marrack, P.; Pierce,
      C.W.; Hood, L.: Three T cell hybridomas do not contain detectable heavy chain variable
      gene transcripts. J. Exp. Med. 158: 210 (1983).
189   Kobori, J.A.; Winoto, A.; McNicholas, J., Hood, L.: Molecular characterization of the
      recombination region of six murine major histocompatibility complex I-region recombi-
      nants. J. Mol. Cell. Immunol. 661: 125–131 (1984).
190   Schrader, J.W.: Nature of the T cell receptor. Scand. J. Immunol. 10: 387–393 (1979).
191   Murphy, D.B.: Commentary on the genetic basis for control of I-J determinants. J.
      Immunol. 135: 1543–1547 (1985).
192   Uracz, W.; Asano, Y.; Abe, R.; Tada, T.: I-J epitopes are adaptively acquired by T cell
      differentiated in the chimaeric condition. Nature 316: 741–743 (1985).
193   Flood, P.M.; Benoist, C.; Mathis, D.; Murphy, D.B.: Altered I-J phenotype in Ea
      transgenic mice. Proc. Natl Acad. Sci. USA 83: 8308–8312 (1986).
194   Hayes, C.E.; Klyczek, K.K.; Krum, D.P.; Whitcomb, R.M.; Hullett, D.A.; Cantor, H.:
      Chromosome 4 Jt gene controls murine T cell surface I-J expression. Science 223:
      559–563 (1984).
195   Engler, P.; Haasch, D.; Pinkert, C.A.; Doglio, L.; Glymour, M.; Brinster, R.; Storb, U.:
      A strain-specific modifier on mouse chromosome 4 controls the methylation of indepen-
      dent transgenic loci. Cell 65: 939–947 (1991).
196   Jardieu, P.; Akasai, M.; Ishizaka, K.: Association of I-J determinants with lipomodulin/
      macrocortin. Proc. Natl Acad. Sci. USA 83: 160–164 (1986).
197   Steele, J.K.; Kuchroo, V.K.; Kawasaki, H.; Jayaraman, S.; Iwata, M.; Ishizaka, K.;
      Dorf, M.E.: A monoclonal antibody raised to lipomodulin recognizes T suppressor
      factors in two independent hapten-specific suppressor networks. J. Immunol. 142:
      2213–2220 (1989).
198   Nakayama, T.; Kubo, R.T.; Kishimoto, H.; Asano, Y.; Tada, T.: Biochemical identifica-
      tion of I-J as a novel dimeric surface molecule on mouse helper and suppressor T cell
      clones. Int. Immunol. 1: 50–58 (1989).

199 Heuer, J.; Degwert, J.; Pauels, H.-G., Kolsch, E.: T cell receptor α and β gene expression in a murine antigen-specific T suppressor lymphocyte clone with cytolytic potential. J. Immunol. *146:* 775–782 (1991).

200 Imai, K.; Kanno, M.; Kimoto, H.; Shigemoto, K.; Yamamoto, S.; Taniguchi, M.: Sequence and expression of transcripts of the T-cell antigen receptor α-chain gene in a functional, antigen-specific suppressor-T-cell hybridoma. Proc. Natl Acad. Sci. USA *83:* 8708 (1986).

201 Morimoto, C.; Matsuyama, T,; Rudd, C.E.; Forsgren, A.; Letvin, N.L.; Schlossman, S.F.: Role of the 2H4 molecule in the activation of suppressor inducer function. Eur. J. Immunol. *18:* 731–737 (1988).

202 Takeuchi, T., DiMaggio, M.; Levine, H.; Schlossman, S.F.; Morimoto, C.: CD11 molecule defines two types of suppressor cells within the T8+ population. Cell. Immunol. *111:* 398–409 (1988).

203 Takeuchi, T.; Rudd, C.E.; Tanaka, S.; Rothstein, D.M.; Schlossman, S.F.; Morimoto, C.: Functional characterization of the CD45R (2H4) molecule on CD8 (T8) cells in the autologous mixed lymphocyte reaction system. Eur. J. Immunol. *19:* 747–755 (1989).

204 Sohen, S.; Rothstein, D.M.; Tallman, T.; Gaudette, D.; Schlossman, S.F.; Morimoto, C.: The functional heterogeneity of CD8+ cells defined by anti-CD45RA (2H4) and anti-CD29 (4B4) antibodies. Cell. Immunol. *128:* 314–328 (1990).

205 Matsuyama, T.; Yamada, A.; Rothstein, D.M.; Anderson, K.C.; Schlossman, S.F.; Morimoto, C.: CD45 isoforms associated with distinct functions of CD4 cells derived from unusual healthy donors lacking CD45RA-T lymphocytes. Cell. Immunol. *137:* 406–419 (1991).

206 Nagoya, S.; Kikuchi, K., Uede, T.: Monoclonal antibody to a structure expressed on a subpopulation of rat CD8 T cell subsets. Microbiol. Immunol. *35:* 895–911 (1991).

207 Ferguson, T.A.; Iverson, G.M.; Flood, P.M.: Infectious and noninfectious tolerance are blocked by a monoclonal antibody to T-suppressor factor. Cell. Immunol. *115:* 403–412 (1988).

208 Ferguson, T.A.; Ptak, W.; Iverson, G.M.; Flood, P.: The role of suppression in immunoregulation: In vivo analysis using a monoclonal antibody to T suppressor factors. Eur. J. Immunol. *18:* 1179–1185 (1988).

209 Devens, B.H.: Sorensen, C.M.; Kapp, J.A.; Pierce, C.W.; Webb, D.R.: Indirect evidence of the nature of the antigen receptor on antigen-specific suppressor cells: Monoclonal antibodies against GAT-specific factors can alter the capacity of H-2-specific suppressor cells to regulate the response of CTL to allogeneic target cells; in Molecular Basis of the Immune Response. Ann. N.Y. Acad. Sci. *546:* 204–206 (1988).

210 Devens, B.H.; Koontz, A.W.; Kapp, J.A.; Pierce, C.W.; Webb, D.R.: Involvement of two distinct cell populations in the antigen-specific suppression of cytolytic T cell generation. J. Immunol. *146:* 1394–1401 (1991).

211 Hoskin, D.W.; Brooks-Kaiser, J.C.; Kaiser, M.; Murgita, R.A.: Reactivity of mono-clonal antibody 1E5.B5 with a novel phenotypic marker expressed on a murine natural suppressor cell subset. Hybridoma *11:* 203–215 (1992).

212 Freedman, M.S.; Ruijs, T.C.; Blain, M.; Antel, J.P.: Phenotypic and functional charac-teristics of activated CD8+ cells: A CD11b−CD28− subset mediates noncytolytic functional suppression. Clin. Immunol. Immunopathol. *60:* 254–267 (1991).

213 Yowell, T.L.; Araneo, B.A.; Sercarz, E.: Amputation of a suppressor determinant on lysozyme reveals underlying T-cell reactivity to other determinants. Nature *279:* 70–71 (1979).

214 Shivakumar, S.; Sercarz, E.E.; Krzych, U.: The molecular context of determinants within the priming antigen establishes a hierarchy of T cell induction: T cell specificities

induced by peptides of beta galactosidase vs. the whole antigen. Eur. J. Immunol. *19:* 681–687 (1989).

215 Shaked, A.; Sperber, K.; Mayer, L.: Stimulation of distinct T cell subsets in MLR using human macrophage hybridomas differentially expressing class II antigens. Transplantation *52:* 1068–1072 (1991).

216 Chang, M.D.; Jaureguiberry, B.; Garrido, E.; Diamond, B.: A murine macrophage line of the H-2d/f haplotype can activate H-2k suppressor T cells. Proc. Natl Acad. Sci. USA *87:* 2501–2505 (1990).

217 Newcomb, J.R.; Lin, Y.S.; Rogers, T.J.: Requirement for accessory cells in suppression of MOPC-315 IgA secretion by staphylococcal enterotoxin B-induced T-suppressor cells. Cell. Immunol. *129:* 528–537 (1990).

218 Green, D.R.; Chue, B.; Zheng, H.G.; Ferguson, T.A.; Beaman, K.D.; Flood, P.M.: A helper T cell clone produces an antigen-specific molecule (T-ABM) which functions in the induction of suppression. J. Mol. Cell. Immunol. *3:* 95–108 (1987).

219 Kuchroo, V.K.; Minami, M.; Diamond, B.; Dorf, M.E.: Functional analysis of cloned macrophage hybridomas. VI. Differential ability to induce immunity or suppression. J. Immunol. *141:* 10–16 (1988).

220 Ishikura, H.; Jayaraman, S.; Kuchroo, V.; Diamond, B.; Saito, S.; Dorf, M.E.: Functional analysis of cloned macrophage hybridomas. VII. Modulation of suppressor T cell-inducing activity. J. Immunol. *143:* 414–419 (1989).

221 Jayaraman, S.; Mensi, N.; Webb, D.R.; Dorf, M.E.: Involvement of protein kinase C in competence induction of macrophages to generate T suppressor cells. J. Immunol. *146:* 4085–4091 (1991).

222 Ishikura, H.; Dorf, M.E.: Thymic stromal cells induce hapten-specific, genetically restricted effector suppressor cells in vivo. Immunobiology *182:* 11–21 (1990).

223 Lahat, N.; Sheinfeld, M.; Sobel, E.; Baron, E.; Kraiem, Z.: Class II HLA-DR antigens on non-autoimmune human thyroid cells stimulate autologous T cells with high suppressor activity. Autoimmunity *8:* 125–133 (1990).

224 Oksenberg, J.R.; Mor-Yosef, S.; Ezra, Y.; Brautbar, C.; Antigen presenting cells in human decidual tissue. III. Role of accessory cells in the activation of suppressor cells. Am. J. Reprod. Immunol. Microbiol. *16:* 151–158 (1988).

225 Rich, R.R.; Pierce, C.W.: Biological expressions of lymphocyte activation. II. Generation of a population of thymus-derived suppressor lymphocytes. J. Exp. Med. *137:* 649 (1973).

226 Dutton, R.W.: Suppressor T cells. Transplant. Rev. *26:* 39 (1975).

227 Jegasothy, B.V.; Namba, Y.; Waksman, B.H.: Regulatory substances produced by lymphocytes. IV. IDS (inhibitor of DNA synthesis) inhibits stimulated lymphocyte proliferation by activation of membrane adenylate cyclase at a restriction point in late G1. Immunochemistry *15:* 551 (1978).

228 Green, W.C.; Fleisher, T.A.; Waldmann, T. A.: Soluble suppressor supernatants elaborated by concanavalin A-activated human mononuclear cells. I. Characterization of a soluble suppressor of T cell proliferation. J. Immunol. *126:* 1185 (1981).

229 Stutman, O.: Immunodepression and malignancy. Adv. Cancer Res. *22:* 261 (1975).

230 Goodwin, J.S.; Webb, D.R.: Regulation of the immune response by prostaglandins. Clin. Immunol. Immunopathol. *15:* 106 (1980).

231 Gualde, N.; Atluru, D.; Goodwin, J.S.: Effect of lipoxygenase metabolites of arachidonic acid on proliferation of human T cells and T cell subsets. J. Immunol. *134:* 1125 (1985).

232 Webb, D.R.; Nowowiejski, I.; Healy, C.; Rogers, T.J.: Immunosuppressive properties of leukotriene D4 and E4 in vitro. Biochem. Biophys. Res. Commun. *104:* 1617 (1982).

233 Zimecki, M.; Webb, D.R.: The regulation of the immune response to T-independent antigens by prostaglandins and B cells. J. Immunol. *117:* 2158–2164 (1976).

234 Zimecki, M.; Webb, D.R.: The role of prostaglandins in the control of the immune response to the autologous red blood cell antigen (Hb). Clin. Immunol. Immunopathol. *8:* 420–429 (1977).

235 Zimecki, M.; Webb, D.R.: The influence of molecular weight on immunogenicity and suppressor cells in the immune response to polyvinyl-pyrrolidone. Clin. Immunol. Immunopathol. *9:* 75–79 (1978).

236 Webb, D.R.; Nowowiejski, I.: The role of prostaglandins in the control of the primary 19S immune response to sRBC. Cell. Immunol. *33:* 1–10 (1977).

237 Webb, D.R.: Prostaglandins and the immune response. Prostaglandins Therapeutics *4:* 1 (1978).

238 Webb, D.R.; Rogers, T.J.; Nowowiejski, I.: Endogenous prostaglandin synthesis and the control of lymphocyte function; in Subcellular Factors in Immunity. Proc. N.Y. Acad. Sci. USA *332:* 262–270 (1979).

239 Aune, T.M.; Pierce, C.W.: Monoclonal soluble immune response suppressor (SIRS) derived from T-cell hybridomas; in Hammerling, G.J.; Hammerling, U.; Kearney, J.F. (eds): Monoclonal Antibodies and T-Cell Hybridomas. Amsterdam, Elsevier/North-Holland, 1981, p. 516.

240 Rich, R.; Pierce, C.W.: Biological expressions of lymphocyte activation. III. Suppression of plaque-forming cell responses in vitro by supernatant fluids from concanavalin A-activated spleen cell cultures. J. Immunol. *112:* 1360 (1974).

241 Aune, T.M.; Pierce, C.W.: Identification and initial characterization of a nonspecific suppressor factor (MO-SF) produced by soluble immune response suppressor (SIRS)-treated macrophages. J. Immunol. *127:* 1828 (1981).

242 Aune, T.M.; Pierce, C.W.: Conversion of soluble immune response suppressor to macrophage-derived suppressor factor by peroxide. Proc. Natl Acad. Sci. USA *75:* 5099–6033 (1981).

243 Aune, T.M.; Pierce, C.W.: Mechanism of SIRS action at the cellular and biochemical level; in Pick, E. (ed.): Lymphokines. New York, Academic Press, 1984, vol. 9, p. 257.

244 Aune, T. M.; Webb, D.R.; Pierce, C.W.: Purification and initial characterization of the lymphokine soluble immune response suppressor. J. Immunol. *131:* 2848 (1983).

245 Webb, D.R.; Mason, K.; Semenuk, G.; Aune, T.M.; Pierce, C.W.: Purification and analysis of isoforms of soluble immune response suppressor. J. Immunol. *135:* 3238 (1985).

246 Webb, D.R.; Mensi, N.; Freire-Moar, J.; Schnaper, H.W.; Lewis, R.V.; Semenuk, G.; Devens, B.H.; Koontz, A.; Danho, W.; Pan, Y.-C.; Wesselschmidt, R.; Aune, T.M.; Pierce, C.W.: Putative amino terminal sequence of murine soluble immune response suppressor: Significant homology with short neurotoxin 1. Int. Immunol. *2:* 765–774 (1990).

247 Schnaper, H.W.; Pierce, C.W.; Aune, T.M.: Identification and initial characterization of concanavalin A- and interferon-induced human suppressor factors: Evidence for a human equivalent of murine soluble immune response suppressor. J. Immunol. *132:* 2429 (1984).

248 Schnaper, H.W.; Aune, T.M.: Identification of the lymphokine-soluble immune response suppressor in urine of nephrotic children. J. Clin. Invest. *76:* 341 (1985).

249 Moore, S.C.; Theus, S.A.; Barnett, J.B.; Soderberg, L.S.: Cytokine regulation of bone marrow natural suppressor cell activity in the suppression of lymphocyte function. Cell. Immunol. *141:* 398–408 (1992).

250 Sykes, M.; Sachs, D.H.: Mechanisms of suppression in mixed allogeneic chimeras. Transplantation *46*(suppl): 135S–142S (1988).

251 Sykes, M.; Eisenthal, A.; Sachs, D.H.: Mechanism of protection from graft-vs.-host disease in murine mixed allogeneic chimeras. I. Development of a null cell population suppressive of cell-mediated lympholysis responses and derived from the syngeneic bone marrow component. J. Immunol. *140*: 2903–2911 (1988).

252 Sykes, M.; Sharabi, Y.; Sachs, D.H.: Natural suppressor cells in spleens of irradiated, bone marrow-reconstituted mice and normal bone marrow: Lack of Sca-1 expression and enrichment by depletion of Mac1-positive cells. Cell. Immunol. *127*: 260–274 (1990).

253 Holds, J.H.: LPS activation of bone marrow natural suppressor cells. Cell. Immunol. *141*: 518–527 (1992).

254 Sagiura, K.; Ikehara, S.; Inaba, M.; Haraguchi, S.; Ogata, H.; Sardina, E.E.; Sugawara, M.; Ohta, Y.; Good, R.A.: Enrichment of murine bone marrow natural suppressor activity in the fraction of hematopoietic progenitors with interleukin-3 receptor-associated antigen. Exp. Hematol. *20*: 256–263 (1992).

255 Yoshida, T.; Norihisa, Y.; Habu, S.; Kobayashi, N.; Takei, M.; Kanehira, N.; Shimamura, T.: Proliferation of natural suppressor cells in long-term cultures of spleen cells from normal adult mice. J. Immunol. *147*: 4136–4139 (1991).

256 Saffran, D.C.; Singhal, S.K.: Suppression of mixed lymphocyte reactivity by murine bone-marrow-derived suppressor factor – Inhibition of proliferation due to a deficit in IL-2 production. Transplantation *52*: 685–690 (1991).

257 Strober, S.; Niki, T.; Van Vlasselaer, P.: A novel immunosuppressive cytokine derived from CD4 – CD8 – $\alpha\beta$ T cells. J. Cell. Biochem. *17B*: 54 (1993).

258 Hertel-Wulff, B.; Strober, S.: Immunosuppressive lymphokine derived from natural suppressor cells. J. Immunol. *140*: 2633–2638 (1988).

259 Miller, R.G.; Derry, H.: A cell population in nu/nu spleen can prevent generation of cytotoxic lymphocytes by normal spleen cells against self antigens of the nu/nu spleen. J. Immunol. *122*: 1502–1509 (1979).

260 Miller, R.G.: The veto phenomenon and T cell regulation. Immunol. Today *7*: 112–114 (1986).

261 Rammensee, H.; Bevan, M.J.; Fink, P.J.: Antigen-specific suppression of T cell responses – The veto concept. Immunol. Today *6*: 41–44 (1985).

262 Fink, P.J.; Shimonkevitz, R.P.; Bevan, M.J.: Veto cells. Annu. Rev. Immunol. *6*: 115–137 (1988).

263 Muraoka, S.; Ehman, D.L.; Miller, R.G.: Irreversible inactivation of activated cytotoxic T lymphocyte precursoe cells by 'anti-self' suppressor cells present in murine bone marrow T cell colonies. Eur. J. Immunol. *14*: 1010–1016 (1984).

264 Nakamura, H.; Gress, R.E.: Interleukin-2 enhancement of veto suppressor cell function in T-cell-depleted bone marrow in vitro and in vivo. Transplantation *49*: 931–937 (1990).

David R. Webb, Institute of Immunology and Biological Science, Syntex Discovery Research, Hillview Ave., Palo Alto, CA 94303 (USA)

Topics of Clinical Interest

Granstein RD (ed): Mechanisms of Immune Regulation.
Chem Immunol. Basel, Karger, 1994, vol 58, pp 193–205

T Cell Tolerance: Models for Clinical Application to Allergy and Autoimmunity

Victoria C. Schad

ImmuLogic Pharmaceutical Corp., Waltham, Mass., USA

T cells are central players in the immune response. They recognize and respond to protein antigens in the form of peptides displayed by the histocompatibility leukocyte antigens (HLA) on the surface of antigen-presenting cells (APC). This review will focus on those peptides which are bound by the class II HLA on the surface of myelomonocytic cells and B cells and the consequences of T cell recognition and response. The way in which the recognition event occurs is critical to the outcome. Paradoxically, the T cell may be either activated or inactivated by this recognition event, and an activated T cell may also elicit profoundly different types of immunity. This review will try to provide a synthesis of what is known about these events and how immunologically active peptides may be used as therapies to regulate immunologically based disorders such as allergy and autoimmunity.

The Recognition Triad

The T cell antigen receptor (TCR) binds to a complex ligand, the antigenic peptide and the HLA molecule. The identification of a peptide-binding groove from the crystal structure of HLA-A2 (and subsequently two other class I HLA molecules) has provided a model for the identification of critical contact residues, both for the peptide and HLA and the peptide and TCR [1–3]. Although a crystal structure of a class II molecule has yet to be solved, the amino acid sequence similarities between class I and class II provide a basis for a model of this ternary complex as well. Given the structural constraints of the HLA-binding groove, certain peptides will have a greater affinity for a given HLA than other peptides. Empirical findings demonstrate that a given HLA may bind a number of different peptides, however certain peptides, some of which conform to the structural restriction models of the binding site, are stronger binders than others [4]. Antigens recognized by T cells in the context of class II molecules are, in general, antigens which have been taken up by APC, proteolytically cleaved or 'processed', and the peptides expressed on the cell surface bound to class II. The limitations on what peptides a TCR can 'see' are partially due to what peptides are bound to class II and partially due to the ligand-binding constraints of the TCR. Certain TCR critical residues can be identified in peptides and empirically it has been determined that protein antigens contain dominant peptides recognized by T cells [5]. Thus, the immunologically relevant peptides are relatively limited compared to the world of peptides which may constitute an antigen.

In addition to the recognition triad other receptor-ligand interactions between the T cell and APC are required for T cell activation to occur. Some of these interactions may serve to enhance the interaction between the T cell and the APC, and act as 'adhesion' molecules, such as the LFA-1/ICAM-1 interaction [6]. Other molecules, such as CD4, can function as a coreceptor with the TCR for class II-restricted T cells [7]. In addition, some 'adhesion' molecules expressed by APC which interact with receptors on the T cell can provide additional 'second signals' required for immunocompetent activation. One such interaction is between CD28 expressed on the surface of T cells and the B7 molecule on the surface of APC [8].

The Differential Outcome of Class II-Restricted T Cell Activation

After activation through the ternary complex of TCR/peptide/HLA, and additional second signal interactions, the T cell proliferates and secretes an array of cytokines which play a large part in class II-restricted T cell effector function. There are two poles in T cell reactivity as described using murine T cell clones, Th1 and Th2. Th1 cells are likened to the T cell responsible for classic delayed-type hypersensitivity (DTH) and they secrete IL-2, IL-3, IFN-γ and TNF. In contrast, Th2 cells are thought to be the classic T helper cell which provide 'help' for B cell antibody production, and they secrete, among other cytokines, IL-4, IL-5, IL-6 and IL-10 [9]. Although human T cell clones may have such extreme phenotypes, they generally are capable of producing a mixture of these cytokines [10, 11]. Interestingly, in the mouse, it has been shown that a single peptide can generate a Th1-like or Th2-like response depending on which class II allele is expressed, suggesting that the composition of the ternary complex may also effect the lymphokine profile a T cell produces [12].

The pleiotrophic cytokines produced by activated T cells have profound effects on many cell types. Some cytokines are directly implicated in the control of the T cell response. For example, some cytokines produced by Th1 cells down-regulate Th2 cell cytokine responsiveness. IFN-γ will decrease the ability of Th2 cells to proliferate to IL-4 [13]. The converse is also true, some Th2 cytokines down-regulate Th1 cytokines. IL-10 has negative effects on IL-3 and IFN-γ synthesis [14]. Thus, the activated T cell may provide a microenvironment conducive to one type of response by down-regulating other types.

Th1-like T cells and Th2-like T cells also elicit help for antibody production. Th1 cells tend to provide help for the production of IgG1/IgG2a and Th2 cells mediate the isotype switch to the production of IgE [13]. A practical example of the importance of T cell cytokine balance is in the context of allergy, an IgE-mediated disease. The allergic cascade is characterized by the allergen-driven production of IgE. This allergen-specific IgE binds to the surface of mast cells in tissues and basophils in the blood. Upon subsequent exposure to allergen, surface IgE is cross-linked and granules containing a host of reactive compounds, including histamine, are released leading to allergic symptoms [15]. It has been determined that IL-4 is necessary for the production of IgE [16]. IFN-γ inhibits the production of IgE [17]. Therefore, it has been postulated that Th2 responses increase IgE production and Th1 responses decrease IgE production. Thus, if a T cell response is shifted from Th2 to Th1, allergic symptoms should be alleviated.

The Mechanisms of T Cell Unresponsiveness

There are three general ways in which mature, peripheral T cells are functionally inactivated: anergy, in which reactive cells are present but incapable of responding to antigen; deletion, or cell death which may be mediated by an active process termed apoptosis; and suppression, in which a reactive T cell is actively kept from function by another cell.

As mentioned above, T cells require additional non-TCR-mediated interactions with costimulatory molecules provided by the APC to be functionally activated. In fact, when the TCR is occupied in the absence of costimulation a long-lasting state of antigenic unresponsiveness or 'anergy' occurs. One adhesion molecule which can mediate this second signal is CD28, a 44-kD homodimeric glycoprotein expressed on the surface of T cells [18]. The murine CD28 has also been described and acts in a similar fashion [19]. If T cells are stimulated through the TCR and CD28, activation occurs as assessed by proliferation and lymphokine production. TCR engagement leads to a classic Ig gene superfamily activation pathway involving phospholipase C activation, the generation of the second messen-

gers diacylglycerol and inositol triphosphate, the activation and translocation of protein kinase C from the cytosol to the plasma membrane, the
mobilization of calcium from intracellular stores, and the activation of
src-family protein-tyrosine kinases. This pathway eventually leads to the
transfer of signals to the nucleus and the activation of lymphokine gene
transcription. Interestingly, engagement of CD28, which like TCR is an Ig
superfamily member, does not lead to proliferation or lymphokine secretion.
Signals generated by CD28, presumably through a protein-tyrosine kinase-
mediated pathway, may act by stabilizing the message for IL-2 or other
cytokines thereby leading to enhanced lymphokine production in TCR-stim-
ulated cells [18, 20]. CD28 signalling is also associated with DNA-binding
protein (transcription factor) activity in the IL-2 gene enhancer [21]. The
resultant enhanced lymphokine production may be necessary for paracrine
rather than only autocrine lymphokine-mediated stimulation. In effect, this
costimulatory signal allows the antigen-driven system to get past a threshold,
or maintenance level, of lymphokine production required for the expansion
of an immune response. If the interaction of CD28 with its ligand B7 is
blocked, or if B7 is not present on the APC, anergy is induced. In addition,
signals generated by CD28 can block the induction of anergy [19]. Other
molecules which can function as costimulatory molecules in antigen-primed
T cells are LFA-3, ICAM-1 and VCAM-1 [22]. The role that these molecules
may play in the induction of T cell anergy is unknown. One can envision
that the relative amounts of costimulatory molecules expressed by different
APC or the receptors for such molecules on T cells could be variably
regulated during the course of an immune response to allow for either
activation during the initiation and expansion of the immune response or
inactivation when a response is no longer required.

A significant proportion of the deletion of immature thymocytes after
engagement of the TCR on elements of the thymic epithelium has been
shown to be due to an active process termed apoptosis or programmed cell
death [23]. This process has been well studied and is characterized by
phenotypic cytoplasmic changes, protein synthesis, chromatin condensation, and nucleosome length DNA fragmentation [24]. It is also possible
that peripheral, mature T cells could be signalled to undergo cell death.
Such 'activation-driven' cell death has been demonstrated in transformed T
cell hybridomas and in nontransformed T cell lines in a number of different
ways in vitro. Interestingly, there appears to be a role for the induction of
proto-oncogenes in this process as antisense c-myc, c-myb, and bcl-2 can
block apoptosis [25, 26]. Peripheral T cell apoptosis has also been demonstrated in vivo using athymic mice made tolerant to the superantigen
staphylococcal enterotoxin B (SEB) [27]. Other examples of peripheral
deletion in the Mls system (subsequently determined to be an MMTV-en-

coded superantigen) may also involve apoptosis [28–31]. Superantigen-me-
diated deletion of mature peripheral cells may involve signalling distinct
from that generated by a 'normal' antigen as superantigens bind to an area
of class II which does not mediate peptide binding [32].

Active suppression of T cell responses may also occur, although the
ability to transfer T cell tolerance to other animals using select cell
populations has failed in most experimental systems and the molecular
mechanisms have not been delineated. A better understanding of T cell
subsets may help explain immune suppression phenomena by anti-idiotypic
T cell specificities and the elaboration of suppressive cytokines by Th cells
as well as the different activation requirements such cells may possess [33].

Models of T Cell Tolerance

The mechanisms of T cell unresponsiveness described above have been
delineated in a number of experimental systems. Each system may contain
elements of anergy, deletion or suppression. The insights gained from these
systems are now being used to study intervention in a number of immunologically based disease states, and the remainder of this review will focus on
some well-characterized models of T cell tolerance and their practical
application to the treatment of allergy and autoimmunity.

Tolerance to Protein Antigens

The generation of T cell tolerance in adult animals by the intravenous
injection of soluble protein antigens has been well documented. These
models typically use deaggregated serum proteins, such as human γ-globu-
lin (HGG) or various serum albumins conjugated to a hapten such that the
antigen specificity can be examined. With the growing understanding of the
role that T cell subsets and their cytokines play in the regulation of class
II-restricted immune responses, these systems have been re-examined to
explore the mechanisms of T cell tolerance at a finer level. A number of
investigators have suggested that there is preferential T cell anergy induced
in Th1-type cells when animals are tolerized to deaggregated immunoglobulin or ovalbumin [34, 35]. Anergy in the Th1 subset was reflected by the
finding that both IFN-γ production and antigen-specific IgG2a were decreased. However, if one looks in vitro at Th1 and Th2 T cell clones
specific for HGG, another story emerges. When HGG-pulsed spleen cells
are fixed and allowed to present antigen to HGG-specific T cell clones, Th1
cell proliferation and their ability to provide T cell help for IgG2a

production are decreased and Th2 cell-mediated help for antibody production is also decreased although their proliferation is not affected [36]. Thus, tolerogenic presentation of HGG, i.e. in the absence of costimulatory signals by use of fixed APC, affects both Th1 and Th2 cells.

Such observations highlight the importance of APC signals in the generation of T cell activation and tolerance. It is possible that delivering soluble proteins by intravenous injection allows certain types of APC to preferentially process and present antigen in a tolerogenic fashion to Th1 cells. In addition, since many T cells in the animal are present in a resting Th0-like state (i.e. capable of producing both Th1-like and Th2-like cytokines) the regulatory cytokine balance may be a result of different activation requirements [37]. Notably, different APC vary in their ability to stimulate T cells. This has been clearly demonstrated in models of in vitro T cell tolerance.

In vitro, if a murine Th1 clone is stimulated through the TCR complex by immobilized anti-CD3, by peptide-pulsed, chemically fixed APC, or by peptide/class II complexes present on planar membranes, a long-lasting state of antigen unresponsiveness occurs [38–40]. Thus, TCR engagement in the absence of costimulation by APC elicits T cell anergy. This appears to be related to a defect in the anergized T cell's ability to produce IL-2, which may be due to altered levels of src-family protein-tyrosine kinases in anergic cells [41]. Importantly, T cell anergy has also been induced in murine Th0-like clones [42]. One other way of inducing unresponsiveness in vitro is by incubating human T cell clones, in the absence of APC, with high doses of peptide [43]. This effect is peptide dose-dependent and may be due to qualitative differences in T cell signalling when human T cells, which in contrast to murine T cells express class II antigens, act as APC rather than other 'professional' APC [44].

T Cell Tolerance in the Experimental Allergic
Encephalomyelitis (EAE) Model

An example of the application of experimental T cell tolerance induction using protein antigen, and subsequently T cell epitope containing peptides, to human disease is in the experimental mouse model for human multiple sclerosis (MS) termed EAE. Human MS may take many forms but is characterized as a demyelinating disease of the central nervous system in which autoreactive T cells specific for myelin basic protein (MBP) are implicated [45]. While the course of the experimentally induced form is somewhat different then the human disease, so much has been learned about the T cell basis of the disease in the model that therapies ranging from T cell 'vaccination' using MBP-specific T cell clones, the administration of blocking TCR-derived peptides, to oral delivery of MBP are currently being evaluated in MS patients [46].

EAE is induced in susceptible strains of mice (H-2^u and H-2^s) by subcutaneous injection of either MBP or MBP-derived peptides emulsified in adjuvant [47, 48]. In H-2^u mice, a dominant, disease-inducing T cell epitope is contained in amino acids 1–9 of MBP and in H-2^s mice the dominant T cell epitope is contained within amino acids 87–104 [47, 48]. Interestingly, a high frequency of T cell lines derived from humans with MS are reactive with the 84–102 peptide and there is a HLA DR-2, and DQw1 association [49]. In addition to a restricted peptide/HLA profile, MS may also be associated with limited TCR Vβ specificities [50]. A TCR Vβ association has been found in the murine model as well [51]. Thus, much is known about the nature of the ternary complex required for T cell responsiveness to MBP in mouse and man. One potential way to intervene in the disease process is to inactivate, or tolerize the MBP-reactive T cells which are implicated in the pathology. Recently, it has been demonstrated in H-2^u mice that delivery of two peptides, 1–11 and 35–47, in a tolerogenic fashion (peptide in the absence of complete adjuvant) could block subsequent disease induction by whole MBP, and could alleviate symptoms in ongoing disease [52]. In addition, there was evidence that peptide-induced T cell tolerance was mediated by the induction of anergy in MBP-reactive T cells.

This study suggests that T cell epitope-containing peptides which elicit T cell activation also elicit T cell inactivation. The concept of T cell epitope dominance where certain antigenic peptides dominate immune responsiveness has been demonstrated in other systems [53]. In the λ repressor system, T cell responses in BALB/c mice are predominantly directed to the amino acids 12–26 [54]. When this dominant epitope is synthetically linked to a weaker epitope, the T cell response is still dominated by 12–26. In addition, if the synthetically linked peptide is used to tolerize mice, T cell reactivity can be recalled to the weaker, but not the dominant epitope [55]. Together these findings suggest that T cell epitopes which mediate immune responsiveness may also elicit T cell anergy. Thus, by identification of dominant areas of T cell reactivity to protein antigens, one may be able to induce T cell anergy by the delivery of dominant T cell epitopes in a tolerogenic manner, in the absence of 'costimulatory' adjuvants.

The Model of Allergic Disease

The findings that the same T cell epitopes may be used to activate or inactivate T cells, that both Th1- and Th2-type responses may be anergized, and that ongoing responses may be effectively halted by tolerogenic delivery of T cell epitope-containing peptides suggests that the tolerogenic delivery of allergen-specific T cell epitopes should abrogate allergic symptoms. Allergy, as described previously, is an IgE-mediated disease. This disorder can allow the examination of the concepts of T cell epitope dominance and the effect of T cell epitope tolerization on the Th1-like/Th2-like lymphokine balance which may be critical for the generation of symptom-mediating allergen-specific IgE. Allergy may be the ideal system to test these concepts as it is a non-life-threatening condition with a clear immunologic outcome. Current treatments, excluding immunotherapy for certain allergens, do not effect a cure. In addition, in contrast to immunotherapy, which uses whole allergen extracts capable of inducing life-threatening anaphylaxis, T cell epitope containing peptides will not bind IgE and thereby avoid such side effects.

The allergic immune response, like the immune responses described above, is initiated by T cell recognition of peptidic antigen/HLA complex. The protein antigens responsible for the induction of allergy in atopic, or sensitive individuals, are termed allergens. Many of the common allergens have been identified, cloned and sequenced, including the major ragweed allergen, *Amb a* I, the cat allergen, *Fel d* I, and house dust mite allergens, *Der p* I and *Der p* II [56]. In addition, T cell reactivity to many cloned or purified allergens has been examined, and hierarchies of areas of reactivity have been identified [57, 58]. This suggests that T cell epitope dominance may occur in the immune response to allergens. Notably, T cells reactive to allergens may be identified in nonatopic individuals, suggesting that T cell recognition of allergens is similar in atopics and nonatopics, but the nature of the T cell response to allergens is different in atopic individuals [59]. This is reflected in the fact that T cells from nonatopics can provide help for IgG antibody to allergens, but help elicited by T cells from atopic individuals leads to much greater levels of IgE [60].

If T cell recognition of allergens is similar in sensitive and nonsensitive individuals, how is the generation of allergen-specific IgE in atopics regulated? As described above, T cell recognition of antigens can lead to different immunologic outcomes based on the balance of lymphokines produced. There is evidence that house dust mite-specific T cell clones from atopic individuals produce IL-4 whereas those from nonatopics fail to produce IL-4 and that the IL-4-producing cells can elicit an isotype switch to IgE production [60]. In addition, there appears to be a genetic predisposition to atopy which may be reflected in selective use of HLA antigens to present allergen [61]. This may be significant, for as discussed previously, different T cells may have different activation requirements, and restricted expression of class II on different types of APC or allergenic peptide/HLA on APC with varying costimulatory capabilities may alter the type of immune response, Th1-like or Th2-like, generated.

There is evidence that T cell anergy can be induced in allergen-reactive T cells both in vitro and in vivo. Conventional immunotherapy has been shown to lead to a decrease in proliferation of leukocytes to the targeted allergen [62]. In addition, ragweed allergics who received ragweed allergoid immunotherapy had a marked decrease in T cell proliferation to recombinant *Amb a* I.1, the major ragweed allergen, and to immunodominant T cell epitope-containing peptides relative to allergics who had not undergone immunotherapy [63]. There has also been a report that house dust mite-specific T cells can be rendered anergic in vitro using specific *Der p* I peptides [64]. Together, these data suggest that allergen-specific T cells, regardless of which array of lymphokines they may produce, may be rendered anergic in a similar fashion asd has been documented for Th1 T cell clones. The data also suggest that T cell anergy, as induced by immunotherapy, may be effective in an ongoing allergic response.

Conclusion

The current level of understanding of T cell biology is allowing new treatment approaches for mild to severe immunologically based disorders. The ability to inactivate T cells by the use of peptides which are responsible for activation may ameliorate the debilitating autoimmune disorders. Conversely, the ability to selectively activate T cells may lead to new avenues in prophylactic vaccines for infectious diseases. In addition, a greater understanding of the regulation of T cell subsets in vivo may provide an even finer target for immune intervention.

References

1   Bjorkman, P.J.; Saper, M.A.; Samraoui, B.; Bennett, W.S.; Strominger, J.L.; Wiley, D.C.: Structure of the human class I histocompatibility antigen, HLA-A2. Nature *329:* 506–512 (1987).
2   Garrett, T.P.; Saper, M.A.; Bjorkman, P.J.; Strominger, J.L.; Wiley, D.C.: Specificity pockets for the side chains of peptide antigens in HLA-Aw68. Nature *342:* 692–696 (1989).
3   Madden, D.R.; Gorga, J.C.; Strominger, J.L.; Wiley, D.C.: The structure of HLA-B27 reveals nonamer self-peptides bound in an extended conformation. Nature *353:* 321–325 (1991).
4   Rothbard, J.B.; Gefter, M.L.: Interactions between peptides and MHC proteins. Annu. Rev. Immunol. *9:* 527–565 (1991).

5   Perkins, D.L.; Berriz, G.; Kamradt, T.; Smith, J.A.; Gefter, M.L.: Immunodominance: Intramolecular competition between T cell epitopes. J. Immunol. *146:* 2137–2144 (1991).

6   Springer, T.A.; Dustin, M.L.; Kishimoto, T.K.; Marlin, S.D.: The lymphocyte function-associated LFA-1, CD2, and LFA-3 molecules: Cell adhesion receptors of the immune system. Annu. Rev. Immunol. *5:* 223–252 (1987).

7   Janeway, C.A., Jr.; Rojo, J.; Saizawa, K.; Dianzani, U.; Portoles, P.; Tite, J.; Hague, S.; Jones, B.: The co-receptor function of murine CD4. Immunol. Rev. *109:* 77–92 (1989).

8   Jenkins, M.K.; Taylor, P.S.; Norton, S.D.; Urdahl, K.B.: CD28 delivers a costimulatory signal involved in antigen-specific IL-2 production by human T cells. J. Immunol. *147:* 2461–2466 (1991).

9   Mosmann, T.R.; Coffman, R.L.: TH1 and TH2 cells: Different patterns of lymphokine secretion lead to different functional properties. Annu. Rev. Immunol. *7:* 145–174 (1989).

10 Paliard, X.; de Waal Malefijt, R.; Yssel, J.; Banchard, D.; Chretien, I.; Abrams, J.; De Vries, J.; Spits, H.: Simultaneous production of IL-2, IL-4, and IFN-γ by activated human CD4$^+$ and CD8$^+$ T cell clones. J. Immunol. *141:* 849–855 (1988).

11 Romagnani, S.: Induction of TH1 and TH2 responses: A key role for the 'natural' immune response? Immunol. Today *13:* 379–381 (1992).

12 Murray, J.S.; Pfeiffer, C.; Madri, J.; Bottomly, K.: Major histocompatibility complex (MHC) control of CD4 T cell subset activation. II. A single peptide induces either humoral or cell-mediated responses in mice of distinct MHC genotype. Eur. J. Immunol. *22:* 559–565 (1992).

13 Gajewski, T.F.; Fitch, F.W.: Anti-proliferative effect of IFN-γ in immune regulation. I. IFN-γ inhibits the proliferation of TH2 but not TH1 murine HTL clones. J. Immunol. *140:* 4245–4252 (1988).

14 Howard, M.; O'Garra, A.; Ishida, H.; De Waal Malefijt, R.; De Vries, J.: Biological properties of IL-10. J. Clin. Immunol. *12:* 239–247 (1992).

15 Kaliner, M.; Lemanske, R.: Rhinitis and asthma. JAMA *268:* 2807–2829 (1992).

16 Vercelli, D.; Jabara, H.H.; Arai, K.-I.; Geha, R.S.: Induction of human IgE synthesis requires interleukin-4 and T/B interactions involving the T cell receptor/CD3 complex and MHC class II antigens. J. Exp. Med. *169:* 1295–1307 (1989).

17 Vercelli, D.; Jabara, H.H.; Lauener, R.P.; Geha, R.S.: IL-4 inhibits the synthesis of IFN-γ and induces the synthesis of IgE in human mixed lymphocyte cultures. J. Immunol. *144:* 570–573 (1990).

18 June, C.H.; Ledbetter, J.A.; Linsley, P.S.; Thompson, C.B.: Role of the CD28 receptor in T-cell activation. Immunol. Today *11:* 211–216 (1990).

19 Harding, F.A.; McArthur, J.G.; Gross, J.A.; Raulet, D.H.; Allison, J.P.: CD28-mediated signalling co-stimulates murine T cells and prevents induction of anergy in T-cell clones. Nature *356:* 607–609 (1992).

20 Lu, Y.; Granelli-Piperno, A.; Bjorndahl, J.M.; Phillips, C.A.; Trevillyan, J. M.: CD28-induced T cell activation: Evidence for a protein-tyrosine kinase signal transduction pathway. J. Immunol. *149:* 24–29 (1992).

21 Fraser, J.D.; Irving, B.A.; Crabtree, G.R.; Weiss, A.: Regulation of IL-2 gene enhancer activity by the T cell accessory molecule CD28. Science *251:* 313–316 (1991).

22 Damle, N.K.; Klussman, K.; Linsley, P.; Aruffo, A.: Differential costimulatory effects of adhesion molecules B7, ICAM-1, LFA-3, and VCAM-1 on resting and antigen-primed CD4$^+$ T lymphocytes. J. Immunol. *148:* 1985–1992 (1992).

23 Smith, C.A.; Williams, G.T.; Kingston, R.; Jenkinson, E.J.; Owen, J.T.: Antibodies to CD3/T-cell receptor complex induce death by apoptosis in immature T cells in thymic cultures. Nature *337:* 181–184 (1989).

24   Ucker, D.S.; Ashwell, J.D.; Nickas, G.: Activation-driven T cell death. I. Requirements
      for de novo transcription and translation and association with genome fragmentation. J.
      Immunol. *143:* 3461–3469 (1989).
25   Venturelli, D.; Travali, S.; Calabretta, B.: Inhibition of T-cell proliferation by a MYB
      antisense oligomer is accompanied by selective down-regulation of DNA polymerase α
      expression. Proc. Natl Acad. Sci. USA *87:* 5963–5967 (1990).
26   Shi, Y.; Glynn, J.M.; Guilbert, L.J.; Cotter, T.G.; Bissonnette, R.P.; Green, D.R.: Role
      for c-myc in activation-induced apoptotic cell death in T cell hybridomas. Science *257:*
      212–214 (1992).
27   Kawabe, Y.; Ochi, A.: Programmed cell death and extrathymic reduction of V beta 8$^+$
      CD4$^+$ T cells in mice tolerant to *Staphylococcus aureus* enterotoxin B. Nature *349:*
      245–248 (1991).
28   Webb, S.; Morris, C.; Sprent, J.: Extrathymic tolerance of mature T cells: Clonal
      elinimation as a consequence of immunity. Cell *63:* 1249–1256 (1990).
29   Woodland, D.L.; Happ, M.P.; Gollub, K.J.; Palmer, E.: An endogenous retrovirus
      mediating deletion of αβ T cells? Nature *349:* 529–530 (1991).
30   Frankel, W.N.; Rudy, C.; Coffin, J.M.; Huber, B.T.: Linkage of Mls genes to endoge-
      nous mammary tumor viruses of inbred mice. Nature *349:* 526–528 (1991).
31   Dyson, P.J.; Knight, A.M.; Fairchild, S.; Simpson, E.; Tomonari, K.: Genes encoding
      ligands for deletion of Vβ 11 T cells cosegregate with mammary tumor virus genomes.
      Nature *349:* 531–532 (1991).
32   Dellabona, P.; Peccoud, J.; Kappler, J.; Marrack, P.; Benoist, C.; Mathis, D.: Superanti-
      gens interact with MHC class II molecules outside of the antigen groove. Cell *62:*
      1115–1121 (1990).
33   Bloom, B.R.; Salgame, P.; Diamond, B.: Revisiting and revising suppressor T cells.
      Immunol. Today *13:* 131–135 (1992).
34   Burstein, H.J.; Shea, C.M.; Abbas, A.K.: Aqueous antigens induce in vivo tolerance
      selectivity in IL-2- and IFN-γ-producing (TH1) cells. J. Immunol. *148:* 3687–3691
      (1992).
35   De Wit, D.; Van Mechelen, M.; Ryelandt, M.; Figueiredo, A.C.; Abramowicz, D.;
      Goldman, M.; Bazin, H.; Urbain, J.; Leo, O.: The injection of deaggregated gamma
      globulins in adult mice induces antigen-specific unresponsiveness of T helper type 1 but
      not type 2 lymphocytes. J. Exp. Med. *175:* 9–14 (1992).
36   Gilbert, K.M.; Hoang, K.D.; Weigle, W.O.: Th1 and Th2 clones differ in their response
      to a tolerogenic signal. J. Immunol. *144:* 2063–2071 (1990).
37   Miller, J.F.; Morahan, G.: Peripheral T cell tolerance. Annu. Rev. Immunol. *10:* 51–69
      (1992).
38   Jenkins, M.K.; Schwartz, R.H.: Antigen presentation by chemically modified spleno-
      cytes induces antigen-specific T cell unresponsiveness in vitro and in vivo. J. Exp. Med.
      165: 302–319 (1987).
39   Quill, H.; Schwartz, R.H.: Stimulation of normal inducer T cell clones with antigen
      presented by purified Ia molecules in planar lipid membranes: Specific induction of
      a long-lived state of proliferative nonresponsiveness. J. Immunol. *138:* 3704–3712
      (1987).
40   Jenkins, M.K.; Chen, C.; Jung, G.; Mueller, D.L.; Schwartz, R.H.: Inhibition of
      antigen-specific proliferation of type 1 murine T cell clones after stimulation with
      immobilized anti-CD3 monoclonal antibody. J. Immunol. *144:* 16–22 (1990).
41   Quill, H.; Riley, M.P.; Cho, E.A.; Casnellie, J.E.; Reed, J.C.; Torigoe, T.: Anergic Th1
      cells express altered levels of the protein tyrosine kinases p56^{lck} and p59^{fyn}. J. Immunol.
      149: 2887–2893 (1992).

42 Mueller, D.L.; Chiodetti, L.; Bacon, P.A.; Schwartz, R.H.: Clonal anergy blocks the
    response to IL-4, as well as the production of IL-2, in dual-producing T helper cell
    clones. J. Immunol. *147:* 4118–4125 (1991).
43 Lamb, J.R.; Skidmore, B.J.; Green, N.; Chiller, J.M.; Feldman, M.: Induction of
    tolerance in influenza virus-immune T lymphocyte clones with synthetic peptides of
    influenza hemagglutinin. J. Exp. Med. *154:* 1434–1447 (1983).
44 Hewitt, C.; Feldman, M.: Human T cell clones present antigen. J. Immunol. *142:*
    1429–1436 (1989).
45 Hafler, D.A.; Weiner, H.L.: MS: A CNS and systemic autoimmune disease. Immunol.
    Today *10:* 104–107 (1989).
46 Richert, J.R.: Neurologic and muscular disease; in Proceedings of Immunosuppression
    in the Treatment of Disease. National Institutes of Health, National Institute of Allergy
    and Infectious Diseases, Oct. 23, 1992.
47 Zamvil, S.S.; Mitchell, D.J.; Moore, A.C.; Kitamura, K.; Steinman, L.; Rothbard, J.B.:
    T cell epitope of the autoantigen myelin basic protein that induces encephalomyelitis.
    Nature *324:* 258–260 (1986).
48 Kono, D.H.; Urbain, J.L.; Horvath, S.J.; Ando, D.G.; Saavedra, R.A.; Hood, L.: Two
    minor determinants of myelin basic protein induce experimental allergic encephalomyeli-
    tis in SJL/J mice. J. Exp. Med. *168:* 213–228 (1988).
49 Ota, K.; Matsui, M.; Milford, E.L.; Mackin, G.A.; Weiner, H.L.; Hafler, D.A.: T-cell
    recognition of an immuno-dominant myelin basic protein epitope in multiple sclerosis.
    Nature *346:* 183–187 (1990).
50 Hafler, D.A.; Duby, A.D.; Lee, S.J.; Benjamin, D.; Seidman, J.G.; Weiner, H.L.:
    Oligoclonal T lymphocytes in the cerebrospinal fluid of patients with multiple sclerosis.
    J. Exp. Med. *167:* 1313–1322 (1988).
51 Zamvil, S.S.; Mitchell, D.J.; Lee, N.E.; Moore, A.C.; Waldor, M.K.; Sakai, K.;
    Rothbard, J.B.; McDevitt, H.O.; Steinman, L.; Acha-Orbea, H.: Predominant expres-
    sion of a T cell receptor Vβ gene subfamily in autoimmune encephalomyelitis. J. Exp.
    Med. *167:* 1586–1596 (1988).
52 Gaur, A.; Wiers, B.; Liu, A.; Rothbard, J.; Fathman, C.G: Amelioration of autoimmune
    encephalomyelitis by myelin basic protein synthetic-peptide induced anergy. Science *258:*
    1491–1494 (1992).
53 Schad, V.C.; Garman, R.D.; Greenstein, J.L.: The potential use of T cell epitopes to
    alter the immune response. Semin. Immunol. *3:* 217–224 (1991).
54 Lai, M.Z.; Ross, D.T.; Guillet, J.-G.; Briner, T.J.; Gefter, M.L.; Smith, J.A.: T
    lymphocyte response to bacteriophage λ repressor cI protein. J. Immunol. *139:* 3973–
    3980 (1987).
55 Ria, F.; Chan, B.M.C.; Scherer, M.T.; Smith, J.A.; Gefter, M.L.: Immunological
    activity of covalently linked T-cell epitopes. Nature *343:* 381–383 (1990).
56 Scheiner, O.: Recombinant allergens: Biological, immunological and practical aspects.
    Int. Arch. Allergy Immunol. *98:* 93-96 (1992).
57 Dhillon, M.; Roberts, C.; Nunn, T.; Kuo, M.: Mapping the T cell epitopes on
    phospholipase A_2: The major bee-venom allergen. J. Allergy Clin. Immunol. *90:* 42–51
    (1992).
58 Yssel, H.; Johnson, K.E.; Schneider, P.V.; Wideman, J.; Terr, A.; Kasterlein, R.; De
    Vries, J.E.: T cell activation-inducing epitopes of the house dust mite allergen *Der p I*.
    Proliferation and lymphokine production patterns by *Der p I*-specific CD4[+] T cell
    clones. J. Immunol. *148:* 738–745 (1992).
59 O'Hehir, R.E.; Garman, R.D.; Greenstein, J.L.; Lamb, J.R.: The specificity and
    regulation of T-cell responsiveness to allergens. Annu. Rev. Immunol. *9:* 67–95 (1991).

60 Chambers, C.A.; Zimmerman, B.; Hozumi, N.: Functional heterogeneity of human T cell clones from atopic and non-atopic donors. Clin. Exp. Immunol. *88:* 149–156 (1992).

61 O'Hehir, R.E.; Bal, V.; Quint, D.; Moqbel, R.; Kay, A.B.; Zanders, E.D.; Lamb, J.R.: An in vitro model of allergen-dependent IgE synthesis by human B cells: Comparison of the response of an atopic and non-atopic individual to *Dermatophagoides* spp. Immunology *66:* 499–504 (1989).

62 Evans, R.; Pence, H.; Kaplan, H.; Rocklin, E.: The effect of immunotherapy on humoral and cellular responses in ragweed hay fever. J. Clin. Invest. *57:* 1378–1380 (1976).

63 Greenstein, J.L.; Morgenstern, J.P.; LaRaia, J.; Counsell, C.M.; Goodwin, W.H.; Lussier, A.; Creticos, P.S.; Norman, P.S.; Garman, R.D.: Ragweed immunotherapy decreases T cell reactivity to recombinant *Amb a* I.1. J. Allergy Clin. Immunol. *89:* 322 (1992).

64 Higgins, J.A.; Lamb, J.R.; Marsh, S.G.E.; Tonks, S.; Hayball, J.D.; Rosen-Bronson, S.; Bodmer, J.G.; O'Hehir, R.E.: Peptide-induced nonresponsiveness of HLA-DP restricted human T cells reactive with *Dermatophagoides* spp. (house dust mite). J. Allergy Clin. Immunol. *90:* 749–756 (1992).

Victoria C. Schad, ImmuLogic Pharmaceutical Corp., 610 Lincoln Street, Waltham, MA 02154 (USA)

Granstein RD (ed): Mechanisms of Immune Regulation.
Chem Immunol. Basel, Karger, 1994, vol 58, pp 206–235

Towards T Cell Vaccination in Rheumatoid Arthritis

Jacob M. van Laar[a], *André M.M. Miltenburg*[a],
Ferdinand C. Breedveld[a], *Irun R. Cohen*[c], *René R.P. de Vries*[b]

[a]Department of Rheumatology, [b]Immunohaematology and Bloodbank,
University Hospital, Leiden, The Netherlands; [c]Department of Cell Biology,
Weizmann Institute of Science, Rehovot, Israel

Introduction

Two centuries have elapsed since an English peasant inoculated his son with innocuous cowpox as a prophylaxis against lethal infection with smallpox, the first account of a process later called 'vaccination' (vacca = cow). At present it is realized that the beneficial manipulation of the immune system underlying vaccination may not only be useful in preventing infectious diseases but also in preventing autoimmune diseases. In the case of autoimmunity, disease results from an uncontrolled immune response towards an unknown autoantigen which in most cases has not yet been defined. Autoaggressive T lymphocytes have been implicated in a variety of experimental and clinical autoimmune diseases and have been used experimentally to redirect the immune system away from the autoaggressive response by a procedure known as T cell vaccination (TCV), analogous to microbial vaccinations. In the past decade, TCV has been exploited in several animal models of autoimmune disease and nonautoimmune disease (table 1). TCV was shown not only to be effective in the prevention but also in the treatment of autoimmune disease. In general, TCV may be a treatment of choice in those disease conditions where T cells play a central role in the pathogenesis either as effector cells or as regulatory cells. The successful results obtained in the animal studies prompted us to evaluate the applicability of TCV in patients with rheumatoid arthritis (RA), an entity thought to be an autoimmune disease characterized by dysregulated T cell function. In this review the lessons learned from TCV studies in animal models as well as the considerations that led us to investigate TCV in RA patients will be discussed. Specifically, three animal models of TCV will be reviewed: experimental allergic

Table 1. Animal models of experimental antigen-mediated (auto)immune disease where TCV has been studied

Disease	Species	Antigen	Reference
Experimental allergic encephalomyelitis	rat, guinea pig	myelin basic protein	5, 12–19, 25, 26, 29, 30
Adjuvant arthritis	rat	*M. tuberculosis*	36, 38, 39, 41
Collagen arthritis	mouse	collagen type II	53
Experimental allergic thyroiditis	mouse	thyreoglobulin	107
Myasthenia gravis	rat	acetylcholine receptor	52
Insulin-dependent diabetes mellitus	mouse	65-kD heat-shock protein	44

encephalomyelitis (EAE), the best studied animal model with regard to TCV; adjuvant arthritis (AA) as a model for RA, and insulin-dependent diabetes mellitus (IDDM) serving as a model for spontaneous autoimmune disease. The immunological basis and the clinical effects of TCV in these animal models will receive particular attention. Finally, future perspectives of TCV as a potential immunotherapy in man will be discussed.

Experimental Allergic Encephalomyelitis

Role of T Cells in the Pathogenesis of EAE
EAE can be actively induced in susceptible animals by the injection of myelin basic protein (MBP) emulsified in an adjuvant such as complete Freund's adjuvant (CFA) or passively by the injection of lymph node cells or activated anti-MBP T lymphocyte lines and clones from MBP/CFA immunized donors [1, 2]. In Lewis rats, the animal commonly used for TCV studies, the disease is characterized clinically by paralysis that is most marked in the tail and hind limbs starting usually ≈12 days after the injection of MBP/CFA or ≈4 days after passive transfer. The central nervous system (CNS) parenchyma is found to contain a perivascular mononuclear cell infiltrate, indicative of a local antigen-driven immune response. That the T cells form the principle cell population in the induction of EAE comes from the finding that EAE actively induced in irradiated rats correlates with the presence of IL-2 receptor-positive T cells in the CNS [3]. Transfer experiments show that T cells, activated in vitro with MBP or mitogen, accumulate in the brain, thymus, spleen and liver [4, 5]. The MBP-reactive T cells that home in the brain are probably involved in the induction of the paralytic lesions that are characteristic of EAE by reacting cytotoxically with MBP-presenting cells such as astrocytes [6], endothelial cells [7] or other MHC class II-bearing cells [8]. T-cell derived cytokines with cytotoxic effects may also play an important role as

it has been shown in mice that encephalitogenicity of MBP-specific T cell clones correlates with lymphotoxin and tumor necrosis factor-α production [9].

Role of T Cells in Acquired Immunity to EAE

Rats that have recovered from EAE remain resistant to further attempts to induce disease. It is thought that T cells not only play a key role in the induction of disease but also in this subsequent resistance to renewed active challenge. The strongest argument for this comes from the observation that selected T cell populations transfer protection against EAE. The putative suppressor T cells responsible for this phenomenon may be antigen-specific, directed at suppressor epitopes on MBP [10, 11], anti-idiotypic, directed at the T cell receptor (TcR) of the disease-inducing T cells or antiergotypic, directed at activation molecules of T cells. The latter two are the focus of TCV.

Anti-idiotypic T cells were first implicated in EAE when it was observed that while injection of in vitro activated cells of an MBP-reactive T cell line, Z1a, induced EAE, injection of the same cells after attenuation by irradiation or mitomycin C prevented actively induced disease in the majority of animals and delayed onset of disease in the remainder of animals [12, 13]. Since T cells reactive against irrelevant control antigens did not induce protection it was reasoned that the TcR played a key role in TCV.

For T cells to be effective in TCV the cells had to be activated either by MBP or by the mitogen concanavalin A (ConA). This latter observation argues against carry-over of tolerogenic doses of antigen within the injected cell preparation as the explanation of the beneficial effect. The initial experiments using T cells that were attenuated by irradiation or treatment with mitomycin C failed to demonstrate protection against passive EAE. When the T cells were treated with hydrostatic pressure instead, protection against actively induced disease was 80% and against passively induced disease 100% [14]. It was suggested that treatment with hydrostatic pressure caused aggregation of surface molecules such as the TcR, thereby enchancing immunogenicity. Thus, both activation and attenuation were concluded to be prerequisites for efficient TCV. That the TcR was the target of protective immune responses was also deduced from experiments indicating the determinant specificity of TCV. TCV with the Z1a T cell clone specific for the encephalitogenic peptide (EP) determinant of guinea pig MBP (GP-MBP) (AA68–88) induced resistance against encephalomyelitis induced with GP-MBP/CFA but not with bovine MBP (B-MBP) in CFA. On the other hand, the T cell line B1, reactive against the EP determinant of B-MBP, protected against encephalomyelitis induced with B-MBP/CFA but not with GP-MBP/CFA [15]. This indicated that acquired immunity to EAE was directed by the receptor specificity of the virulent anti-MBP T cells. Furthermore, in another study it was shown that several days after induction of TCV with subcutaneously injected Z1a cells, a proliferative response against Z1a but not to an unrelated T cell clone could be detected in the draining lymph node and later in the cervical lymph node. Removal of the draining lymph node at the time of its response against Z1a abrogated the protection in the donor whereas transfer of these cells after mitogen activation in vitro endowed the recipients with protection. CD4+ and CD8+ clones were isolated from the draining lymph node showing the anti-Z1a response that influenced the in vitro proliferation of Z1a [16]. The CD4+ T cell clones enhanced the in vitro proliferation of Z1a irrespective of the presence of GP-MBP while the CD8+ T cell clones down-regulated the in vitro proliferation of Z1a only in the presence of GP-MBP. The CD4+ and CD8+ T cell clones neither recognized the antigen itself nor influenced the in vitro proliferation of irrelevant control clones. From this it was concluded that both subsets recognized a clonotypic determinant on Z1a such as the TcR which would define them as anti-idio-typic T cell clones. When the Z1a cells were injected intravenously instead of subcutaneously, Z1a-reactive cells could be isolated from the thymus and the spleen that upon transfer induced protection [17, 18]. Interestingly, the intravenously injected irradiated Z1a cells themselves have been shown to accumulate in thymus, spleen and liver [5]. The finding that both Z1a and anti-Z1a cells could be isolated from the same site further underscored the in vivo significance of the anti-idiotypic T cells after TCV.

For the anti-idiotypic cells to be protective these cells have to either down-regulate or eliminate the virulent T cells after TCV. That the anti-idiotypic cells did not completely eliminate but rather suppressed the virulent T cells was concluded from elegant experiments showing that encephalitogenic T cells with female karyotype could be recovered 2 months after inoculation in syngeneic male rats, at a time the male rats had well recovered from EAE. After in vitro activation with MBP these T cells with female karyotype were still able to mediate EAE [4]. The notion that the virulent T cells remain in the body in a suppressed or nonactivated state [19] also comes from the finding that unvaccinated healthy rats harbor poten-tially virulent T cells. An MBP-reactive T cell line was propagated from healthy GP-MBP-immunized PVG rats, resistant against active EAE induc-tion, that was capable of mediating passive EAE in syngeneic rats [20].

Other mechanisms may also be involved. From Lewis rats that were vaccinated previously with an irradiated encephalitogenic T cell line, S1, a CD8+ T cell line, was prepared by in vitro incubation of spleen cells with S1. This cell line proliferated selectively against S1, was cytotoxic in vitro and was effective in preventing EAE induction [17]. The cytotoxic capacity of the anti-S1 T cell line did not eliminate all S1 cells in vivo since it was reported that the S1 cell line survived for 2 months in rats injected with both S1 and anti-S1 cells, in the absence of manifest disease. The precise nature of the determinant recognized by the anti-S1 CD8+ T cell line was not identified but the absence of reactivity of anti-S1 with other T cells and MBP suggested it was related to a clonotypic determinant, possibly the TcR.

With the identification of the encephalitogenic epitopes of MBP and T cell receptors involved in the recognition of these epitopes, it became possible to study in more detail the role of the TcR in TCV. The response against the EP was shown to be dictated by T cells displaying a limited TcR heterogeneity [21]. Both in mice and rats, TcR $V\alpha2$-$J\alpha$TA39 and $V\beta8.2$ were predominant among relevant T cell clones [22]. Synthetic peptides analogous to elements of these TcR were constructed and used to vaccinate against actively induced disease. Immunization with 50 μg of the r-VDJ2$_9$ peptide in CFA 2 weeks prior to challenge with MBP/CFA prevented EAE induction and histologic abnormalities in 100% of rats [23]. Similar clinical effects were demonstrated after immunization with 100 μg of a peptide representing the $V\beta8$-CDR2 region in CFA [24]. The finding of antibodies and a delayed-type hypersensitivity (DTH) skin reaction against the peptide proved the immunogenicity of the peptide. Experiments were undertaken to find an immunological basis for the beneficial effect observed by immunizing rats with the peptide and GP-MBP simultaneously and preparing lines against either antigen from the draining lymph node 20 days after immunization. A T cell line reactive against GP-MBP showed decreased activity against EP but not to other antigenic determinants present on MBP. This decreased reactivity was not due to a deletion of EP-reactive T cell clones since the line could still mediate EAE after in vitro culture and activation with MBP. Rather, the activity of the encephalitogenic clones appeared to be suppressed, presumably because T cells reactive with the TcR of the encephalitogenic clones were elicited by the peptide immunization. Indeed, the T cell line that was made against the $V\beta8$ peptide prevented induction of active disease when coadministered with GP-MBP/CFA and was shown to home to the draining lymph node. Furthermore, the anti-$V\beta8$ T cell line not only proliferated against the peptide but also against $V\beta8$+ T cells in vitro, underlining the in vivo significance of anti-idiotypic cells.

In addition to inducing or enhancing an anti-idiotypic network, TCV may induce a response directed against other structures of activated T cells. This response was referred to as being antiergotypic. The concept of anti-ergotypic immunity stems from the observation that adoptive EAE was not lethal in rats that were vaccinated with an attenuated, activated clone of unrelated antigen specificity in comparison with the lethal course in rats that were not vaccinated at all. Moreover, Lewis rats vaccinated with activated, glutaraldehyde-fixed cells of the encephalitogenic D9 T cell clone developed a DTH skin reaction not only towards the injected cells but also towards unrelated activated T cells. An antiergotypic T cell response was thought to have mediated the nonspecific effects. The antiergotypic cells were shown to be T cells both of the CD4 and CD8 phenotype that recognized a structural cellular component of activated T cells other than the IL-2 receptor or MHC class I or II molecules. Upon in vivo injection the antiergotypic cells mediated protection against EAE. Transfer of viable antiergotypic cells obtained from draining lymph nodes of rats that had beed injected with an activated T cell clone prevented EAE when activated D9 cells were injected simultaneously but at a different site and mitigated EAE by MBP/CFA [25].

Clinical Effectiveness of TCV in EAE

Relevant to the applicability of TCV in man is knowledge of the immunological and clinical effectiveness of various protocols of TCV in EAE. Table 2 summarizes the clinical effectiveness of vaccination with (a) attenuated T cells, (b) TCR peptide constructs, and (c) subencephalitogenic doses of T cells, designated low-dose TCV. Table 2A shows the results from TCV studies with attenuated T cells. The mode of attenuation appeared to be a critical factor for effective TCV. Irradiation, hydrostatic pressure, or mitomycin C have been used for attenuation. In one study, pressure treatment was superior to irradiation as a mode of attenuation [14]. In another study, pressure treatment did not fully attenuate the virulent T cells and additional irradiation was necessary [26]. As the protocols used were different, the results of the various studies are difficult to compare. Incidence of EAE after active and passive challenge was reduced up to 100% following TCV depending on the protocol used. In those studies where reduction of incidence was less than 100% a decrease of disease activity, disease duration and delay of onset of disease was observed in the remaining animals. Vaccination with TcR peptides has been tested on a small scale with promising results (table 2A). A 100% prophylaxis against active EAE was found in two studies [23, 24]. The composition of the peptide chosen was critical since minor alterations in peptide structure significantly decreased the protective effects of vaccination. In another study, vaccination with a TcR peptide actually led to a severe chronic relapsing disease [27]. This observation may be related to the recent finding of a suppressor T cell clone that utilizes the same TcR as the pathogenic T cells [28]. It is conceivable that an anti-idiotypic response directed against the TcR of the pathogenic T cells also down-regulated the suppressor cells.

Table 2A. Prophylactic effects of TCV with attenuated T cells and TCR peptide vaccination on actively and passively induced EAE

Species	Cell No.	Activation	Attenuation[1]	Route[2]
Lewis rat	10^7 Z1a	GP-MBP	R or mitomycin C	i.v.
Lewis rat	$2 \cdot 10^7$ Z1a	GP-MBP or ConA	R P	i.p. in IFA
	$2 \cdot 10^7$ D9	GP-MBP	R P	i.p. in IFA
Lewis rat	$2 \cdot 10^7$ Z1a	GP-MBP	R	i.v.
	$2 \cdot 10^7$ B1	B-MBP	R	i.v.
Lewis rat	$3 \times 2 \cdot 10^7$ BP1	GP-MBP	R	i.p.
	$1 \times 2 \cdot 10^7$ D9	GP-MBP	P + R	i.p.
	$3 \times 2 \cdot 10^7$ D9	GP-MBP	P + R	i.p.
	$3 \times 2 \cdot 10^7$ BP20	GP-MBP	P + R	i.p.
Lewis rat	*Peptide* rVDJ2$_9$, 50 µg in CFA			i.c.
Lewis rat	Vβ8-CDR2, 100 µg			s.c.

[1] Attenuation by irradiation (R), treatment with hydrostatic pressure (P) or mitomycin C.
[2] i.v. = Intravenously; i.p. = intraperitoneally; IFA = incomplete Freud's adjuvant; i.c. = intracutaneous: s.c. = subcutaneous.
[3] Active EAE induced by subcutaneous injection of GP-MBP/CFA unless indicated otherwise. Protective effect of TCV expressed as % decreased incidence of EAE in TCV-treated animals

Lastly, low-dose TCV was shown in several studies to be successful in the prevention of both active and passive EAE (table 2B), although one other study found no benefit [26]. The administration of as few as 10^3 D9 cells prevented active and passive EAE completely. In the remaining animals, disease activity and disease duration diminished significantly. The route of administration did not appear to be critical as intravenously, intraperitoneally and subcutaneously injected animals also acquired similar protection [29]. Attenuation of the injected cells abolished the effects

Active EAE[3]	Passive EAE[4]	Reference
70% + onset d12 → d15	Z1a: 0%	5, 12, 13
70%	Z1a: 0%	14
80%	Z1a: 100%	
0%	Z1a: 0%	
80%	Z1a: 100%	
GP-MBP/CFA: 78%	ND	15
B-MBP/CFA: 20%		
GP-MBP/CFA: 0%		
B-MBP/CFA: 80%		
80% + dis.act 1.3/3	BP1: 0%	26
33% + dis.act. 2/3	D9: 67% + dis.act.1.3/2.9	
	BP20: 0% + dis.act. 4.0/2.6	
100% + dis.act. 0.3/3	D9: 100% + dis.act. 0/2.9	
	BP20: 0% + dis.act. 3.3/2.6	
100% + dis.act. 0.8/3	D9: 17% + dis.act. 2.4/2.9	
	BP20: 11% + dis.act. 2.6/2.6	
100%	ND	23
100%	ND	24

versus control animals. If investigated, disease activity (dis.act.) of TCV-treated versus control animals is shown as ../..
[4] Passive EAE induced by the indicated cell lines/clones. Protective effects of TCV against passive EAE expressed as described for active EAE.

suggesting that in vivo proliferation or specific homing of the injected cells contributed to the effect.

Two studies addressed the issue of duration of protection. In guinea pigs the duration of protection against active EAE was at least 60 days [19], while in Lewis rats protection against passive EAE was present from day 7 onwards and lasted for 125 days after low-dose TCV [29].

EAE as an autoantigen-defined autoimmune disease offers the opportunity to examine not only the clinical effectiveness but also the im-

Table 2B. Prophylactic effects of TCV with subencephalitogenic doses of T cells on actively and passively induced EAE[1]

Species	Cell No.	Activation	Route	Active EAE	Passive EAE	Reference
Guinea pig	$0.1-1.0 \cdot 10^7$ LNC	GP-MBP	i.p.	GP-MBP/CFA 90–100% +dis. act.1.4/7.7 CNS/CFA 0% ($0.5-2 \cdot 10^6$ cells) 55% ($0.5-1 \cdot 10^7$ cells)	ND	19
Lewis rat	10^4 Z1a	GP-MBP	i.v.	35% +dis.act. 1.3/3.5 +dis.dur. 4.8/7.1	Z1a: 55% +dis.act. 1/3.5 +dis.dur. 3/4.5	18, 29
	10^4 Z1a	ConA	i.v., ip., s.c.	ND	Z1a: 80–100%	
	10^2 D9			ND	D9: 40%	
	10^3 D9			ND	D9: 100%	
Lewis rat	10^4 D9, BP20	GP-MBP	i.v. or i.p.	0%	D9 or BP20: 0%	26

[1] Legend as in table 2A.

munomodulatory potential of TCV. In one study it was shown that the response against BP of pooled lymph nodes in EAE induced with Z1a cells was inhibited after low-dose TCV with Z1a cells. This type of TCV also inhibited the DTH response to BP in actively induced EAE but not to a control antigen [18]. However, it is conceivable that the mechanism whereby low-dose TCV exerts its effects may involve both the up-regulation of putative anti-idiotypic T cells and also the induction of antigen-specific suppressor cells. In various studies both the degree of EAE and effectiveness of low-dose TCV depended on the number of injected T cells [29], suggesting that T cells reactive with suppressor determinants of MBP or other antigens exposed during the inflammation were also involved. MBP-specific CD4+ T suppressor cells have indeed been isolated and shown to be effective in the prevention of adoptive EAE [11].

Taken together, the TCV studies in EAE show that TCV induces a cytotoxic or suppressive anti-T cell response probably directed at the TcR and activation molecules of the injected T cells. Depending on the protocol, protection could be obtained against both actively and passively induced EAE.

Adjuvant Arthritis

Role of T Cells in Pathogenesis of AA

AA can be actively induced in genetically susceptible Lewis rats by a single injection of CFA which is a dispension of heat-killed *Mycobacterium tuberculosis* in oil, or adoptively by the injection of spleen and lymph node cells from animals immunized previously with CFA. AA is a commonly used experimental model for human RA sharing several clinical and histopathological aspects. In contrast to EAE, AA can only be successfully transferred to irradiated recipient rats [30]. Irradiation may facilitate the action of arthritogenic cells by down-regulating suppressor cells or by inducing stress proteins in the joint [31]. It is thought that T cells triggered by mycobacterial antigens cross-reactive with autologous epitopes mediate the disease. A T cell clone, A2b, reactive against a nine amino acid epitope of mycobacterial 65-kD heat-shock protein (hsp) (residues 180–188), transferred disease when injected in syngeneic irradiated rats [32]. This CD4+ T cell clone also recognized a fragment of proteoglycans in vitro [33]. The epitope recognized and the mechanism by which this clone exerts its arthritogenic effect in vivo remains to be elucidated. Surprisingly, immunization with the 65-kD hsp or the nonapeptide of 65-kD hsp, recognized by A2b, did not induce but rather protected against AA [34, 35]. Thus, the 65-kD hsp not only elicits arthritogenic but also protective T cells in vivo [36]. It is therefore conceivable that mycobacterial antigens other than the 65-kD hsp also play a role in AA. Moreover, the finding that experimental arthritis can also be induced by synthetic adjuvants implies that mycobacterial antigens are not a prerequisite for AA and that other (self) antigens may also be involved [37].

Role of T Cells in Acquired Immunity to AA

As in EAE, virulent T cells have been used to protect against AA. Injection of rats with a T cell line, designated A2, isolated from the lymph node of a rat immunized previously with CFA, prevented AA by CFA, depending on the number of T cells injected (table 3) [38, 39]. While the A2 line was only arthritogenic in irradiated rats, it was effective for TCV both in irradiated and normal rats. A2 attenuated by irradiation was even more effective in nonirradiated rats. Full protection against AA occurred when the A2 line was injected at least 16 days before the injection of CFA. The protective effect of TCV with A2 lasted up to 180 days after the injection of this line. Two clones, A2b and A2c, were isolated from the A2 line with the same antigen specificity but different in vivo characteristics. It was

Table 3. Prophylactic and therapeutic effects of TCV on actively induced AA

Species	Cell No.	Activation[1]	Attenuation[2]	Route[3]
Lewis rat	$0.1 \cdot 10^7$ A2	MT, ConA	–	i.v.
	$0.5 \cdot 10^7$ A2	MT	–	i.v.
	$1 \cdot 10^7$ A2	MT	–	i.v.
	$2 \cdot 10^7$ A2	MT	–	i.v.
	$2 \cdot 10^7$ A2	MT	R	i.v.
Lewis rat	$3 \times 2 \cdot 10^7$ A2b	MT	–	s.c. (i.v. or i.p.)
			P/colchicine/ cytochalasin B/ glutaraldehyde	
	$3 \times 2 \cdot 10^7$ LNC	ConA	P	s.c.

[1] Activation: with *M. tuberculosis* (MT) or ConA in the presence of accessory cells.
[2] Attenuation: not performed (–), by irradiation (R), treatment with hydrostatic pressure (P) or as indicated.
[3] i.v. = Intravenously; s.c. = subcutaneously; i.p. = intraperitoneal.

reported that the A2c clone was not arthritogenic but shared the protective capacity with the line [40]. The arthritogenic A2b T cell clone could also be used for TCV if attenuated by pressure treatment or chimical cross-linkers [41]. TCV with A2c or attenuated A2b was effective in both the prevention and treatment of AA. Not only cloned cells, but also crude preparations of lymph node cells from CFA-immunized animals, activated and attenuated, were effective in the prevention and treatment of AA. Since the lymph node populations probably only contained a minority of relevant T cells, this was interpreted as an indication of the potency of TCV in strengthening a preformed network of anti-idiotypic cells. The observation that TCV induces remission of existing disease within 2 days after treatment supports the notion that a memory response mediates the effects of TCV. The principle underlying these phenomena would be that the immune system is focussed on a few immunodominant antigens and on the T cells recognizing these antigens. Recent data support this idea. Within 4 days after immunization of Lewis rats with CFA, a response could be detected in the

Active AA[4]	Therapeutic effect	Reference
0% + dis.act. 12/13.6 + onset d17/d14	ND	38, 39
0% + dis.act. 7.4/13.6 + onset d17.4/14		
80% + dis.act. 2/13.6 + onset d17/14		
100% (both in R and non-R rat) 25% (R rat) 56% (non-R rat)		
0% 80–85% + dis.act. 10–20/80 dis.dur. 10–12/40	no yes, with $1 \times 2 \cdot 10^7$ P-treated A2b	41
ND	yes	

[4] AA induced by immunization with CFA. Protective effect of TCV expressed as % decreased incidence of AA in TCV-treated animals versus control animals. If investigated, disease activity (dis.act.), disease duration (dis.dur.) or onset of disease of TCV-treated animals versus control animals is shown as ../..

draining lymph node against a T cell clone, termed M1, specific for an epitope on 65-kD hsp, but not against A2b, that preceded the response against the antigen itself [42]. It is not surprising that M1 could also be successfully exploited for TCV. Although the cells that recognize M1 and A2b have not yet been identified, cells capable of transferring the protective effect were isolated from the thymus and spleen of rats that had been vaccinated previously with A2b [41]. Whether these cells are anti-idiotypic or activated by other mechanisms is as yet unknown.

Insulin-Dependent Diabetes mellitus

IDDM differs from the foregoing autoimmune diseases in that it is not an experiment of man but of nature. It is caused by the spontaneous inflammation of the insulin-producing β cells of the pancreas, insulitis, progressing to overt clinical disease once the majority of β cells have been destroyed. Interestingly, in NOD/Lt mice, a strain that spontaneously develops IDDM, IDDM was shown to be associated with T cell reactivity against an epitope on the 65-kD hsp of *M. tuberculosis* [43]. The development of IDDM was preceded by an increase and subsequent decline of T cell reactivity against the 65-kD hsp. This was followed by an increase in serum of the 65-kD hsp antigen and by the appearance of anti-65-kD hsp, insulin and anti-insulin antibodies. The demonstration that IDDM could be transferred with 65-kD hsp-reactive T cells underscores the importance of T cells in IDDM and made these T cells an attractive target for TCV. Indeed, TCV with an irradiated 65-kD hsp-reactive T cell clone or spleen cells prevented IDDM [44]. This was probably due to the down-regulation of the virulent T cells in vivo since it was accompanied by the inhibition of the spontaneous responses to 65-kD hsp characteristic of NOD mice developing IDDM. It was reported that anti-idiotypic T cells were induced that could have mediated the suppressive effects.

EAE, AA and IDDM: A Common Mechanism For TCV?

Introduction

The experimental diseases discussed above differ in etiology and pathogenesis but have in common that T cells and a disease-related antigen are considered to play a pivotal role. The realization that virulent T cells can also be manipulated to prevent or even treat a variety of diseases suggests that a common regulatory mechanism is operative in the different models of experimental disease. A network of T cells recognizing and reacting to other T cells, reminiscent of the networks proposed by Jerne [45], has been suggested to form the basis of TCV. The purpose of TCV is to boost this network to function properly [46]. According to this view, disease is the consequence of the way the immune system deals with foreign or self antigens and with the cells that are involved in the recognition of these antigens [47]. This concept has drawn attention since other traditional postulations for autoimmune disease have been undermined by experimental data. It is now recognized that autoimmune disease cannot be solely accounted for by either the emergence of autoreactive T cells or by the binding of disease inducing T cell epitopes to certain HLA haplotypes since healthy individuals also carry autoreactive T cells and disease-associated HLA haplotypes.

At a cellular level, several issues have been addressed but not yet resolved: (1) the determinants recognized by T cells responding to other T cells; (2) the effector mechanism utilized by such T cells; (3) the contribution of other cells of the immune system, and (4) the extent of the network.

Determinants Involved in T-T Cell Recognition

The question is not only whether or why networks exist but how the cells comprising them interact and what signals they use. From the TCV studies it has been concluded that the TcR is a major recognition element involved in T-T cell interactions. The strongest argument for this comes from the TCV studies using TcR peptides. However, a relatively high dose of peptide was needed to obtain the effect and it is therefore unlikely that this is the only explanation for the effects of TCV with whole cell preparations. In addition, the observation that only activated T cells were effective in TCV suggests that other mechanisms associated with activation are also involved, e.g. activation markers or hsp present on activated T cells [48] or cytokines. That not only the TcR but also cytokines are involved in the cellular interactions is illustrated by data showing that the arthritogenic A2b and protective A2c T cell clones share the same TcR [49] but have different cytokine profiles [50]. It is well appreciated at present that T cells can interact via cytokine networks. T cell clones can be categorized in TH0, TH1, and TH2 subsets according to their cytokine profiles. Not only are these subsets associated with different functional properties but also with different sensitivities to particular cytokines [51]. Thus, the successful outcome of TCV may not only be determined by recognition of the TcR of the injected T cells but also of their cytokine profile.

Effector Mechanisms

With the exception of EAE, little is known about the down-regulating mechanisms utilized by anti-idiotypic T cells. In EAE, cytolytic and suppressive mechanisms have been implicated. With regard to the suppressive effects, it is unclear whether the anti-idiotypic cells down-regulate the virulent T cells directly, e.g. via cytokines, or indirectly, via recruitment of other cells. The characterization of suppressive cytokines and their receptors may provide more insight into possible down-regulatory T-T cell responses.

Contribution of Other Cells

Since the anti-idiotypic cells investigated thus far were all T cells, little attention has been paid to the possible contribution of other cells. The absence of serum antibodies against the injected T cells after TCV argues against a regulatory role of anti-idiotypic antibodies in mediating the effects of TCV. However, this does not exclude a secondary effect on B cells producing relevant antibodies. In the experimental model of myasthenia gravis, TCV induced an increase of pathogenic anti-acetylcholine receptor antibodies indicating an effect of TCV on T-cell-dependent autoreactive B cells [52]. It is tempting to speculate that this was mediated by a TH2 response induced by TCV with a TH1 T cell clone. In collagen-induced arthritis TCV elicited an anti-idiotypic antibody response against anti-collagen antibodies [53].

Extent of the Network

It has been reasoned that analogous to B cell networks, T cell networks may be a vital part of one's immune system [54]. It was concluded that although little is known about TcR connectivities, there is good ground for believing that 10–20% of immunocompetent cells are constantly fixed on each other irrespective of foreign antigenic challenge. V region connectivity was proposed to underlie this phenomenon compatible with the idea of an anti-idiotypic network. That such a network must be powerful comes from the findings in both EAE and AA that TCV with a single virulent T cell clone prevents active disease induction. This implies that all potentially virulent T cell clones present in the recipient are down-regulated, probably because virulent T cells share TcR fragments as has been demonstrated at least in EAE.

Rheumatoid Arthritis

Role of T Cells in RA

RA is a chronic disease primarily affecting the joints but with extra-articular symptoms or systemic complications in a number of patients. The hallmark of the disease is inflammation of the synovial tissue ultimately leading to destruction of the cartilage. Neither etiology nor pathogenesis are fully understood but various findings are compatible with deranged T cell function and lend support to the view that RA is an autoimmune disease:

(1) RA is associated with particular HLA class II molecules. The most striking feature of this association is the high prevalence of HLA-DR4 and HLA-DR1 in RA patients. Two subtypes, Dw4 and Dw14, mainly account for the association of HLA-DR4 with the disease. The genes conferring susceptibility to RA share a nucleotide sequence encompassing a portion of the DRβ molecule that may be functionally critical for T cell recognition of a common autoantigen [55].

(2) Immunohistological [56] and electron microscopical [57] analysis of the synovial tissue show infiltrating T cells, mainly of the CD4+ CD45RO phenotype, in close contact with HLA class II positive dendritic cells [58,59]. The T cells are partially activated as shown by expression of HLA class II [60], LFA-1, VLA-1 [61] and to a lesser extent IL-2 receptors.

The above picture resembles a DTH reaction [62] and, given the regulatory role of T cells in the immune response in general, has stimulated many investigators to examine the role of T cells and the putative autoantigens they respond to in detail. Despite intense research the autoantigen(s) that incites autoreactive T cells is unknown. Collagen type II or other cartilage components [63, 64], chondrocyte membranes [65], and mycobacterial hsp [66] have been implicated but rejected as the primary autoantigen either because healthy individuals or only a minority of RA patients displayed reactivity against the antigens or because reactivity was low. A recent study has raised hope for this field of research by showing reactivity against synovial fluid antigens of IL-2-responsive T cells in 2 out of 4 HLA-DR4Dw4 RA patients [67].

Based on the assumption that recognition of a limited number of autoantigens by T cells leads to an oligoclonal T cell response, a number of studies have addressed the issue whether synovial T cell oligoclonality occurs in RA. Although the first reports indeed suggested predominance of a limited number of T cell clones expanded from synovial tissue as detected by Southern blot analysis of TcR β-chain rearrangements [68], other studies focussing on freshly isolated synovial T cells failed to demonstrate T cell oligoclonality [69, 71]. In an attempt to resolve the controversy, studies in our laboratory [72] and by another group [73] indicated that expansion of 'in vivo activated' T cells from synovial tissue yielded dominant TcR β chain rearrangements while the study of freshly isolated or nonspecifically expanded synovial T cell populations pointed to a lack of oligoclonality. However, these 'in vivo activated' T cells are not evenly distributed through the tissue as TcR rearrangement patterns of T cell populations expanded from different tissue fragments of the same patient showed significant variability [74].

Due to technical limitations of Southern blot analysis of TcR rearrangements, the polymerase chain reaction is at present more frequently employed to assess V-region usage by synovial T cells. The results thus far are inconclusive. In the case of synovial T cells, either restricted V-region usage [75, 77] of different V-gene segments in different studies, or TcR heterogeneity was found [78, and L. Struyk et al., submitted]. Heterogeneity of synovial tissue T cells was also found using mAbs against a few V-region families [79]. The recent finding that the T cell repertoire in man is strongly HLA determined [80] may shed some light on the observed inter- and intraindividual differences in TcR usage. Taken together, the data do not allow a definite conclusion with regard to T cell oligoclonality or restricted V-gene usage. Nevertheless, the results of two recent studies are worth mentioning at this point. In one, it was observed that the synovial TcR repertoire of RA patients with recent onset disease was less heterogeneous than of patients with longlasting disease [81]. In the other, in vivo activated synovial T cells displayed limited TcR β-chain heterogeneity [82]. This has been interpreted to be the result of stimulation of possibly disease-related T cells in the joints by superantigens that have bound to HLA class II molecules in some individuals. In animal models of arthritis the activation of specific T cells by superantigen has been suggested to be a mechanism whereby autoreactive T cells, normally quiescent, are activated and induce disease [83]. However, more studies are needed to determine the role of superantigens in RA.

That T cells play a major role in the pathogenesis of RA also comes from experimental studies in vivo directed at down-regulation of the number or activity of T cells. Total lymphoid irradiation, thoracic duct drainage and lymphopheresis have been relatively successful in ameliorating disease activity [84, 87]. However, such treatment modalities are relatively nonspecific since they are at least targeted at all CD4+ T cells and harbor the potential risk of side effects such as infections and neoplasms. TCV as performed in the animal studies has fulfilled the wish for a nontoxic therapy that selectively suppresses the virulent T cells.

TCV in RA Patients: An Experimental Study

Based on the successful treatment of experimental models of autoimmune disease in animals by means of TCV and the central role of T cells in RA, a Phase I study was undertaken in RA patients [88]. Thirteen RA patients were treated with attenuated, autologous T cells isolated from synovial tissue or synovial fluid, the compartments presumably harboring the virulent T cells. The goal of the study was to determine the technical feasibility, toxicity, and immunomodulatory potential of TCV in RA patients. Furthermore, disease activity was monitored although it must be realized that any Phase I trial only allows for a preliminary impression of clinical effectiveness.

Four patients received an inoculum containing $50 \cdot 10^6$ T cells of an in vivo activated, chemically fixed, CD4+ T cell clone. The T cell clones were obtained by an antigen-driven expansion method using putative autoantigens possibly relevant for the disease or, in the absence of antigen reactivity, by a nonspecific expansion method. In fact, in only 1 out of 4 patients an antigen-reactive clone could be obtained. This T cell clone was reactive against the acetone-precipitable (AP) fraction of *M. tuberculosis*, HLA DR4Dw14 restricted, and was isolated from synovial fluid of a patient suffering from RA for 3 months. In the remaining patients, T cell clones were selected based on their CD4 positivity and in vitro growth characteristics. As it became known during the course of the study that, in animal models, T cell lines containing disease-relevant T cells could be expanded not only with the autoantigen but also a mitogen [89], it was decided to treat another group of RA patients with inocula containing T cell lines. Nine patients received an inoculum containing $50 \cdot 10^6$ T cells of an in vitro activated, irradiated, T cell line since this would more likely contain disease-relevant T cells than nonspecifically expanded T cell populations. Figure 1 shows the technical protocol employed to prepare the inocula. The number of cells injected, the route of administration and modes of attenuation were roughly extrapolated from the TCV protocols used in AA and EAE. However, the activation step was different from that used in the animal studies. Instead of activation of the T cells with irradiated syngeneic antigen-presenting cells in combination with the antigen or ConA, T cells were activated on OKT3 mAb-coated culture plates in the presence of rIL-2 to avoid contamination of the inoculum with feeder cells. The results of the study can be summarized as follows:

Technical Feasibility. RA has the advantage that the lesions are easily accessible for removal of inflamed synovial tissue or fluid thus facilitating the use of synovial fluid T cells for treatment. Obviously, the amount of synovial fluid or synovial tissue, the number of T cells isolated and the growth characteristics during subsequent in vitro expansion until sufficient numbers are obtained all affect the time needed to produce the inoculum. Furthermore, in RA patients intrinsic hyporesponsiveness of synovial T cells to mitogens [90] and deficient accessory cell function of PBMC used as feeder cells [91] may have an unfavorable effect during the expansion of the T cells. Nevertheless, inocula could be prepared in all patients. In general, the mean time required to make the inocula from a T cell clone was longer than from a T cell line (6.3 vs. 4.3 months).

Toxicity. In experimental studies toxicity is a major concern. It was therefore reassuring that in none of the patients adverse effects were reported or demonstrated as determined by laboratory tests. Particular attention was paid to sterile preparation of the cells since in vitro culture of the cells for a relatively long period of time harbors the risk of contamination of the inocula. To minimize the chance of contamination the cells were cultured with autologous serum and feeder cells as well as intensive screening before inoculation.

Clinical Response. Patients were seen in the outpatient clinic. Disease activity was determined by monitoring of Ritchie index, grip strength, morning stiffness, number of swollen joints and laboratory parameters of disease activity. Although as a phase I trial the study was not designed to evaluate the clinical effectiveness of the treatment, a preliminary impression

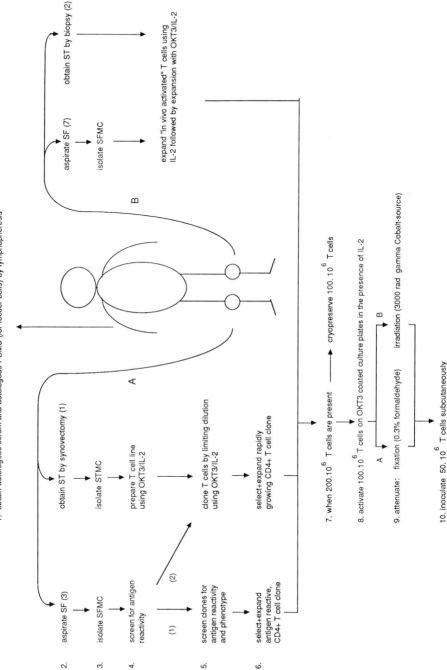

1. obtain autologous serum and autologous PBMC (for feeder cells) by lymphapheresis

obtain ST by biopsy (2)

aspirate SF (7)

isolate SFMC

expand "in vivo activated" T cells using IL-2 followed by expansion with OKT3/IL-2

B

2. aspirate SF (3)

3. isolate SFMC

4. screen for antigen reactivity

(1)

(2)

5. screen clones for antigen reactivity and phenotype

6. select+expand antigen reactive, CD4+ T cell clone

A

obtain ST by synovectomy (1)

isolate STMC

prepare T cell line using OKT3/IL-2

clone T cells by limiting dilution using OKT3/IL-2

select+expand rapidly growing CD4+ T cell clone

7. when 200.10^6 T cells are present

cryopreserve 100.10^6 T cells

8. activate 100.10^6 T cells on OKT3 coated culture plates in the presence of IL-2

9. attenuate: fixation (0.3% formaldehyde)   irradiation (3000 rad gamma Cobalt-source)

A   B

10. inoculate 50.10^6 T cells subcutaneously

of the influence on disease activity could be obtained. Taken as a group, mean values of all parameters indicated an improvement of disease activity during the first 2 months after treatment. However, variability of the responses among the individual patients was high and a remission of disease was not observed in any of the patients.

Immunological Effects. To determine the immunomodulatory potential of TCV, experiments were carried out to detect the development of a cellular or humoral immune response against the injected cells as well as to measure mitogen reactivity and rheumatoid factors (RF) in peripheral blood. In none of the patients investigated did the findings indicate that a classic cellular or humoral immune response by the inoculation with attenuated, activated T cells was elicited. Responses of PBMC against irradiated, activated T cells from the inoculum could be detected but, with the exception of 1 patient (see below), did not differ from pretreatment levels. In addition, patients that received a T cell line were tested for the development of DTH skin reactions by the intracutaneous challenge with cells from the inoculum. Both before and after treatment skin reactions occurred but although the histological aspects showed characteristics of a DTH reaction, the clinical course with rapid onset and minor induration did not provide indubitable eidence of DTH. Furthermore, no serum anti-T cell antibodies were detected. Nevertheless, several findings indicated that the treatment induced altered immune reactivity. Interestingly, this was most pronounced in the above mentioned patient with recent onset disease who was inoculated with a defined T cell clone. Apart from a beneficial clinical response within 2 weeks after treatment and lasting for 3 months, mitogen reactivity of PBMC was suppressed for 3 months and the increased titers of RF of the IgM and IgG isotype returned to normal levels after treatment. The decrease in titers of RF was not merely the result of a decrease of total immunoglobulin content as total IgM and total IgG did not change. In 2 other patients that had received a T cell clone a decrease of serum RF was also measured.

In addition, a shortlasting enhancement of the proliferative response in peripheral blood against the injected cells was measured suggestive of an inoculation effect. This response also occurred against a CD4+ control T cell clone reactive with the same antigen preparation but not against a

---

Fig. 1. Schematic representation of the protocol employed to prepare T cell inocula for the treatment of patients with RA. Patients were either treated with T cell clones (group A) or T cell lines (group B). Numbers between parentheses represent number of patients. SF = Synovial fluid; ST = synovial tissue; MC = mononuclear cells.

CD4+ control T cell clone of unknown specificity. Whether this intriguing finding is related to an idiotypic/anti-idiotypic network in man remains to be determined.

It is tempting to speculate that the observation of beneficial clinical responses and of down-regulated immune reactivity after TCV in a patient with recent onset disease is a reflection of the central role of T cells in early disease. T cell reactivity has been found to be most pronounced in early chronic arthritis [92]. Accordingly, a response elicited against activated T cells may be most effective in early disease. Assuming that the in vitro immune reactivity found in peripheral blood against a few activated T cell clones is representative of the in vivo situation, it can be concluded that the response induced is not entirely generalized but focussed on a limited number of activated T cell clones. However, to obtain formal evidence that the T cell clones are indeed down-regulated in vivo is virtually impossible since it would require the demonstration that the target T cell clone is either suppressed in situ or eliminated.

Differences with Animal Studies of TCV

Treatment with attenuated, autologous T cells as performed in RA patients differs from TCV as employed in the animal studies in several respects.

Selection of T Cells for TCV. Whereas in most animal models of experimental autoimmune disease detailed knowledge of the autoantigen and the T cells involved exists, the pathogenesis of RA is still poorly understood. Even though RA may be a T-cell-mediated disease, the relevant T cells cannot as yet be identified. The recent demonstration in AA and EAE that antigen-reactive T cell clones can be expanded nonspecifically from lymph nodes of rats previously sensitized with the antigen implies that precise knowledge of the autoantigen is not a precondition for successful TCV. Since this was done with ConA the question arises which expansion method will work best with human cells. In our study, patients were treated with T cell lines expanded from the lesion with IL-2 since it was reasoned that these would contain T cells activated in vivo by the putative autoantigen. With the exception of one T cell line, none of the T cell lines showed reactivity against putative autoantigens such as collagen type II, proteoglycans, mycobacterial antigens or autologous synovial fluid. This fits with the idea that most T cells present at chronic inflammatory sites such as the joint are CD45RO memory T cells of unrelated antigen reactivities that have been attracted nonspecifically to the lesion [93, 94] but does not exclude the possibility that a limited number of T cells have initiated the inflammation specifically. Preliminary results obtained in animal models indeed suggest that disease-relevant T cells are predominantly present in the lesion. In rats suffering from actively induced EAE the frequency of MBP-reactive T cell clones was found to be highest in the cerebrospinal fluids as compared with peripheral blood or lymphoid organs [F. Mor et al., unpubl. observations]. Whether such lesional T cells can vice versa successfully be used for TCV remains to be determined as vaccines thus far have been prepared from T cells isolated from draining lymph nodes. If so, it would be of interest to investigate whether such T cells can also be isolated from chronic lesions.

The limited heterogeneity of T cell clones reactive with MBP in EAE has tempted investigators to employ TcR peptides for vaccination purposes in animals. In contrast, from the available data on TcR usage in RA, no role for a subset of T cells bearing specific TcR can yet be concluded. With the study of TcR usage in early disease, possible relevant T cells can be identified but the limited data obtained thus far do not allow the selection of particular TcR-bearing T cells for TCV in RA.

Vaccine Preparation. Although the outlines for the preparation of the vaccine in RA patients were drawn by analogy to the animal studies, the application of TCV in RA required some adjustments. Lack of knowledge of the autoantigen involved forced us to use a mitogen to activate the T cells before inoculation. Activation using OKT-3-coated culture plates was chosen because this obviated the need of autologous feeder cells and kept the final inoculum free from contaminating feeder cells. It must be kept in mind that in the animal studies T cells were activated using a mitogen or antigen and syngeneic, not autologous, feeder cells. Since future studies of TCV in man will employ an activation step that either is accessory cell independent or uses autologous feeder cells, the animal models may be used to examine the influence of these changes in the protocol on the efficacy of TCV. However, not only the method of activation but also the number of cells injected, mode of attenuation, and site of injection, may determine which protocol works best as was recently demonstrated in a pilot TCV study in primates and RA patients [95]. Only from the clinical application of TCV will we be able to obtain this information.

Monitoring the Effects of TCV. In most animal studies to evaluate the efficacy of TCV disease activity was assessed. However, RA differs from the animal models of experimental disease with respect to disease course. RA frequently is a chronic relapsing disease whereas most experimentally induced diseases have a limited disease course. This poses a major problem for evaluation of effects of any experimental treatment of RA as compared with the animal studies. It is unclear whether relapse of disease activity in RA is T cell mediated. Increases in serum titers of soluble CD4 antigens have been found to precede relapses and, conversely, decreases have been noted during remissions [96] while changes in serum titers of soluble CD8 showed an inverse pattern [97]. This argues for a role of CD4+ T cells not only in the initiation of the inflammatory process but also in its maintenance. The isolation of MBP-reactive CD4+ T cell clones that can mediate chronic relapsing EAE substantiates this notion and may provide a tool for studying the effects of TCV on a protracted disease [98]. Furthermore, to conclude that a beneficial clinical response following TCV is due to the treatment, one would like to detect a suppressive immune response against the injected cells. In the animal models cellular responses have been investigated in lymphoid organs but not in peripheral blood nor in the lesion. Whether these responses are generalized or confined to lymphoid organs remains to be elucidated. However, the demonstration of such an immune response is hampered in humans since only peripheral blood is available to measure responses and because it is difficult to design in vitro experiments for monitoring without knowledge of disease-relevant antigens. Furthermore, the use of activated T cells as stimulator cells for in vitro experiments leads to high background responses in peripheral blood possibly masking relevant responses.

Networks. If TCV is to have a chance in man, greater understanding of the interactions between human T cells is needed. While new concepts on networks are emerging form animal studies, the validity of these views on man deserves more attention.

At the cellular level, T-T cell interactions have been described in humans. Anti-idiotypic T cells recognizing an unidentified determinant of an alloreactive T cell clone were capable of down-regulating the allo-response of the stimulator clone [99, 100]. A similar type of study has been done using as stimulator cell a T cell clone reactive with a viral antigen [101]. Down-regulation did not occur via a cytotoxic mechanism. Furthermore, responses against activated T cells have been measured in PBMC representing a variant of the autologous mixed lymphocyte reaction [102]. In addition, at least some T-T cell interactions were shown to be based on cellular interactions involving accessory molecules [103]. Given the multitude of such molecules involved and the complexity of cytokine interactions, responses of PBMC to activated T cells are difficult to dissect. Preliminary results in our laboratory suggest that both accessory cell independent [104] and dependent T-T cell interactions exist [105].

Concluding Remarks

The severe and intractable course of chronic human diseases such as RA, as well as the lack of satisfactory therapies currently available to treat

RA, warrant new therapeutic strategies to be developed. The concept that RA may be a T-cell-mediated autoimmune disease makes T cell vaccination a potential treatment for RA. Its potential clinical benefits need to be further explored and a number of methodological problems have to be solved before TCV can be fully appreciated. Some relate to the different approaches required for the study of TCV in man as compared to the animal studies while others relate to more fundamental questions such as the mechanism underlying TCV. In the coming decade, new techniques such as the polymerase chain reaction and in situ hybridization may provide valuable tools to resolve some of the issues mentioned in this review. In particular, cytokine profiles and TcR usage of early infiltrating T cells should be investigated to further delineate the role of T cells in the initiation and perpetuation of the inflammation. In parallel, the mechanism of TCV in animal studies should be further elucidated. Since of all therapies targeted at T cells TCV is the most specific, TCV as a potential treatment of human T-cell-mediated autoimmune disease deserves thorough scientific attention. History teaches that this may require undaunted efforts as it is recorded that when Benjamin Jesty inoculated his sons with cowpox his sceptical neighbors 'feared their metamorphosis into horned beasts' [106].

References

1 Ben-Nun, A.; Wekerle, H.; Cohen, I.R.: The rapid isolation of clonable antigen-specific T lymphocyte lines capable of mediating autoimmune encephalomyelitis. Eur. J. Immunol. *11:* 195–199 (1981).

2 Ben-Nun, A.; Cohen, I.R.: Experimental autoimmune encephalomyelitis mediated by T cell lines: Process of selection of lines and characterization of the T cells. J. Immunol. *129:* 303–308 (1982).

3 Sedgwick, J.; Brostoff, S.; Mason, D.: Experimental allergic encephalomyelitis in the absence of a classical delayed-type hypersensitivity reaction. J. Exp. Med. *165:* 1058–1075 (1987).

4 Naparstek, Y.; Holoshitz, J.; Eisenstein, S.; Reshef, T.; Rappaport, S.; Chemke, J.; Ben-Nun, A.; Cohen, I.R.: Effector T lymphocyte line cells migrate to the thymus and persist there. Nature *300:* 262–264 (1982).

5 Naparstek, Y.; Ben-Nun, A.; Holoshitz, J.; Reshef, T.; Frenkel, A.; Rosenberg, M.; Cohen, I.R.: T lymphocyte lines producing or vaccinating against autoimmune encephalomyelitis. Functional activation induces peanut agglutinin receptors and accumulation in the brain and thymus of line cells. Eur. J. Immunol. *13:* 418–423 (1983).

6 Sun, D.; Wekerle, H.: Ia-restricted encephalitogenic T lymphocytes mediating EAE lyse autoantigen-presenting astrocytes. Nature *320:* 70–72 (1986).

7 McCarran, R.M.; Kempski, O.; Spatz, M.; McFarlin, D.E.: Presentation of myelin basic protein by murine cerebral vascular endothelial cells. J. Immunol. *134:* 3100–3103 (1985).

8 Hinrichs, D.J.; Wegmann, K.W.; Dietsch, G.N.: Transfer of experimental allergic encephalomyelitis to bone marrow chimeras. Endothelial cells are not a restricting element. J. Exp.Med. *166:* 1906–1911 (1987).

9 Powell, M.B.; Mitchell, D.; Lederman, J.; Buckmeier, J.; Zamvil, S.S.; Graham, M.; Ruddle, N.H.: Lymphotoxin and tumor necrosis factor-alpha production by myelin basic protein-specific T cell clones correlates with encephalitogenicity. Int. Immunol. *2:* 539–544 (1990).

10 Higgins, P.J.; Weiner, H.L.: Suppression of experimental autoimmune encephalomyelitis by oral administration of myelin basic protein and its fragments. J. Immunol. *140:* 440–445 (1988).

11 Ellerman, K.E.; Powers, S.M.; Brostoff, S.W.: A suppressor T-lymphocyte cell line for autoimmune encephalomyelitis. Nature *331:* 265–267 (1988).

12 Ben-Nun, A.; Cohen, I.R.: Vaccination against autoimmune encephalomyelitis (EAE): Attenuated autoimmune T lymphocytes confer resistance to induction of active EAE but not to EAE mediated by the intact T lymphocyte line. Eur. J. Immunol. *11:* 949–952 (1981).

13 Ben-Nun, A.; Wekerle, H.; Cohen, I.R.: Vaccination against autoimmune encephalomyelitis with T lymphocyte line cells reactive against myelin basic protein. Nature *292:* 60–61 (1981).

14 Lider, O.; Shinitzky, M.; Cohen, I.R.: Vaccination against experimental autoimmune diseases using T lymphocytes treated with hydrostatic pressure. Ann. NY. Acad. Sci. *475:* 267–273 (1986).

15 Holoshitz, J.; Frenkel, A.; Ben-Nun, A.; Cohen, I.R.: Autoimmune encephalomyelitis mediated or prevented by T lymphocyte lines directed against diverse antigenic determinants of myelin basic protein. Vaccination is determinant specific. J. Immunol. *131:* 2810–2813 (1983).

16 Lider, O.; Reshef, T.; Beraud, E.; Ben-Nun, A.; Cohen, I.R.: Anti-idiotypic network induced by T cell vaccination against experimental autoimmune encephalomyelitis. Science *239:* 181–183 (1988).

17 Sun, D.; Qin, Y.; Chluba, J.; Epplen, J.; Wekerle, H.: Suppression of experimentally induced autoimmune encephalomyelitis by cytolytic T-T cell interactions. Nature *332:* 843–845 (1988).

18 Lider, O.; Beraud, E.; Reshef, T.; Friedman, A.; Cohen, I.R.: Vaccination against experimental autoimmune encephalomyelitis using a subencephalitogenic dose of autoimmune effector cells. 2. Induction of a protective anti-idiotypic response. J. Autoimmun. *2:* 87–99 (1989).

19 Driscoll, B.F.; Kies, M.W.; Alvord, E.C, Jr.: Suppression of acute experimental allergic encephalomyelitis in guinea pigs by prior transfer of suboptimal numbers of EAE-effector cells: Induction of chronic EAE in whole tissue-sensitized guinea pigs. J. Immunol. *128:* 635–638 (1982).

20 Ben-Nun, A.; Eisenstein, S.; Cohen, I.R.: Experimental autoimmune encephalomyelitis in genetically resistant rats: PVG rats resist active induction of EAE but are susceptible to and can generate EAE effector T cell lines. J. Immunol. *129:* 918–919 (1982).

21 Smilek, D.E.; Lock, C.B.; McDevitt, H.O.: Antigen recognition and peptide-mediated immunotherapy in autoimmune disease. Immunol. Rev. *118:* 37–71 (1990).

22 Kumar, V.; Kono, D.H.; Urban, J.L.; Hood, L.: The T-cell receptor repertoire and autoimmune diseases. Annu. Rev. Immunol. *7:* 657–682 (1989).

23 Howell, M.D.; Winters, S.T.; Olee, T.; Powell, H.C.; Carlo, D.J.; Brostoff, S.W.: Vaccination against experimental allergic encephalomyelitis with T cell receptor peptides. Science *246:* 668–670 (1989).

24 Vandenbark, A.A.; Hashim, G.; Offner, H.: Immunization with a synthetic T-cell receptor V-region peptide protects against experimental autoimmune encephalomyelitis. Nature *341:* 541–544 (1989).

25 Lohse, A.W.; Mor, F.; Karin, N.; Cohen, I.R.: Control of experimental autoimmune encephalomyelitis by T cells responding to activated T cells. Science *244:* 820–822 (1989).

26 Offner, H.; Jones, R.; Celnik, B.; Vandenbark, A.A.: Lymphocyte vaccination against experimental autoimmune encephalomyelitis: Evaluation of vaccination protocols. J. Neuroimmunol. *21:* 13–22 (1989).

27 Desquenne-Clark, L.; Esch, T.R.; Otvos, L.; Heber-Katz, E.: T-cell receptor peptide immunization leads to enhanced and chronic experimental allergic encephalomyelitis. Proc. Natl Acad. Sci. USA *88:* 7219–7223 (1991).

28 Lider, O.; Miller, A.; Miron, S.; Hershkovic, R.; Weiner, H.L.; Zhang, X.; Heber-Katz, E.: Nonencephalitogenic CD4– CD8– Vα2Vβ8.2+ anti-myelin basic protein rat T lymphocytes inhibit disease induction. J. Immunol. *147:* 1208–1213 (1991).

29 Beraud, E.; Lider, O.; Baharav, E.; Reshef, T.; Cohen, I.R.: Vaccination against experimental autoimmune encephalomyelitis using a subencephalitogenic dose of auto-immune effector cells. 1. Characteristics of vaccination. J. Autoimmun. *2:* 75–86 (1989).

30 Cohen, I.R.; Ben-Nun, A.; Holoshitz, J.; Maron, R.; Zerubavel, R.: Vaccination against autoimmune disease with lines of autoimmune T lymphocytes. Immunol. Today *4:* 227–230 (1983).

31 Cohen, I.R.: Autoimmunity to chaperonins in the pathogenesis of arthritis and diabetes. Annu. Rev. Immumol. *9:* 567–589 (1991).

32 Holoshitz, J.; Matitiau, A.; Cohen, I.R.: Arthritis induced in rats by cloned T lympho-cytes responsive to mycobacteria but not to collagen type II. J. Clin. Invest. *73:* 211–215 (1984).

33 Van Eden, W.; Holoshitz, J.; Nevo, Z.; Frenkel, A.; Klajman, A.; Cohen, I.R.: Arthritis induced by a T-lymphocyte clone that responds to *Mycobacterium tuberculosis* and to cartilage proteoglycans. Proc. Natl Acad. Sci. USA *82:* 5117–5120 (1985).

34 Billingham, M.E.J.; Carney, S.; Butler, R.; Colston, M.J.: A mycobacterial 65 kD heat shock protein induces antigen-specific suppression of adjuvant arthritis, but is not itself arthritogenic. J. Exp. Med. *171:* 339–344 (1990).

35 Yang, X. D.; Gasser, J.; Riniker, B.; Feige, U.: Treatment of adjuvant arthritis in rats: Vaccination potential of a synthetic nonapeptide from the 65 kD heat shock protein of mycobacteria. J. Autoimmun. *3:* 11–23 (1990).

36 Cohen, I.R.; Holoshitz, J.; van Eden, W.; Frenkel, A.: T lymphocyte clones illuminate pathogenesis and affect therapy of experimental arthritis. Arthritis Rheum. *28:* 841–845 (1985).

37 Yang, X.D.; Feige, U.: The 65 kD heat shock protein: A key molecule mediating the development of autoimmune arthritis? Autoimmunity *9:* 83–88 (1991).

38 Holoshitz, J.; Naparstek, Y.; Ben-Nun, A.; Cohen, I.R.: Lines of T lymphocytes induce or vaccinate against autoimmune arthritis. Science *219:* 56–58 (1982).

39 Holoshitz, J.; Matitiau, A.; Cohen, I.R: Role of the thymus in induction and transfer of vaccination against adjuvant arthritis with a T lymphocyte line in rats. J. Clin. Invest. *75:* 472–477 (1985).

40 Cohen, I.R.: Regulation of autoimmune disease. Physiological and therapeutic. Im-munol. Rev. *94:* 5–19 (1986).

41 Lider, O.; Karin, N.; Shinitzky, M.; Cohen, I.R.: Therapeutic vaccination against adjuvant arthritis using autoimmune T cells treated with hydrostatic pressure. Proc. Natl Acad. Sci. USA *84:* 4577–4580 (1987).

42 Cohen, I.R.: Physiological basis of T-cell vaccination against autoimmune disease. Cold
    Spring Harb. Symp. Quant. Biol. *154:* 879–884 (1989).
43 Elias, D.; Markovits, D.; Reshef, T.; van der Zee, R.; Cohen, I.R.: Induction and
    therapy of autoimmune diabetes in the non-obese diabetic (NOD/Lt)mouse by a 65 kD
    heat shock protein. Proc. Natl Acad. USA *87:* 1576–1580 (1989).
44 Elias, D.; Reshef, T.; Birk, O.S.; van der Zee, R.; Walker, M.D.; Cohen, I.R.:
    Vaccination against autoimmune mouse diabetes with a T-cell epitope of the human
    65-kD heat shock protein. Proc. Natl Acad. Sci. USA *88:* 3088–3091 (1991).
45 Jerne, N.K.: Towards a network theory of the immune system. Ann. Immunol. (Paris)
    125C: 373–389 (1974).
46 Cohen, I.R.; Atlan, H.: Network regulation of autoimmunity: An automaton model. J.
    Autoimmun. *2:* 613–625 (1989).
47 Cohen, I.R.: The immunological homunculus and autoimmune disease. Mol. Autoim-
    mun., in press.
48 Ferris, D.K.; Harel-Bellan, A.; Morimoto, R.I.; Welch, W.J.; Farrar, W.L.: Mitogen
    and lymphokine stimulation of heat shock proteins in T lymphocytes. Proc. Natl. Acad.
    Sci. USA *85:* 3850–3854 (1988).
49 Broeren, C.P.; Verjans, G.M.; Van Eden, W.; Kusters, J.G.; Lenstra, J.A.; Logtenberg,
    T.: Conserved nucleotide sequences at the 5′ end of T cell receptor variable genes
    facilitate polymerase chain reaction amplification. Eur. J. Immunol. *21:* 569–575 (1991).
50 Jacob, C.O.; Holoshitz, J.; Van der Meide, P.; Strober, S.; McDevitt, H.O: Heterogeneous
    effects of interferon-gamma in adjuvant arthritis. J. Immunol. *141:* 1500–1505 (1989).
51 Peltz, G.: A role for CD4+ T-cell subsets producing a selective pattern of lymphokines
    in the pathogenesis of human chronic inflammatory and allergic diseases. Immunol.
    Rev. *123:* 23–25 (1991).
52 Kahn, C.R.; McIntosh, K.R.; Drachman, D.B.: T-cell vaccination in experimental
    myasthenia gravis: A double-edged sword. J. Autoimmun. *3:* 659–669 (1990).
53 Kakimoto, K.; Katsuki, M.; Hirofuji, T.; Iwata, H.; Koga, T.: Isolation of T cell line cap-
    able of protecting mice against collagen-induced arthritis. J. Immunol. *140:* 78–83 (1988).
54 Pereira, P.; Bandeira, A.; Coutinho, A.: V-region connectivity in T cell repertoires.
    Annu. Rev. Immunol. *7:* 209–249 (1989).
55 Wordsworth, B.P.; Lanchbury, J.S.S.; Sakkas, L.I.; Welsh, K.I.; Panayi, G.S.; Bell, J.I.:
    HLA-DR4 subtype frequency in rheumatoid arthritis indicate that DRB1 is the major
    susceptibility locus within the human leucocyte antigen class II region. Proc. Natl Acad.
    Sci. USA *86:* 10049–10053 (1989).
56 Duke, O.; Panayi, G.S.; Janossy, G.; Poulter, L.W.: An immunohistochemical analysis
    of lymphocyte subpopulations and their microenvironment in the synovial membranes
    of patients with rheumatoid arthritis using monoclonal antibodies. Clin. Exp. Immunol.
    49: 22–30 (1982).
57 Kurosaka, M.; Ziff, M.: Immunoelectron microscopic study of the distribution of T cell
    subsets in rheumatoid synovium. J. Exp. Med. *158:* 1191–1210 (1983).
58 Waalen, K.; Førre, O.; Teigland, J.; Natvig, J.B.: Human rheumatoid synovial and
    normal blood dendritic cells as antigen presenting cells: Comparison with autologous
    monocytes. Clin. Exp. Immunol. *70:* 1–9 (1987).
59 Poulter, L.W.; Janossy, G.: The involvement of dendritic cells in chronic inflammatory
    disease. Scand. J. Immunol. *21:* 401–407 (1985).
60 Laffon, A.; Sánchez-Madrid, F.; Ortiz de Landázuri, M.; Jiménez Cuesta, A.; Ariza, A.;
    Ossorio, C.; Sabando, P.: Very late activation antigen on synovial fluid T cells from
    patients with rheumatoid arthritis and other rheumatic diseases. Arthritis Rheum. *32:*
    386–392 (1988).

61 Cush, J.J.; Lipsky, P.E.: Phenotypic analysis of synovial tissue and peripheral blood lymphocytes isolated from patients with rheumatoid arthritis. Arthritis Rheum. *31:* 1230–1238 (1988).

62 Janossy, G.; Panayi, G.S.; Duke, O.; Bofill, M.; Poulter, L.W.; Goldstein, G.: Rheumatoid arthritis: A disease of T-lymphocyte/macrophage immunoregulation. Lancet *ii:* 839–842 (1981).

63 Stuart, J.M.; Postlethwaite, A.E.; Townes, A.S.; Kang, A.H.: Cell-mediated immunity to collagen and collagen α chains in rheumatoid arthritis and other rheumatic diseases. Am. J. Med. *69:* 13–18 (1980).

64 Sigal, L.H.; Johnston, S.L.; Philips, P.E.: Cellular immune response to cartilage components in rheumatoid arthritis and osteoarthritis: A review and report of a study. Clin. Exp. Rheumatol. *82:* 59–66 (1988).

65 Alsalameh, S.; Mollenhauer, J.; Hain, N.; Stock, K.-P.; Kalden, J.R.; Burmester, G.R.: Cellular immune response toward human articular chondrocytes. Arthritis Rheum. *33:* 1477–1486 (1990).

66 Res, P.; Thole, J.; De Vries, R.: Heat-shock proteins and autoimmunity in humans. Springer Semin. Immunopathol. *13:* 81–97 (1991).

67 Devereux, D.; O'Hehir, R.E.; McGuire, J.; van Schooten, W.C.A.; Lamb, J.: HLA-DR4Dw4-restricted T cell recognition of self antigen(s) in the rheumatoid synovial compartment. Int. Immunol. *3:* 635–640 (1991).

68 Stamenkovic, I.; Stegagno, M.; Wright, K.A.; Krane, S.M.; Amento, E.P.; Colvin, R.B.; Duquesnoy, R.J.; Kurnick, J.T.: Clonal dominance among T lymphocyte infiltrates in arthritis. Proc. Natl Acad. Sci. USA *85:* 1179–1183 (1988).

69 Duby, A.D.; Sinclair, A.K.; Osborne-Lawrence, S.L.; Zeldes, W.; Kan, L.; Fox, D.A.: Clonal heterogeneity of synovial fluid T lymphocyte from patients with rheumatoid arthritis. Proc. Natl Acad. Sci. USA *86:* 6206–6210 (1989).

70 Savill, C.M.; Delves, P.J.; Kioussis, D.; Walker, P.; Lydyard, P.M.; Colaco, B.; Shipley, M.; Roitt, I.M.: A minority of patients with rheumatoid arthritis show a dominant rearrangement of T-cell receptor *β* chain genes in synovial lymphocytes. Scand. J. Immunol. *25:* 629–635 (1987).

71 Keystone, E.C.; Minden, M.; Klock, R.; Poplonski, L.; Zalcberg, J.; Takadera, T.; Mak, T.W.: Structure of T cell antigen receptor *β* chain in synovial fluid cells from patients with rheumatoid arthritis. Arthritis Rheum. *31:* 1555–1557 (1988).

72 Van Laar, J.M.; Miltenburg, A.M.M.; Verdonk, M.J.A.; Daha, M.R.; de Vries, R.R.P.; van den Elsen, P.J.; Breedveld, F.C.: Lack of T cell oligoclonality in enzyme-digested synovial fluid in most patients with rheumatoid arthritis. Clin. Exp. Immunol. *83:* 352–358 (1991).

73 Cooper, S.M.; Dier, D.L.; Roessner, K.D.; Budd, R.C.; Nicklas, J.A: Diversity of rheumatoid synovial tissue T cells by T cell receptor analysis. Oligoclonal expansion in interleukin-2-responsive cells. Arthritis Rheum. *34:* 537–546 (1991).

74 Van Laar, J.M.; Miltenburg, A.M.M.; Verdonk, M.J.A.; Daha, M.R.; de Vries, R.R.P.; van den Elsen, P.J.; Breedveld, F.C.: T-cell receptor *β*-chain gene rearrangements of T-cell populations expanded from multiple sites of synovial tissue obtained from a patient with rheumatoid arthritis. Scand. J. Immunol. *35:* 187–194 (1992).

75 Sottini, A.; Imberti, L.; Gorla, R.; Cattaneo, R.; Primi, D.: Restricted expression of T cell receptor V*β* but not V*α* genes in rheumatoid arthritis. Eur. J. Immunol. *21:* 461–466 (1991).

76 Pluschke, G.; Ricken, G.; Taube, H.; Kroninger, S.; Melchers, I.; Peter H.H.; Eichmann, K.; Krawinkel, U.: Biased T cell receptor V*α* region repertoire in the synovial fluid of rheumatoid arthritis patients. Eur. J. Immunol. *21:* 2749–2754 (1991).

77   Williams, W.V.; Fang, Q.; Demarco, D.; VonFeldt, J.; Zurier, R.B.; Wiener, D.B.: Restricted heterogeneity of T cell receptor transcripts in rheumatoid synovium. J. Clin. Invest. *90:* 326–333 (1992).

78   Uematsu, Y.; Wege, H.; Straus, A.; Ott, M.; Bannwarth, W.; Lanchbury, J.S.; Panayi, G.S.; Steinmetz, M.: The T-cell receptor repertoire in the synovial fluid of a patient with rheumatoid arthritis is polyclonal. Proc. Natl Acad. Sci. USA *88:* 8534–8538 (1991).

79   Brennan, F.M.; Allard, S.; Londei, M.; Savill, C.; Boylston, A.; Carrel, S.; Maini, R.N.; Feldmann, M.: Heterogeneity of T cell receptor idiotypes in rheumatoid arthritis. Clin. Exp. Immunol. *73:* 417–423 (1988).

80   Gulwani-Akolkar, B.; Posnett, D.N.; Janson, C.H.; Grunewald, J.; Wigzell, H.; Akolkar, P.; Gregersen, P.K.; Silver, J.: T cell receptor V-segment frequencies in peripheral blood T cells correlate with human leucocyte antigen type. J. Exp. Med. *174:* 1139–1146 (1991).

81   Bucht, A.; Oksenberg, J.R.; Lindblad, S.; Gronberg, A.; Steinman, L.; Klareskog, L.: Characterization of T-cell receptor alpha/beta repertoire in synovial tissue from different temporal phases of rheumatoid arthritis. Scand. J. Immunol. *35:* 159–165 (1992).

82   Howell, M.D.; Dively, J.P.; Lundeen, K.A.; Esty, A.; Winters, S.T.; Carlo, D.J.; Brostoff, S.W.: Limited T-cell receptor beta-chain heterogeneity among interleukin-2 receptor-positive synovial T cells suggests a role for superantigen in rheumatoid arthritis. Proc. Natl Acad. Sci. USA *88:* 10921–10925 (1991).

83   White, J.; Herman, A.; Pullen, A.M.; Kubo, R.; Kappler, J.W.; Marrack, P.C.: The V beta specific superantigen staphylococcal enterotoxin B: Stimulation of mature T cells and clonal deletion in neonatal mice. Cell *56:* 27–35 (1989).

84   Paulus, H.E.; Machleder, H.I.; Levine, S.; Yu, D.T.Y.; MacDonald, N.S.: Lymphocyte involvement in rheumatoid arthritis: Studies during thoracic duct drainage. Arthritis Rheum. *20:* 1249–1262 (1977).

85   Kotzin, B.L.; Strober, S.; Engleman, E.G.; Calin, A.; Hoppe, R.T.; Kansas, G.S.; Terrell, C.P.; Kaplan, H.S.: Treatment of intractable rheumatoid arthritis with total lymphoid irradiation. N. Engl. J. Med. *305:* 969–976 (1981).

86   Trentham, D.E.; Belli, J.A.; Anderson, R.J.; Buckley, J.A.; Goetzl, E.J.; David, J.R.; Austen, K.F.: Clinical and immunological effects of fractionated total lymphoid irradiation in refractory rheumatoid arthritis. N. Engl. J. Med. *305:* 976–982 (1981).

87   Karsh, J.; Klippel, S.H.; Plotz, P.H.; Decker, J.L.; Wright, D.R.; Flye, M.W.: Lymphapheresis in rheumatoid arthritis: A randomized trial. Arthritis Rheum. *24:* 867–873 (1981).

88   Van Laar, J.M.; Miltenburg, A.M.M.; Verdonk, M.J.A.; Leow, A.; Elferink, B.G.; Daha, M.R.; Cohen, I.R.; de Vries, R.R.P.; Breedveld, F.C.: Effects of inoculation with attenuated autologous T cells in patients with rheumatoid arthritis. J. Autoimmun. *6:* 159–167 (1993).

89   Mor, F.; Lohse, A.W.; Karin, N.; Cohen, I.R.: Clinical modeling of T cell vaccination against autoimmune disease in rats. Selection of antigen-specific T cells using a mitogen. J. Clin. Invest. *85:* 1594–1598 (1990).

90   Lotz, M.; Tsouvas, C.D.; Robinson, C.A.; Dinarello, C.A.; Carson, D.A.; Vaughan, J.H.: Basis for defective response of rheumatoid synovial fluid lymphocytes to anti-CD3 (T3) antibodies. J. Clin. Invest. *78:* 713–721 (1986).

91   Kitas, G.D.; Salmon, M.; Farr, M.; Gaston, J.S.H.; Bacon, P.A.: Deficient interleukin-2 production in rheumatoid arthritis: Association with active disease and systemic complications. Clin. Exp. Immunol. *73:* 242–249 (1988).

92   Res, P.C.M.; Schaar, C.G.; Breedveld, F.C.; van Eden, W.; van Embden, J.D.A.; Cohen, I.R.; de Vries, R.R.P.: Synovial fluid T cell reactivity against 65 kD heat shock protein of mycobacteria in early chronic arthritis. Lancet *ii:* 478–480 (1988).

93 Pitzalis, C.; Kingsley, G.H.; Haskard, D.; Panayi, G.S.: The preferential accumulation of helper-inducer T lymphocytes in inflammatory lesions: Evidence for regulation by selective endothelial and homotypic adhesion. Eur. J. Immunol. *18:* 1397–1404 (1988).

94 Pitzalis, C.; Kingsley, G.H.; Covelli, M.; Meliconi, R.; Markey, A.; Panayi, G.S.: Selective migration of the human helper-inducer memory T cell subset: Confirmation by in vivo cellular kinetic studies. Eur. J. Immunol. *21:* 369–376 (1991).

95 Lohse, A.W.; Bakker, N.P.M.; Hermann, E.; Poralla, T.; Jonker, M.; Meyer zum Buschenfelde, K.H.: Induction of an anti-vaccine response by T cell vaccination in primates and humans. J. Autoimmun. *6:* 121–130 (1993).

96 Symons, J.A.; McCulloch, J.F.; Wood, N.C.; Duff, G.W.: Soluble CD4 in patients with rheumatoid arthritis and osteoarthritis. Clin. Immunol. Immunopathol. *60:* 72–82 (1991).

97 Symons, J.A.; Wood, N.C.; Di Giovine, F.S.; Duff, G.W.: Soluble CD8 in patients with rheumatic diseases. Clin. Exp. Immunol. *80:* 354–359 (1990).

98 Zamvil, S.S.; Nelson, P.; Trotter, J.; Mitchell, D.; Knobler, R.; Fritz, R.; Steinman, L.: T-cell clones specific for myelin basic protein induce chronic relapsing paralysis and demyelination. Nature *317:* 355–358 (1985).

99 Bensussan, A.; Meuer, S.C.; Schlossman, S.F.; Reinherz, E.L.: Delineation of an immunoregulatory amplifier population recognizing autologous Ia molecules. Analysis with human T cell clones. J. Exp. Med. *59:* 559–576 (1984).

100 Mohagheghpour, N.; Damle, N.K.; Takada, S.; Engleman, E.G.: Generation of antigen receptor-specific suppressor T cell clones in man. J. Exp. Med. *164:* 950–955 (1984).

101 Lamb, J.R.; Feldmann, M.: A human suppressor T cell clone which recognizes an autologous helper T cell clone. Nature *300:* 456–458 (1982).

102 Ditzian-Kadanoff, R.; Parks, L.; Evavold, B.; Quintans, J.; Swartz, T.J.: Stimulation of peripheral blood T cells by an activated T-cell line: A novel human autologous T-T lymphocyte reaction. Scand. J. Immunol. *34:* 713–719 (1991).

103 Brod, S.A.; Purvee, M.; Benjamin, D.; Hafler, D.A.: T-T interactions are mediated by adhesion molecules. Eur. J. Immunol. *20:* 2259–2268 (1990).

104 Van Laar, J.M.; Miltenburg, A.M.M.; de Kuiper P.; Daha, M.R.; de Vries, R.R.P.; Breedveld, F.C.: Cellular interactions between synovial T cell clones from a patient with rheumatoid arthritis. Scand. J. Immunol., in press.

105 Van Laar, J.M.; Miltenburg, A.M.M.; Verdonk, M.J.A.; Leow, A.; Elferink, B.G.; Daha, M.R.; de Vries, R.R.P.; Breedveld, F.C.: Activated synovial T cell clones from a patient with rheumatoid arthritis induce proliferation of autologous peripheral blood derived T cells. Cell. Immunol. *146:* 71–79 (1993).

106 Haviland, A.: 'Small-pox'. Lancet, Sept 13, 1862.

107 Maron, R.; Zerubauel, R.; Friedman, A.; Cohen, I.R.: T lymphocyte line specific for thyroglobulin produces or vaccinates against autoimmune thyroiditis in mice. J. Immunol. *131:* 2316–2322 (1983).

Jacob M. van Laar, Department of Rheumatology,
University Hospital, NL-2300 Leiden (The Netherlands)

Granstein RD (ed): Mechanisms of Immune Regulation.
Chem Immunol. Basel, Karger, 1994, vol 58, pp 236–258

Mechanisms of Tolerance to Allografts

Richard S. Lee, Hugh Auchincloss, Jr.

Harvard Medical School and Transplantation Unit, Department of Surgery,
Massachusetts General Hospital, Boston, Mass., USA

Introduction

Immunologic tolerance is the long lasting, specific absence of an immune response. Maintenance of self-tolerance is a fundamental require-ment for the immune system, but in the field of clinical transplantation, the goal of achieving donor-specific tolerance has remained elusive. This review will summarize our understanding of the basic mechanisms of tolerance induction and will describe several strategies to achieve donor-specific tolerance.

Clonal Deletion

T Cell Clonal Deletion in the Thymus

Burnet [1] first suggested in his clonal selection theory that tolerance to self antigens could be achieved by clonal deletion of autoreactive lymphocytes. He proposed that maturing thymocytes expressing high affinity T cell receptors (TCR) for self antigens would be eliminated before reaching the periphery. Evidence supporting this hypothesis was presented by Kappler et al. [2] in 1987, who used antibodies recognizing $V\beta 17^+$ TCR to trace the elimination of such T cells in I-E$^+$ mice ($V\beta 17^+$ TCR T cells are reactive with I-E molecules). Further evidence demonstrating thymic clonal deletion has been obtained from studies using other TCR $V\beta$ elements [3–9]. More recently, TCR transgenic mice eliminate most of the cells expressing a self-reactive transgene [10, 11].

The thymus is the principle site of T cell clonal deletion [2] and deletion of autoreactive cells therefore does not occur in nude mice which lack a thymus [12]. Within the thymus, bone marrow-derived antigen-pre-senting cells (APCs), thymic epithelial cells, and thymocytes themselves have all been shown to contribute to tolerance induction. Bone marrow-derived APCs, found in the thymic medulla and corticomedullary junction, are very potent inducers of clonal deletion for cells reactive with class I and class II molecules [13–15]. Thymic epithelial cells may be only weakly tolerogenic for CD8$^+$ cells [16], but are very tolerogenic for CD4$^+$ cells [16–18]. Thymic epithelial cells can cause clonal deletion in some cases and clonal anergy in others [19–21]. Thymocytes may contribute to achieving tolerance to class I molecules [22]. Evidence indicates that central clonal deletion occurs through apoptosis of autoreactive clones [23–25].

Although thymic clonal deletion is probably the major mechanism for self tolerance [26], other mechanisms may also be required. For example, some self antigens may never be expressed in thymus [27, 28] and occasional autoreactive T cells may escape the thymic deletion process.

B Cell Clonal Deletion

The hypothesis that autoreactive B cells are clonally deleted was included in Burnet's [1] clonal selection theory and later modified by Lederberg [29]. Lederberg proposed that maturing B cells which engage a ligand through their surface receptors would be deleted in the bone marrow before reaching the periphery. The finding that treatment of mice from birth with anti-μ antibodies depletes all B cells is consistent with this hypothesis [30].

The most convincing studies showing B cell deletion have involved transgenic mice. Nemazee and Bürki [31] created transgenic mice producing an IgM antibody specific for class I K^b and K^k. They reported the complete absence of peripheral B cells expressing the transgene and no secretion of IgM antibody in K^b or K^k mice [32]. Goodnow and co-workers [33] have also demonstrated the clonal deletion of autoreactive B cell clones in double transgenic mice which express transgenes for both anitlysozyme Ig and a membrane-bound form of lysozyme. The precise mechanism producing B cell clonal deletion in bone marrow has not yet been determined. Maturation arrest, apoptosis, and veto-like killing have been suggested [34].

Clonal Anergy

T Cell Clonal Anergy

Clonal anergy is the functional inactivation of a cell, without elimination, following its encounter with antigen. T cell anergy results from antigen receptor stimulation in the absence of costimulatory signals. Anergic T cells are characterized by a deficiency in IL-2 secretion and they remain nonresponsive when restimulated with appropriate antigen in the presence of costimulators [35]. Clonal anergy may result from antigen presentation by epithelial cells in the thymus and by contact with other cell populations in the periphery [26]. Several laboratories have addressed the defects in IL-2 production on a molecular level. Defects in transcription of the IL-2 gene [36], perhaps due to deficiencies of several transcriptional factors [37], and a reduced surface expression of the p55 IL-2 receptor have been reported [38].

Bretscher and Cohn [39], in 1970, first proposed the two-signal model of lymphocyte activation as an alternative to Lederberg's [29] temporal model of lymphocyte activation. Discussing B cells, they suggested that antigen recognition by antigen receptor alone (signal one) results in anergy, while antigen recognition plus an undefined second signal results in activation.

More recently, this concept has been extended to T cells as well [40, 41]. Cleveland and Claman [42] showed that intravenous injection of haptenated splenocytes (signal one) induced tolerance, while the simultaneous administration of concanavalin A (signal two) induced activation. Subsequently, Jenkins and Schwartz have demonstrated that Th1 CD4$^+$ T cell clones, exposed to antigen-MHC complexes in planar membranes [35] or on chemically fixed APCs [43], are rendered anergic for at least 7 days. Anergy has also been induced by stimulation of T cells with 'nonprofessional' APCs such as islet cells [44], keratinocytes [45], and T cells themselves [46]. In these situations it seems likely that absence of costimulatory signals during stimulation is responsible for anergy induction.

The precise nature of the second signal is not yet known. Jenkins [47] has proposed that any stimulus promoting cell division can prevent anergy and he has suggested a model whereby the first signal leads to the formation of positive and negative regulators of IL-2 transcription while the second signal inhibits the negative regulators allowing for IL-2 production. Mueller et al. [48] have suggested that the second signal activates a costimulatory pathway, distinct from Ca^{2+} flux and protein kinase C activation, to induce IL-2 transcription. The absence of this biologic signal results in inhibition of IL-2 production and nonresponsiveness.

The second signal need not be provided by only a single molecule or constitute a single pathway and there may be heterogeneity among different T cells with respect to costimulatory requirements [49]. Furthermore, a costimulatory interaction can be difficult to distinguish from a merely physical, adhesive interaction. With these caveats in mind, considerable effort has been made to define the ligand producing the second signal. Roles for ICAM-1/LFA-1 [50], VCAM-1/VLA-4 [50], fibronectin/VLA-4/5 [51], and CD2/LFA-3 [52] have been suggested but may represent the binding effect of these interactions. Janeway and co-workers [53] have demonstrated that heat-stable antigen can be a costimulatory molecule for CD4$^+$ T cells. However, the current leading candidate for the primary costimulator interaction is that between B7 and CD28.

B7 is an activation antigen found on B cells, monocytes, dendritic cells, and on 72-hour cultured Langerhans' cells [54, 55]. CD28 is found on nearly all peripheral T cells in the mouse and is up-regulated upon activation [56]. CD28 and its homologue CTLA-4 bind to B7 [57–61]. The interaction of B7 with CD28 mediates T-B cell adhesion [57] and costimulates CD4$^+$ and CD8$^+$ T cell proliferation [62], IL-2 production [63, 64], T cell alloactivation [65], and CD2- or CD3-dependent cell-mediated cytotoxicity [66]. Blocking this interaction with antibodies to B7 or CD28 inhibits proliferative responses to alloantigen or mitogens [63, 65]. Bluestone and co-workers [67] showed in vivo that antibody to B7 can prevent rejection of human pancreatic islet cells by mice and induce long-term, donor-specific tolerance. Linsley et al. [68] suppressed T cell-dependent antibody responses in mice treated with anti-B7 antibody. On the other hand, some evidence has been reported suggesting that other costimulatory receptors may cooperate during T cell activation. For example, blocking either B7 alone or heat-stable antigen alone has been shown to result in >90% inhibition of CD4$^+$ T cell proliferation to anti-CD3 antibody, while blocking both costimulators together was synergistic [69].

The CD28/B7 interaction may explain the findings of several groups that small, resting B cells, are poor APCs [70–72] and can induce anergy [73, 74], since B7 is weakly expressed on resting B cells but elevated on activated B cells [65, 74, 75]. On the other hand, Fuchs and Matzinger [76] have argued that both resting and activated B cells induce anergy in native T cells while activated B cells can stimulate primed T cells.

Anergic T cells can regain their reactivity, especially in the presence of IL-2. High dose IL-2 can reverse anergy in human T cell clones [77] and it can reactivate self-reactive T cells in neonatally thymectomized mice [78]. In addition, in vivo treatment of athymic BALB/c (Mls-1^b, I-E$^+$) nu/nu mice with IL-2 can activate self-reactive Vβ3$^+$ and Vβ11$^+$ T cells, eventually leading to autoimmunity [79]. The reversibility of anergy has also been shown by Ramsdell and Fowlkes [80] who transferred anergic T cells to an animal or to an in vitro culture system lacking the tolerizing antigen and restored responsiveness to the appropriate antigen.

Anergy to Antigens Expressed Extrathymically in Transgenic Mice

Miller and co-workers [81–83] have demonstrated that peripheral T cell tolerance to class I antigens expressed in selected tissues in transgenic mice can occur through a nondeletional mechanism in addition to a deletional one. They directed class I K^b expression to pancreatic β cells or to hepatocytes and found reduced peripheral anti-K^b cytotoxic T lymphocyte (CTL) activity that was restored after in vitro culture with IL-2. Forman and co-workers [84] have found similar results in mice expressing the nonclassical class I gene, Q10, in hepatocytes. Several reports have also shown evidence for T cell anergy to I-E antigens expressed on pancreatic β cells in transgenic mice. While reduced proliferative responses to I-E were demonstrated in vitro [85, 86], Vβ17a$^+$ T cells, which are reactive to I-E, were not deleted. Using double transgenic mice that express H-2K^b in pancreatic β cells and anti-K^b TCR, Miller and co-workers [87] have also demonstrated that peripheral T cell tolerance can be maintained without down-regulation of TCR or CD8. This report differs from the findings of Arnold and co-workers [88] who demonstrated that peripheral tolerance to K^b expressed under the control of glial fibrillary acidic protein promoter, which restricts expression to cells of neuroectodemal origin, is achieved through down-regulation of anti-K^b TCR and CD8 in double transgenic mice. Anergy has also been reported in systems where Mls antigens or I-E antigens were expressed on thymic cortical epithelial cells. Here, T cells expressing Vβ genes that should be reactive to Mls or I-E were nonresponsive in culture unless exogenous IL-2 was added [12, 89, 90].

B Cell Clonal Anergy

Nossal and Pike [91] have examined the induction of B cell anergy. They observed the persistence of antigen-binding B cells in tolerant mice which were nonresponsive to the antigen or mitogens. Similar results suggesting a nondeletional mechanism of B cell tolerance were found in studies using fluorescein or anti-Ig as tolerogens [92, 93]. In concordance with Bretscher and Cohn's [39] two-signal model of activation, Nossal [94] suggested that anergy induction in B cells results from binding of surface antigen receptors in the absence of second signals. He noted that the susceptibility to negative signaling decreases as the B cell matures from the pre-B cell to the immature B cell to the mature B cell. A peripheral tolerizing mechanism for B cells may be especially valuable since self-reactive B cells may develop during VDJ recombination or Ig hypermutation [95]. Moreover, autoreactive, low-affinity CD5$^+$ B cells have been detected in the periphery [96, 97].

Further evidence for B cell anergy has been provided from studies by Goodnow et al. [98] using lysozyme/antilysozyme double-transgenic mice. They have found that when endogenous lysozyme is expressed in sufficient amounts, antilysozyme antibody secretion stops without any significant change in the frequency of lysozyme-specific B cells [99, 100]. They further demonstrated a 90–98% decrease in surface IgM expression in the tolerant double-transgenic B cells versus nontolerant B cells without any change in surface IgD expression [99]. Moreover, these B cells show deficiencies in differentiation into antibody-secreting cells. Transferring these B cells into nontransgenic mice for 10 days leads to recovery of surface IgM expression but not a spontaneous recovery of antibody response to antigen with T cell help. Eventually, in the absence of tolerizing antigen, the B cells do regain the ability to secrete antibody [101]. Goodnow et al. [102] proposed a model of B cell tolerance which involves antigen receptor down-regulation and desensitization following receptor binding to a threshold amount of tolerogen.

Suppression

Specific Suppression

Active suppression is a third potential mechanism to achieve tolerance. Suppression has been attributed primarily to T cells (Ts) but also to non-T cells. The first studies implicating the presence of suppressor T cells were reported by Gershon [103–105]. He showed that tolerance to sheep erythrocytes was 'infectious' as it could be transferred by T cells to recipient mice. Dorsch and Roser [106, 107] then reported evidence for a suppressor mechanism involved in neonatal transplantation tolerance to MHC antigens based on adoptive transfer studies. Evidence for CD8+ T cell suppressors has been reported by several groups studying lepromatous leprosy [108, 109]. The existence of antigen-specific suppressor cells has been reported in various transplantation models as well [110–113]. However, much difficulty has been encountered in trying to clone suppressor cells and no suppressor-specific surface marker has been identified.

Studies of suppressor cells have suggested that a variety of T cell phenotypes and mechanisms may be involved. For example, Th2 CD4+ T cells can inhibit Th1 proliferation through the action of IL-10 and thus act as a type of 'suppressor' [114–116]. In studies of unresponsiveness to haptens in mice, three types of T cells and three types of suppressive factors (TsF) have been reported [117, 118]. Ts1 cells are antigen-specific CD4+ T suppressor inducers. Ts2 cells are anti-idiotypic CD8+ or CD4+ cells, while Ts3 cells are non-MHC-restricted CD8+ suppressors. Current models of suppression portray these three Ts cells interacting in a cascade where each cell secretes soluble suppressive factors [119, 120]. Engleman and co-workers [121] have reported CD8+ cells, specific for idiotypes on CD4+ cells, which can suppress in vitro proliferative responses to antigen in a MHC-restricted manner. Koide and Engleman [122] have also shown that CD8+ CD28– Ts inhibit MLR responses by a noncytolytic mechanism.

Natural Suppressor (NS) Cells

NS cells mediate suppression in an antigen-nonspecific and non-MHC-restricted fashion [123–126]. They are distinct from natural killer (NK) cells [123]. NS cells may play a role in the suppression of graft-versus-host disease (GVHD) [123] and thus may be important for tolerance induction in the setting of bone marrow transplantation, especially after total lymphoid irradiation (TLI) [127]. NS cells are abundant in neonatal spleen and adult mice after TLI [125, 128]. Their effects are varied, as they can suppress response to mitogens [129], antibody responses [130], T cell cytotoxic and proliferative responses [123, 131], and IL-2 production [132]. Secreted soluble factors may mediate the action of NS cells [133].

Veto Cells

A veto cell is a cell that can produce inactivation of a T cell upon recognition by that T cell [134]. Veto activity has been found among spleen cells from nude mice [135], bone marrow-derived T cell colonies [136], precultured spleen cells [137], Thy-1$^-$ bone marrow cells [136], CD8$^+$ cytotoxic T cells [138], activated CD4$^+$ T cells [139], activated bone marrow cells [140], and NK- or LAK-like bone marrow cells [141]. A mechanism of veto activity, suggested by Sambhara and Miller [142], is that a T cell precursor undergoes apoptosis when the α-3 domain of its class I MHC is bound by the CD8 receptor of a veto cell. Possible suppression of CTL responses by the veto mechanism has been demonstrated in vivo in the cases of class I-specific CTLs [143] and minor H antigen-specific CTLs [144]. An in vivo veto phenomenon resulting in clonal deletion of CD4$^+$ T helper precursors that recognize class II MHC has been suggested in mice receiving an intravenous injection of semiallogeneic lymphocytes [145]. Veto activity may play a role in graft-versus-host tolerance [146], in bone marrow acceptance [140], and in tolerance induced by donor-specific transfusions [147].

Strategies to Achieve Tolerance to Allografts

The mechanisms of tolerance induction discussed above include deletion, anergy, and suppression. Strategies have been developed to try to recapitulate each of these mechanisms in the hopes of achieving donor-specific tolerance.

Bone Marrow Transplantation (BMT)

Allogeneic BMT represents a reliable means of achieving transplantation tolerance in animal models. The concept underlying BMT is to create a chimera with donor-derived lymphocytes educated in the host thymus and therefore displaying tolerance to host and donor antigens. In essence, BMT attempts to recapitulate nature's clonal deletion mechanism.

The requirements for BMT to be successful include (1) the elimination of the recipient's already mature immune response and (2) the provision of 'space'. Several components of the immune system resist allogeneic bone marrow engraftment. Host T cells can prevent bone marrow engraftment [148–150] and radioresistant T cells that can mediate rejection of bone marrow have been found in lethally irradiated mice [151]. In addition, Dennert and co-workers [152] have found evidence that a subpopulation of T cells that are NK-1.1$^+$ can cause acute rejection of bone marrow. Both athymic, T cell-deficient nude mice and SCID mice acutely reject marrow grafts and depletion of NK cells with ^{89}Sr [153] or anti-asialo-GM1 serum improves marrow acceptance [154]. These results suggest that NK cells can effectively mediate rejection of marrow grafts without any T or B cell contribution [155]. However, B cells can also contribute to marrow rejection, since natural antibodies that resist engraftment have been detected [156].

In addition to immune-mediated rejection, nonimmunologic factors may resist even syngeneic engraftment, a phenomenon referred to as lymphohematopoietic 'space' [157]. The concept of space may represent the physical capacity of the marrow microenvironment, which would be increased by eliminating host hematopoietic cells, or the availability of essential cytokines, which might also be increased by treatments affecting host T cells.

Several strategies have been used to overcome these barriers to bone marrow engraftment. Billingham et al. [158] introduced allogeneic cells into fetal mice in utero at a time when their own immune system was not yet mature. To achieve this result in adult animals requires ablation of the already mature immune system, often achieved by whole-body irradiation of the recipient [159].

Three major problems have been encountered in the use of BMT for transplantation tolerance. First the presence of mature T cells in the donor graft causes GVHD. While T cell depletion of donor bone marrow prevents GVHD, it is associated with an increase in graft failure [160]. This may be due to increased susceptibility to rejection by the host [161] or to loss of T cell-derived factors involved in lymphopoiesis [162]. Second, fully allogeneic bone marrow chimeras have been shown to be seriously immunocompromised. Donor T cells are positively selected by the host-type thymic epithelial cells and thus their repertoire is host-restricted [163]. However, the APCs in the periphery are derived from bone marrow cells and these are donor-type. Therefore, fully allogeneic chimeras have T cells which are poorly selected to recognize foreign pathogens in their own environment. To overcome this problem, T cell-depleted allogeneic marrow has been administered along with T cell-depleted syngeneic marrow to create a mixed allogeneic chimera. These mixed chimeras demonstrate donor-specific tolerance without GVHD and better immunocompetence than animals receiving allogeneic marrow alone [164–166]. Third, the technique of whole-body irradiation used to eliminate the mature immune system is too toxic for application in large animals except under dire circumstances. Therefore, other protocols that are immunosuppressive and myelotoxic, but not myeloblative, have been used to promote BMT. Since such regimens allow some host hemopoietic stem cells to survive, a mixed allogeneic chimerism results after infusion of donor cells. However, the persistence of host T cells which can reject the marrow grafts necessitates the infusion of large numbers of donor cells [167, 168]. Nonmyeloablative protocols include total lymphoid irradiation [169], sublethal whole-body irradiation [167], cyclophosphamide [170], and monoclonal antibodies directed against T cell subsets with [148, 171] and without [149, 172, 171] other regimens.

While thymic clonal deletion may represent the major mechanism for tolerance induction in bone marrow chimeras, other mechanisms, such as suppression and clonal anergy, have not been entirely ruled out. Peripheral mechanisms may be especially relevant in the setting of nonmyeloablative regimes. Suppression, veto, and peripheral clonal anergy have been suggested as possible mechanisms involved in tolerance induction after fractionated sublethal whole-body irradiation and BMT [173]. Antigen-specific, donor-type suppressor T cells have been implicated in achieving tolerance to skin grafts in mice treated with antilymphocyte serum and BMT [113]. When nondepleting anti-T cell monoclonal antibodies have been used to induce tolerance to minor histocompatibility antigens in the setting of BMT, clonal anergy may be involved in achieving tolerance [171, 174].

Intrathymic Injection

Since the concept underlying BMT is to achieve tolerance by thymic clonal deletion, a new approach is to inject donor cells directly into thymus. Since this strategy is directed only at new T cells, the mature immune system must be eliminated to allow the new T cells exposed to foreign antigens in the thymus to repopulate the periphery. Several early reports have demonstrated the effectiveness of intrathymic injection of foreign cells in tolerance induction [175–177]. Naji and co-workers [179] first reported successful use of this approach with islet transplants and tolerance to autoimmune diabetes has also been achieved in mice by intrathymic injection of islets. Significant prolongation of class I-disparate skin grafts has been demonstrated following intrathymic injection of donor-type splenocytes [180]. The particular advantage of this strategy in transplantation might be either that localization of donor antigen in the thymus (the site of clonal deletion) can be more efficiently achieved by direct injection to this site than by BMT or that the thymus is a 'privileged' site for transplantation.

Chimerism Following Organ Transplantation

Recently, there has been an interest in determining whether transplanted organs that have survived for long periods with exogenous immunosuppression might generate a state of microchimerism in the recipient. Immunohistochemistry and PCR analysis have been used to demonstrate the persistence of donor cells outside the transplanted organ in the lymph nodes, skin, blood, and other tissues [181]. This microchimerism may cause donor-specific tolerance in these recipients [182]. However, it is also possible that microchimerism is the result, not the cause, of long-term graft survival. No definitive study has been reported showing that microchimerism is predictive of either long-term graft survival or the ability to withdraw exogenous immunosuppression.

Peripheral Tolerance Induction by Intravenous Injection of Donor Cells

The most common clinical approach to transplantation tolerance has involved strategies which are directed at achieving clonal anergy. Improved renal graft survival in human recipients receiving multiple blood transfusions prior to renal transplantation has been noted repeatedly [183, 184]. The mechanisms for this benefit are probably multiple, but a specific down-regulation of the immune response to donor antigens is probably involved. Experimentally, intravenous injection of mice with allogeneic lymphoid cells can induce donor-specific tolerance as evidenced by reduced cytotoxic lymphocyte and Th responses [185, 143, 145, 147].

While anergy is probably the primary mechanism involved when donor cells are injected intravenously, some evidence has suggested that veto cells may contribute to tolerance induction. Recently, Sheng-Tanner and Miller [186] have reported a correlation between donor-specific tolerance after donor cell infusion and donor cell recirculation that is opposed by NK reactivity. A requirement for a shared HLA-DR antigen in the donor blood with the recipient has also been shown in some situations [187].

APC Depletion and Modification

Another strategy to induce tolerance by clonal anergy involves manipulating donor APCs. The elimination of donor APCs prior to engraftment has produced prolonged survival in some types of grafts in murine models [188–192]. However, efforts to achieve solid organ transplantation and cellular transplantation in larger animals by APC depletion have met little success. Although anergy may be involved in tolerance induction resulting from APC-depleted grafts, several recent studies have suggested that other mechanisms, including suppression, may play a role [193]. An alternative approach is to modify the function of the immunogenic donor APCs and thus convert them into tolerizing APCs. One example is the treatment of APCs with UV irradiation [194, 195] which appears to induce donor-specific immunosuppression in some cases.

Monoclonal Antibodies

Since APC depletion alone has not been successful in larger animals, additional strategies have been explored to induce anergy by blocking the 'second signal'. Monoclonal antibodies provide a powerful tool for this purpose. Waldmann and co-workers have reported tolerance induction to skin grafts in mice that differ in multiple minor antigens [196] or certain MHC antigens [197] after treatment with anti-CD4 and anti-CD8 antibodies. They have also recently reported induction of tolerance to MHC-incompatible heart allo- and xenografts using a brief treatment with anti-CD4 or anti-CD8 antibodies or both [198]. Anti-CD4 antibody treatment has also produced indefinite survival of islet [199] and heart allografts [200] and anti-CD4 treatment has produced clonal anergy to I-E^k islets in diabetic I-E$^-$ recipients [201]. The use of anti-CD4 antibodies to induce tolerance presumably involves elimination of adequate IL-2 production. Additional strategies to induce anergy through the use of monoclonal antibodies have employed antibodies against various accessory molecules. In mice, combination therapy with anti-ICAM-1 and anti-LFA-1 has achieved donor-specific tolerance to cardiac allografts [202]. In addition, anti-B7 and inhibitors of CD28 and CTLA-4 have been used to induce anergy in some circimstances. The aim is to block costimulatory signals.

Conclusion

The numerous strategies that are currently being investigated to induce donor-specific tolerance suggest that no regimen has achieved adequate clinical success. In general, techniques to achieve central tolerance require unacceptable levels of myeloablation. Strategies to induce peripheral tolerance have become the central focus for much of current research. However, while many of these strategies have proven successful in rodents, none has achieved equivalent success in larger animals. Conceptually, one major problem with all current peripheral tolerance strategies is that they cannot prevent the activation of recipient T cells via the indirect pathway, i.e. presentation of donor antigens in association with MHC antigens on the recipient APCs. While the cells of the graft may be able to tolerize recipient effector cells, the APCs of the host should be able to activate effector cells through indirect recognition of donor antigens. This persistent antigenic stimulation can only be permanently eliminated by techniques that induce central tolerance. Thus, it seems likely that the achievement of clinical transplantation tolerance will require a combination of central and peripheral tolerance-inducing strategies.

References

1 Burnet, F.M.: The Clonal Selection Theory of Acquired Immunity (Cambridge University Press, New York 1959).
2 Kappler, J.W.; Roehm, N.; Marrack, P.: T cell tolerance by clonal elimination in the thymus. Cell 49: 273–280 (1987).
3 Kappler, J.W.; Staerz, U.; White, J.; Marrack, P.C.: Self-tolerance eliminates T cells specific for Mls-modified products of the major histocompatibility complex. Nature 332: 35–40 (1988).
4 MacDonald, H.R.; Schneider, R.; Lees, R.K.; Howe, R.C.; Acha-Orbea, H.; Festenstein, H.; Zinkernagel, R.M.; Hengartner, H.: T-cell receptor V-beta use predicts reactivity and tolerance to Mlsa-encoded antigens. Nature 332: 40–45 (1988).
5 Pullen, A.M.; Marrack, P.; Kappler, J.W.: The T-cell repertoire is heavily influenced by tolerance to polymorphic self-antigens. Nature 335: 796–801 (1988).
6 Abe, R.; Vacchio, M.S.; Fox, B.; Hodes, R.J.: Preferential expression of the T-cell receptor Vβ3 gene by Mls c reactive T cells. Nature 335: 827–830 (1988).
7 Fry, A.M.; Matis, L.A.: Self-tolerance alters T-cell receptor expression in an antigen-specific MHC-restricted immune response. Nature 335: 830–832 (1988).
8 Bill, J.; Kanagawa, O.; Woodland, D.L.; Palmer, E.: The MHC molecule I-E is necessary but not sufficient for the clonal deletion of V-beta-11-bearing T cells. J. Exp. Med. 169: 1405–1419 (1989).
9 White, J.; Herman, A.; Pullen, A.M.; Kubo, R.; Kappler, J.W.; Marrack, P.: The V-beta-specific superantigen staphylococcal enterotoxin B: Simulation of mature T cells and clonal deletion in neonatal mice. Cell 56: 27–35 (1989).
10 Sha, W.C.; Nelson, C.A.; Newberry, R.D.; Kranz, D.M.; Russel, J.H.; Loh, D.Y.: Positive and negative selection of antigen receptor on T cells in transgenic mice. Nature 336: 73–76 (1988).
11 Von Boehmer, H.; Kisielow, P.: Self-nonself discrimination by T cell. Science 248: 1369–1373 (1990).
12 Fry, A.M.; Jones, L.A.; Kruisbeek, A.M.; Matis, L.A.: Thymic requirement for clonal deletion during T cell development. Science 246: 1044–1046 (1989).
13 Sprent, J.; Lo, D.; Gao, E.-K.; Ron, Y.: T cell selection in the thymus. Immunol. Rev. 101: 173–190 (1988).
14 Jenkinson, E.J.; Jhittay, P.; Kingston, R.; Owen J.J.T.: Studies of the role of the thymic environment in the induction of tolerance to MHC antigens. Transplantation 39: 331–333 (1985).

15 Inaba, M.; Inaba, K.; Hosono, M.; Kumamoto, T.; Ishida, T.; Muramatsu, S.; Masuda, T.; Ikehara, S.: Distinct mechanisms of neonatal tolerance induced by dendritic cells and thymic B cell. J. Exp. Med. *173:* 549 (1991).

16 Webb, S.R.; Sprent, J.: Tolerogenicity of thymic epithelium. Eur. J. Immunol. *20:* 2525–2528 (1990).

17 Gao, E.-K.; Lo, D.; Sprent, J.: Strong T cell tolerance in parent → F₁ bone marrow chimeras prepared with supralethal irradiation. Evidence for clonal deletion and anergy. J. Exp. Med. *171:* 1101–1121 (1990).

18 Roberts, A.R.; Sharron, S.O.; Singer, A.: Clonal deletion and clonal anergy in the thymus induced by cellular elements with different radiation sensitivities. J. Exp. Med. *171:* 935–940 (1990).

19 Ready, A.R.; Jenkinson, E.J.; Kingston, R.; Owen, J.J.T.: Successful transplantation across major histocompatibility barrier of deoxyguanosine-treated embryonic thymus expressing class II antigens. Nature *310:* 231 (1984).

20 Salaun, J.; Bandeira, A.; Khazaal, I.; Calman, F.; Coltey, M.; Coutinho, A.; LeDouarin, N.M.: Thymic epithelium tolerizes for histocompatibility antigens. Science *247:* 1471 (1990).

21 Sprent, J.; Gao, E.-K.; Webb, S.R.: T cell reactivity to MHC molecules: Immunity versus tolerance. Science *248:* 13757–1363 (1990).

22 Shimonkevitz, R.P.; Bevan, M.J.: Split tolerance induced by the intrathymic adoptive transfer of thymocyte stem cells. J. Exp. Med. *168:* 143–156 (1988).

23 Smith, C.A.; Williams, G.T.; Kingston, R.; Jenkinson, E.J.; Owen, J.J.T.: Antibodies to CD3/T-cell receptor complex induce death by apoptosis in immature T cells in thymic cultures. Nature *337:* 181–184 (1989).

24 Murphy, K.M.; Heimberger, A.B.; Loh, D.Y.: Induction by antigen of intrathymic apoptosis of CD4+ CD8+ TCRlo thymocytes in vivo. Science *250:* 1720–1723 (1990).

25 Swat, W.; Ignatowicz, L.; von Boehmer, H.; Kisielow, P.: Clonal deletion of immature CD4+8+ thymocytes in suspension culture by extra-thymic antigen-presenting cells. Nature *351:*150–153 (1990).

26 Ramsdell, F.; Fowlkes, B.J.: Clonal deletion versus clonal anergy: The role of the thymus in inducing self tolerance. Science *248:* 1342 (1990).

27 Ohashi, P.S.; Oehen, S.; Bürki, K.; Pricher, H.; Ohashi, C.T.; Odermatt, B.; Malissen, B.; Zinkernagel, R.M.; Hengartner, H.: Ablation of 'tolerance' and induction of diabetes by virus infection in viral antigen transgenic mice. Cell *65:* 305 (1991).

28 Miller, J.; Daitch, L.; Rath, S.; Selsing, E.: Tissue-specific expression of allogeneic class II MHC molecules induces neither tissue rejection nor clonal inactivation of alloreactive T cell. J. Immunol. *144:* 334 (1990).

29 Lederberg, J.: Genes and antibodies. Science *129:* 1649–1653 (1959).

30 Lawton, A.R.; Asofsky, R.; Hylton, M.B.; Copper, M.D.: Suppression of immunoglobulin class synthesis in mice. I. Effects of treatment with antibody to μ chain. J. Exp. Med. *135:* 277 (1972).

31 Nemazee, D.A.; Bürki, K.: Clonal deletion of B lymphocytes in a transgenic mouse bearing anti-MHC class I antibody genes. Nature *337:* 562–566 (1989).

32 Nemazee, D.; Burki, K.: Clonal deletion of autoreactive B lymphocytes in bone marrow chimeras. Proc. Natl Acad. Sci. USA *86:* 8039–8043 (1989).

33 Hartely, S.B.; Crosbie, J.; Brink, R.A.; Kantor, A.B.; Basten, A.; Goodnow, C.C.: Elimination from peripheral lymphoid tissues of self-reactive B lymphocytes recognizing membrane-bound antigens. Nature *353:* 765–769 (1991).

34 Goodnow, C.C.: Transgenic mice and analysis of B-cell tolerance. Annu. Rev. Immunol. *10:* 489–518 (1992).

35 Quill, H.; Schwartz, R.H.: Stimulation of normal inducer T cell clones with antigen presented by purified Ia molecules in planar lipid membranes: Specific induction of a long-lived state of proliferative nonresponsiveness. J. Immunol. *138:* 3704–3712 (1987).

36 Go, C.; Miller, J.: Differential induction of transcription factors that regulate the interleukin-2 gene during anergy induction and restimulation. J. Exp. Med. *175:* 1327–1336 (1992).

37 Kang, S.-M.; Berverly, B.; Tran, A.C.; Brorson, K.; Schwartz R.H.; Lenardo, M.J.: Transactivation by AP-1 is a molecular target of T cell clonal anergy. Science *257:* 1134–1138 (1992).

38 Dallman, M.J.; Shiho, D.; Page, T.M.; Wood, K.J.; Morris, P.J.: Peripheral tolerance to alloantigen results from altered regulation of the interleukin-2 pathway. J. Exp. Med. *173:* 79–87 (1991).

39 Bretscher, P.; Cohn, M.: A theory of self-nonself discrimination. Science *169:* 1042–1049 (1970).

40 Lafferty, K.J.; Woolnough, J.: The origin and the mechanism of allograft rejection. Immunol. Rev. *35:* 231–262 (1977).

41 Lafferty, K.J.; Prowse, S.J.; Simeonovic, C.J.: Immunobiology of tissue transplantation: A return to the passenger leukocyte concept. Annu. Rev. Immunol. *1:* 143–173 (1983).

42 Cleveland, R.P.; Claman, H.N.: T cell signals: Tolerance to DNFB is converted to sensitization by a separate nonspecific second signal. J. Immunol. *124:* 474–480 (1980).

43 Jenkins, M.K.; Schwartz, R.H.: Antigen presentation by chemically modified splenocytes induces antigen-specific T cell unresponsiveness in vitro and in vivo. J. Exp. Med. *165:* 302–319 (1987).

44 Markmann, J.; Lo, D.; Naji, A.; Palmiter, R.D.; Brinster, R.L.; Heber-Katz, E.: Antigen presenting function of class II MHC expressing pancreatic beta cells. Nature *336:* 476–479 (1988).

45 Gaspari, A.A.; Jenkins, M.K.; Katz, S.I.: Class II MHC-bearing keratinocytes induce antigen-specific unresponsiveness in hapten-specific Th1 clones. J. Immunol. *141:* 2216–2220 (1988).

46 Sidhu, S.; Deacock, S.; Bal, J.R.; Batchelor, J.R.; Lombardi, G.; Lechler, R.I.: Human T cells cannot act as autonomous antigen-presenting cells, but induce tolerance in antigen-specific and alloreactive responder cells. J. Exp. Med. *176:* 875–880 (1992).

47 Jenkins, M.K.: The role of cell division in the induction of clonal anergy. Immunol. Today *13:* 69–73 (1992).

48 Mueller, D.L.; Jenkins, M.K.; Schwartz, R.H.: Clonal expansion versus functional clonal inactivation: A costimulatory signalling pathway determines the outcome of T cell antigen receptor occupancy. Annu. Rev. Immunol. *7:* 445–480 (1989).

49 Damle, N.K.; Klussman, K.; Linsley, P.S.; Aruffo, A.: Differential costimulatory effects of adhesion molecules B7, ICAM-1, LFA-3, and VACM-1 on resting and antigen-primed CD4+ T lymphocytes. J. Immunol. *148:* 1985–1992 (1992).

50 Van Seventer, G.A.; Newman, W.; Shimizu, Y.; Nutman, T.B.; Tanaka, Y.; Horgan, J.J.; Gopal, T.V.; Ennis, E.; O'Sullivan, D.; Grey H.; Shaw, S.: Analysis of T cell stimulation by superantigen plus major histocompatibility complex class II molecules or by CD3 monoclonal antibody: Costimulation by purified adhesion ligands VCAM-1, ICAM-1, but not ELAM-1. J. Exp. Med. *174:* 901–913 (1991).

51 Shimizu, Y.; van Seventer, G.A.; Horgan, K.J.; Shaw, S.: Costimulation of proliferative responses of resting CD4+ T cells by the interaction of VLA-4 and VLA-5 with fibronectin or VLA-6 with laminin. J. Immunol. *145:* 59–67 (1990).

52 Bierer, B.E.; Peterson, A.; Gorga, J.C.; Herrmann, S.H.; Burakoff, S.J.: Synergistic T cell activation via the physiological ligands for CD2 and the T cell receptor. J. Exp. Med. *168:* 1145–1156 (1988).

53 Liu, Y.; Jones, B.; Aruffo, A.; Sullivan, K.M.; Linsley, P.S.; Janeway, C.A., Jr.:
    Heat-stable antigen is a costimulatory molecule for CD4 T cell growth. J. Exp. Med.
    175: 437–445 (1992).

54 Freeman, G.J.; Freedman, A.S.; Segil, J.M.; Lee, G.; Whitman, J.F.; Nadler, L.M.: B7,
    a new member of the Ig superfamily with unique expression on activated and neoplastic
    B cells. J. Immunol. *143:* 2714 (1989).

55 Larsen, C.P.; Ritchies, S.C.; Pearson, T.C.; Linsley, P.S.; Lowry, R.P.: Functional
    expression of the costimulatory molecule, B7/BB1 on murine dendritic populations. J.
    Exp. Med. *176:* 1215–1220 (1992).

56 Gross, J.A.; Callas, E.; Allison, J.P.: Identification and distribution of the costimulatory
    receptor CD28 in the mouse. J. Immunol. *149:* 380–388 (1992).

57 Linsley, P.; Clark, E.; Ledbetter, J.: T-cell antigen CD28 mediates adhesion with B cells
    by interacting with activation antigen B7/BB1. Proc. Natl Acad. Sci. USA *87:* 5031
    (1990).

58 Linsley, P.S.; Brady, W.; Urnes, M.; Grosmaire, L.S.; Damle, N.K.; Ledbetter, J.A.:
    CTLA-4 is a second receptor for the B cell activation antigen B7. J. Exp. Med. *174:* 625
    (1991).

59 Hara, T.; Fu, S.M.; Hansen, J.A.: Human T cell activation. II. A new activation
    pathway used by a major T cell population via a disulfide-bound dimer of a 44-kilodal-
    ton polypeptide (9.3 antigen). J. Exp. Med. *161:* 1513–1524 (1985).

60 Martin, P.J.; Ledbetter, J.A.; Morishita, Y.; June, C.H.; Beatty, P.G.; Hansen, J.A.: A
    44-kilodalton cell surface homodimer regulates interleukin-2 production by activated
    human T lymphocytes. J. Immunol. *136:* 3282–3287 (1986).

61 Yang, S.Y.; Denning, S.M.; Mizuno, S.; Dupont, B.; Haynes, B.F.: A novel activation
    pathway for mature thymocytes. Costimulation of CD2 (T, p50) and CD28 (T, p44)
    induces autocrine interleukin-2/interleukin-2 receptor-mediated cell proliferation. J. Exp.
    Med. *168:* 1457–1468 (1988).

62 Harding, F.A.; McArthur, J.G.; Gross, J.A.; Raulet, D.H.; Allison, J.P.: CD28-medi-
    ated signalling co-stimulates murine T cells and prevents induction of anergy in T-cell
    clones. Nature *356:* 607–609 (1992).

63 Linsley, P.S.; Brady, W.; Grosmaire, L.; Aruffo, A.; Damle, N.K.; Ledbetter, J.A.:
    Binding of the β cell activation antigen B7 to CD28 costimulates T cell proliferation and
    interleukin-2 mRNA accumulation. J. Exp. Med. *173:* 721–730 (1991).

64 Tan, R.; Teh, S.-J.; Ledbetter, J.A.; Linsley, P.S.; Teh, H.-S.: B7 costimulates prolifera-
    tion of CD4–CD8+ T lymphocytes but is not required for the deletion of immature
    CD4 + CD8+ thymocytes. J. Immunol. *149:* 3217–3224 (1992).

65 Koulova, L.; Clark, E.A.; Shu, G.; Dupont, B.: The CD28 ligand B7/BB1 provides
    costimulatory signal for alloactivation of CD4+ T cells. J. Exp. Med. *173:* 759–762
    (1991).

66 Azuma, M.; Cayabyab, M.; Buck, D.; Phillips, J.H.; Lanier, L.L.: CD28 interaction with
    B7 costimulates primary allogeneic proliferative responses and cytotoxicity mediated by
    small, resting T lymphocytes. J. Exp. Med. *175:* 353–360 (1992).

67 Lenshow, D.J.; Zeng, Y.; Thistlewaite, J.R.; Montag, A.; Brady, W.; Gibson, M.G.;
    Linsley, P.S.; Bluestone, J.A.: Long-term survival of xenogeneic pancreatic islet grafts
    induced by CTLA-4 Ig. Science *257:* 789–792 (1992).

68 Linsley, P.S.; Wallace, P.M.; Johnson, J.; Gibson, M.G.; Greene, J.L.; Ledbetter, J.A.;
    Singh, C.; Tepper, M.A.: Immunosuppression in vivo by a soluble form of the CTLA-4
    T cell activation molecule. Science *257:* 792 (1992).

69 Liu, Y.; Jones, B.; Brady, W.; Janeway, C.A., Jr.; Linsley, P.S.: Co-stimulation of
    murine CD4 T cell growth: Cooperation between B7 and heat-stable antigen. Eur. J.
    Immunol. *22:* 2855–2859 (1992).

70 Metlay, J.P.; Pure, E.; Steinman, R.M.: Distinct features of dendritic cells and anti-Ig
    activated B cells as stimulators of the primary mixed leukocyte reaction. J. Exp. Med.
    169: 239–254 (1989).
71 Sprent, J.: Features of cells controlling H-2-restricted presentation of antigen to T helper
    cells in vivo. J. Immunol. *125:* 2089–2096 (1980).
72 Lassila, O.; Vainio, O.; Matzinger, P.: Can B cells turn on virgin T cells? Nature *334:*
    253–255 (1988).
73 Eynon, E.E.; Parker, D.C.: Small B cells as antigen-presenting cells in the induction of
    tolerance to soluble protein antigens. J. Exp. Med. *175:* 131–138 (1992).
74 Freedman, A.S.; Freeman, G.; Horowitz, J.C.; Daley, J.; Nadler, L.M.: B7, a B-cell-
    restricted antigen that identifies preactivated B cells. J. Immunol. *139:* 3260–3267
    (1987).
75 Yokochi, T.; Holly, R.D.; Clark, E.A.: B lymphoblast antigen (BB-1) expressed on
    Epstein-Barr virus-activated B cell blasts, B lymphoblastoid cell lines, and Burkitt's
    lymphoma. J. Immunol. *128:* 823–827 (1982).
76 Fuchs, E.; Matzinger, P.: B cells turn off virgin but not memory T cells. Science *258:*
    1156–1159 (1992).
77 Essery, G.; Feldmann, M.; Lamb, J.R.: Interleukin-2 can prevent and reverse antigen-
    induced unresponsiveness in cloned human T lymphocytes. Immunology *64:* 413–417
    (1988).
78 Andreu-Sanchez, J.L.; Moreno de Alboran, N.; Marcos, M.A.R.; Sanchez-Mivilla, A.;
    Martinez, A.C.; Kroemer, G.: Interleukin-2 abrogates the nonresponsive state of T cells
    expressing a forbidden T cell receptor repertoire and induces autoimmune disease in
    neonatally thymectomized mice. J. Exp. Med. *173:* 1323–1329 (1991).
79 Guitierrez-Ramos, J.C.; Moreno de Alboran, I.; Martinez, A.C.: In vivo administration
    of interleukin-2 turns on anergic self-reactive T cells and leads to autoimmune disease.
    Eur. J. Immunol. *22:* 2867–2872 (1992).
80 Ramsdell, F.; Fowlkes, B.J.: Maintenance of in vivo tolerance by persistence of antigen.
    Science *257:* 1130–1134 (1992).
81 Allison, J.; Campbell, I.L.; Morahan, G.; Mandel, T.E.; Harrison, L.C.; Miller, J.F.:
    Diabetes in transgenic mice resulting from over-expression of class I histocompatibility
    molecules in pancreatic beta cells. Nature *333:* 529–533 (1988).
82 Morahan, G.; Allison, J.; Miller, J.F.: Tolerance of class I histocompatibility antigens
    expressed extrathymically. Nature *339:* 622–624 (1989).
83 Morahan, G.; Brennan, F.E.; Bhatal, P.S.; Allison, J.; Cox, K.O.; Miller, J.F.A.P.:
    Expression of transgenic mice of class I histocompatibility antigens controlled by the
    metallothionein promoter. Proc. Natl Acad. Sci. USA *86:* 3782–3786 (1989).
84 Wieties, K.; Hammer, R.E.; Jones-Youngblood, S.; Forman, J.: Peripheral tolerance in
    mice expressing a liver-specific class I molecule: Inactivation/deletion of a T-cell subpop-
    ulation. Proc. Natl Acad. Sci. USA *87:* 6604–6608 (1990).
85 Lo, D.; Burkly, L.C.; Widera, G.; Cowing, C.; Flavell, R.A.; Palmiter, R.D.; Brinster,
    R.L.: Diabetes and tolerance in transgenic mice expressing class II MHC molecules in
    pancreatic beta cells. Cell *53:* 159–168 (1988).
86 Burkly, L.C.; Lo, D.; Kanagawa, O.; Brinster, R.L.; Flavell, R.A.: T-cell tolerance by
    clonal anergy in transgenic mice with nonlymphoid expression of MHC class II I-E.
    Nature *342:* 564–566 (1989).
87 Morahan, G.; Hoffmann, M.W.; Miller, J.F.A.P.: A nondeletional mechanism of
    peripheral tolerance in T-cell receptor transgenic mice. Proc. Natl Acad. Sci. USA *88:*
    11421–11425 (1991).
88 Schonrich, G.; Kalinke, U.; Momburg, F.; Malissen, M.; Schmitt-Verhulst, A.M.;

Malissen, B.; Hammerling, G.J.; Arnold, B.: Down-regulation of T cell receptors on self-reactive T cells as a novel mechanism for extrathymic tolerance induction. Cell *65:* 293–304 (1991).

89 Ramsdell, F.; Lantz, T.; Fowlkes, B.J.: A nondeletional mechanism of thymic self tolerance. Science *246:* 1038–1041 (1989).

90 Hodes, R.J.; Sharrow, S.O.; Solomon, A.: Failure of T cell receptor V-beta-negative selection in an athymic environment. Science *246:* 1041–1044 (1989).

91 Nossal, G.J.V.; Pike, B.I.: Clonal anergy: Persistence in tolerant mice of antigen-binding B lymphocytes incapable of responding to antigen or mitogen. Proc. Natl Acad. Sci. USA *77:* 1602–1606 (1980).

92 Pike, B.L.; Abrams, J.; Nossal, G.J.V.: Clonal anergy: Inhibition of antigen-driven proliferation among single B lymphocytes from tolerance animals and partial breakage of anergy my mitogens. Eur. J. Immunol. *13:* 214–220 (1983).

93 Pike, B.L.; Boyd, A.W.; Nossal, G.J.V.: Clonal anergy: The universally anergic B lymphocyte. Proc. Natl Acad. Sci. USA *79:* 2013–2017 (1982).

94 Nossal, G.J.V.: Cellular mechanisms of immunologic tolerance. Annu. Rev. Immunol. *1:* 33–62 (1983).

95 Schlomchik, M.J.; Marshak-Rothstein, A.; Wolfowicz, C.B.; Rothstein, T.L.; Weigert, M.G.: The role of clonal selection and somatic mutation in autoimmunity. Nature *328:* 805 (1987).

96 Primi, D.; Hammarstrom, L.; Smith, C.I.E.; Moller, G.: Characterization of self-reactive B cells by polyclonal B cell activators. J. Exp. Med. *145:* 21 (1977).

97 Holmberg, D.; Freitas, A.A.; Portnoi, D.; Jacquemart, F.; Avrameas, S.; Coutinho, A.: Autoantibody repertoires of normal BALB/c mice: B lymphocyte populations defined by state of activation. Immunol. Rev. *93:* 147 (1986).

98 Goodnow, C.C.: Transgenic mice and analysis of B-cell tolerance. Annu. Rev. Immunol. *10:* 489–518 (1992).

99 Goodnow, C.C.; Crosbie, J.; Adelstein, S.; Lavoie, T.B.; Smith-Gill, S.J.; Brink, R.A.; Pritchard-Briscoe, H.; Wotherspoon, J.S.; Loblay, R.H.; Raphael, K.; Trent, R.J.; Basten, A.: Altered immunoglobulin expression and functional silencing of self-reactive B lymphocytes in transgenic mice. Nature *334:* 676–682 (1988).

100 Goodnow, C.C.; Crosbie, J.; Adelstein, S.; Lavoie, T.B.; Smith-Gill, S.J.; Mason, D.Y.; Jorgensen, H.; Brink, R.A.; Pritchard-Brsicoe, H.; Loughnan, M.; Loblay, R.H.; Trent, R.J.; Basten, A.: Clonal silencing of self-reactive B lymphocytes in a transgenic mouse model. Cold Spring Harb. Symp. Quant. Biol. *54:* 907–920 (1989).

101 Goodnow, C.C.; Brink, R.A.; Adams, E.: Breakdown of self-tolerance in anergic B lymphocytes. Nature *352:* 532–536 (1991).

102 Goodnow, C.C.; Adelstein, S.; Basten, A.: The need for central and peripheral tolerance in the B cell repertoire. Science *248:* 1373–1379 (1990).

103 Gershon, R.K.; Kondo, K.: Infectious immunological tolerance. Immunology *21:* 903 (1971).

104 Gershon, R.K.: Activation of suppressor T cells by tumour cells and specific antibody. Nature *240:* 594 (1974).

105 Gershon, R.K.: A disquisition on suppressor T cells. Immunol. Rev. *26:* 170 (1975).

106 Dorsch, S.E.; Roser, B.: The clonal nature of alloantigen-sensitive small lymphocytes in the recirculating pool of normal rats. Aust. J. Exp. Biol. Med. Sci. *52:* 45 (1974).

107 Dorsch, S.E.; Roser, B.: T cells mediate transplantation tolerance. Nature *258:* 174 (1975).

108 Mehra, V.; Convit, J.; Rubinstein, A.; Bloom, B.R.: Activated suppressor T cells in leprosy. J. Immunol. *129:* 1946 (1982).

109 Nelson, E.E.; Wong, L.; Uyemura, K.; Rea, T.H.; Modlin, R.L.: Lepromin-induced suppressor cells in lepromatous leprosy. Cell. Immunol. *14:* 99 (1987).

110 Roser, B.J.: Cellular mechanisms in neonatal and adult tolerance. Immunol. Rev. *107:* 179–202 (1989).

111 Tutschka, R.J.; Ki, P.F.; Beschorner, W.E.; Hes, A.D.; Santos, G.W.: Suppressor cells in transplantation tolerance. II. Maturation of suppressor cells in the bone marrow chimera. Transplantation *32:* 321–325 (1981).

112 Lancaster, F.; Chui, Y.L.; Batchelor, J.R.: Anti-idiotypic T cells suppress rejection of renal allografts in rats. Nature *315:* 336–337 (1985).

113 Maki, T.; Gottshalk, R.; Wood, M.L.; Monaco, A.P.: Specific unresponsiveness to skin allografts in anti-lymphocyte serum-treated, marrow-injected mice: Participation of donor marrow-derived suppressor T cells. J. Immunol. *127:* 1433–1437 (1981).

114 Horowitz, J.B.; Kaye, J.; Conrad, P.J.; Katz, M.E.; Janeway, C.A.: Autocrine growth inhibition of a cloned line of helper T cells. Proc. Natl Acad. Sci. USA *83:* 1886 (1989).

115 Fiorentino, D.F.; Bond, M.W.; Mosmann, T.R.: Two types of mouse helper T cell. IV. Th2 clones secrete a factor that inhibits cytokine production by Th1 clones. J. Exp. Med. *170:* 2081 (1989).

116 Fiorentino, D.F.; Zlotnik, A.; Viera, P.; Mosmann, T.R.; Howard, M.; Moore, K.W.; O'Garra, A.: IL-10 acts on the antigen-presenting cell to inhibit cytokine production by Th1 cells. J. Immunol. *146:* 3444–3451 (1991).

117 Klein, J.: Function in Natural History of the Major Histocompatibility Complex, pp. 423–608 (Wiley, New York 1986).

118 Germain, R.N.; Benacerraf, B.: A single major pathway of T-lymphocyte interactions in antigen-specific immune suppression. Scand. J. Immunol. *13:* 1 (1981).

119 Bloom, B.R.; Modlin, R.L.; Salgame, P.: Stigma variations: Observations on suppressor T cells and leprosy. Annu. Rev. Immunol. *10:* 453–488 (1992).

120 Dorf, M.E.; Benacerraf, B.: Suppressor cells and immunoregulation. Annu. Rev. Immunol. *2:* 127–158 (1984).

121 Damle, N.K.; Mohagheghpour, N.; Engleman, E.G.: Soluble antigen-primed inducer T cells activate antigen-specific suppressor T cells in the absence of antigen-pulsed accessory cells: Phenotypic definition of suppressor-inducer and suppressor effector cells. J. Immunol. *132:* 644 (1984).

122 Koide, J.; Engleman, E.G.: Differences in surface phenotype and mechanism of action between alloantigen-specific CD8+ cytotoxic and suppressor T cell clones. J. Immunol. *144:* 32–40 (1990).

123 Sykes, M.; Eisenthal, A.; Sachs, D.H.: Mechanisms of protection from graft-vs.-host disease in murine mixed allogeneic chimeras. I. Development of a null population suppressive of cell-mediated lympholysis responses and derived from the syngeneic bone marrow component. J. Immunol. *140:* 2903–2911 (1988).

124 Hertel-Wulff, B.; Okada, S.; Oseroff, A.; Strober, S.: In vitro propagation and cloning of murine natural suppressor cells. J. Immunol. *133:* 2791–2796 (1984).

125 Strober, S.: Natural suppressor cells, neonatal tolerance, and total lymphoid irradiation: Exploring obscure relationships. Annu. Rev. Immunol. *2:* 219–231 (1984).

126 Oseroff, A.; Okada, S.; Strober, S.: Natural suppressor cells found in the spleen of neonatal mice and adult mice given total lymphoid irradiation express the null surface phenotype. J. Immunol. *132:* 101–110 (1984).

127 Strober, S.; Palathumpat, V.; Schwardon, R.; Hertel-Wulff, B.: Cloned natural suppressor cells prevent lethal graft-vs.-host disease. J. Immunol. *138:* 699 (1987).

128 Schwardon, R.B.; Gangour, D.M.; Strober, S.: Cloned natural suppressor cell lines derived from the spleens of neonatal mice. J. Exp. Med. *162:* 297 (1985).

129 Holda, J.H.; Maier, T.; Claman, H.N.: Murine graft-vs.-host disease across minor barriers: Immunosuppressive aspects of natural suppressor cells. Immunol. Rev. *88:* 87–105 (1985).

130 Merluzzi, V.J.; Levy, E.M.; Kumar, V.; Bennet, N.; Cooperband, S.R.: In vitro activation of suppressor cells from spleens of mice treated with radioactive strontium. J. Immunol. *121:* 505–508 (1978).

131 King, D.P.; Stober, S.; Kaplan, H.S.: Suppression of the mixed leukocyte response and graft-vs.-host disease by spleen cells following total lymphoid irradiation. J. Immunol. *126:* 1140–1144 (1981).

132 Field, E.H.; Becher, G.C.: Blocking of mixed lymphocyte reaction by spleen cells from total lymphoid-irradiated mice involves interruption of the IL-2 pathway. J. Immunol. *148:* 354–359 (1992).

133 Maes, L.Y.; York, J.L.; Soderberg, L.S.F.: A soluble factor produced by bone marrow natural suppressor cells blocks interleukin-2 production and activity. Cell. Immunol. *116:* 35–43 (1988).

134 Miller, R.G.: Lymphoid suppressor cells; In Sercarz, E.E.; Cunningham, A.J. (eds): Strategies of Immune Regulation, p. 507 (Academic Press, New York 1980).

135 Miller, R.G.; Derry, H.: A cell population in nu/nu spleen can prevent generation of cytotoxic lymphocytes by normal spleen cells against self antigens of the nu/nu spleen. J. Immunol. *122:* 1502 (1979).

136 Muraoka, S.; Miller, R.G.: Cells in bone marrow and in T cell colonies grown from bone marrow can suppress generation of cytotoxic T lymphocytes directed against their self antigens. J. Exp. Med. *152:* 54 (1980).

137 Rammensee, H.G.; Nagy, Z.A.; Klein, J.: Suppression of cell-mediated lymphocytotoxicity strong origin. Eur. J. Immunol. *12:* 930 (1982).

138 Fink, P.J.; Rammensee, H.G.; Bevan, M.J.: Cloned cytolytic T cells can suppress primary cytotoxic responses directed against them. J. Immunol. *133:* 1775 (1984).

139 Cassell, D.; Forman, J.: Two roles for CD4 cells in the control of the generation of cytotoxic T lymphocytes. J. Immunol. *146:* 3–10 (1991).

140 Nakamura, H.; Gress, R.E.: Interleukin-2 enhancement of veto suppressor cell function in T-cell-depleted bone marrow in vitro and in vivo. Transplantation *49:* 931–937 (1990).

141 Azuma, E.; Kaplan, J.: Role of lymphokine-activated killer cells as mediators of veto and natural suppression. J. Immunol. *141:* 2601–2606 (1988).

142 Sambhara, S.R.; Miller, R.G.: Programmed cell death of T cells signaled by the T cell receptor and the alpha-3 domain of class I MHC. Science *252:* 1424–1427 (1991).

143 Rammensee, H.G.; Fink, P.J.; Bevan, M.J.: Functional clonal deletion of class I-specific cytotoxic T lymphocytes by veto cells that express antigen. J. Immunol. *133:* 2390 (1984).

144 Rammensee, H.G.; Nagy, Z.A.; Klein, J.: Suppression of cell-mediated lymphocytotoxicity against minor histocompatibility antigens mediated by Lyt-1+ Lyt-2+ T cells of stimulator strain origin. Eur. J. Immunol. *12:* 930 (1982).

145 Kiziroglu, F.; Muller, R.G.: In vivo functional clonal deletion of recipient CD4+ T helper precursor cells that recognize class II MHC on injected donor lymphoid cells. J. Immunol. *146:* 1104–1112 (1991).

146 Azuma, E.; Yamamoto, H.; Kaplan, J.: Use of lymphokine-activated killer cells to prevent bone marrow graft rejection and lethal graft-vs.-host disease. J. Immunol. *143:* 1524–1529 (1989).

147 Heeg, K.; Wagner, H.: Induction of peripheral tolerance to class I major histocompatibility complex (MHC) alloantigens in adult mice: Transfused class I MHC-incompatible

splenocytes veto clonal response of antigen-reactive Lyt-2+ T cells. J. Exp. Med. *172:* 719–725 (1990).

148   Sharabi, Y.; Sachs, D.H.: Mixed chimerism and permanent specific transplantation tolerance induced by a non-lethal preparative regime. J. Exp. Med. *169:* 493–502 (1989).

149   Cobbold, S.P.; Martin, G.; Qin, S.; Waldmann, H.: Monoclonal antibodies to promote marrow engraftment and tissue graft tolerance. Nature *323:* 164–165 (1986).

150   Schwartz, E.; Lapidot, T.; Gozes, D.; Singer, T.S.; Reisner, Y.: Abrogation of bone marrow allograft resistance in mice by increased total body irradiation correlates with eradication of host clonable T cells and alloreactive cytotoxic precursors. J. Immunol. *138:* 460–465 (1987).

151   Dennert, G.; Anderson, C.G.; Warner, J.: T killer cells play a role in allogeneic marrow graft rejection but not in hybrid resistance. J. Immunol. *135:* 3729–3734 (1985).

152   Yankelvich, B.; Knoblock, C.; Nowicki, M.; Dennert, G.: A novel cell type responsible for marrow graft rejection in mice. T cells with NK phenotype cause acute rejection of marrow grafts. J. Immunol. *142:* 3423–3440 (1989).

153   Bennett, M.: Biology and genetics of hybrid resistance. Adv. Immunol. *41:* 333–445 (1987).

154   Ferrar, J.L.M.; Mauch, P.; van Dijken, P.J.; Crosier, K.E.; Michaelson, J.; Burakoff, S.J.: Evidence that anti-asialo-GM1 in vivo improves engraftment of T cell-depleted bone marrow in hybrid recipients. Transplantation *49:* 134–137 (1990).

155   Murphy, W.J.; Kumar, V.; Bennett, M.: Rejection of bone marrow allografts by mice with severe combined immune deficiency. Evidence that natural killer cells can mediate the specificity of marrow graft rejection. J. Exp. Med. *165:* 1212–1217 (1987).

156   Barge, A.J.; Johnson, G.; Wotherspoon, R.; Torok-Storb B.: Antibody-mediated marrow failure after allogeneic bone marrow transplantation. Blood *74:* 1477–1480 (1989).

157   Voralia, M.; Semeluk, A.; Wegmann, T.G.: Facilitation of syngeneic stem cell engraftment by anti-class I monoclonal antibody pretreatment of unirradiated recipients. Transplantation *44:* 487–494 (1987).

158   Billingham, R.E.; Brent, L.; Medawar, P.B.: 'Actively acquired tolerance' of foreign cell. Nature *172:* 603–606 (1953).

159   Rappaport, F.T.: Immunologic tolerance: Irradiation and bone marrow transplantation induction of canine allogeneic unresponsiveness. Transplant. Proc. *9:* 891 (1977).

160   Martin, P.J.; Hansen, J.A.; Storb, R.; Thomas, E.D.: Human marrow transplantation: An immunological perspective. Adv. Immunol. *40:* 379–438 (1987).

161   Blazar, B.R.; Hirsch, R.; Gress, R.E.; Carroll, S.F.; Vallera, D.A.: In vivo administration of monoclonal antibodies or immunotoxins in murine recipients of allogeneic T-cell-depleted marrow for the promotion of engraftment. J. Immunol. *147:* 1492–1503 (1991).

162   Vallera, D.A.; Blazar, B.R.: T cell depletion for graft-versus-host disease prophylaxis: A perspective on engraftment in mice and humans. Transplantation *47:* 751–760 (1989).

163   Bradley, S.M.; Kruisbeek, A.M.; Singer, A.: Cytotoxic T lymphocyte response in allogeneic radiation bone marrow chimers. The chimeric host strictly dictates the self-repertoire of Ia-restricted T cells but not K/D-restricted T cells. J. Exp. Med. *156:* 1650–1656 (1982).

164   Ildstad, S.T.; Wren, S.M.; Bluestone, J.A.; Barbieri, S.A.; Sachs, D.H.: Characterization of mixed allogeneic chimeras. Immunocompetence, in vitro reactivity, and genetic specificity of tolerance. J. Exp. Med. *162:* 231–244 (1985).

165   Ildstad, S.T.; Sachs, D.H.: Reconstitution with syngeneic plus allogeneic or xenogeneic bone marrow leads to specific acceptance of allografts or xenografts. Nature *307:* 168–170 (1984).

166 Singer, A.; Hathcock, K.S.; Hodes, R.J.: Self-recognition in allogeneic radiation chimeras. A radiation host element dictates the self-specificity and immune response gene phenotype of T-helper cells. J. Exp. Med. *153:* 1286–1301 (1981).

167 Pierce, G.E.: Allogeneic versus semiallogeneic F1 bone marrow transplantation into sublethally irradiated MHC-disparate hosts. Effects on mixed lymphoid chimerism, skin graft tolerance, host survival, and alloreactivity. Transplantation *49:* 138–144 (1990).

168 Soderling, C.C.B.; Song, C.W.; Blazar, B.R.; Vallera, D.A.: A correlation between conditioning and engraftment in recipients of MHC-mismatched T cell-depleted murine bone marrow transplants. J. Immunol. *135:* 941–945 (1985).

169 Slavin, S.; Strober, S.; Fuks, Z.; Kaplan, H.S.: Induction of specific tissue transplantation tolerance using fractionated total lymphoid irradiation in adult mice: Long-term survival of allogeneic bone marrow and skin grafts. J. Exp. Med. *146:* 34–48 (1977).

170 Mayumi, H.; Good, R.A.: Long-lasting skin allograft tolerance in adult mice induced across fully allogeneic (multimajor H-2 plus multiminor histocompatibility) antigen barriers by a tolerance-inducing method using cyclophosphamide. J. Exp. Med. *169:* 213–238 (1989).

171 Qin, S.; Cobbold, S.; Benjamin, R.; Waldmann, H.: Induction of classical transplantation tolerance in the adults. J. Exp. Med. *169:* 779–794 (1989).

172 Wood, M.L.; Monaco, A.P.; Gozzo, J.J.; Liegois, A.: Use of homozygous allogeneic bone marrow for induction of tolerance with anti-lymphocyte serum: Dose and timing. Transplant. Proc. *3:* 676–678 (1970).

173 Pierce, G.E.; Watts, L.M.: Effects of Thy-1+ cell depletion on the capacity of donor lymphoid cells to induce tolerance across an entire MHC disparity in sublethally irradiated adult hosts. Transplantation *48:* 289–296 (1989).

174 Cobbold, S.P.; Waldmann, H.: Reprogramming the immune system for tolerance with monoclonal antibodies. Semin. Immunol. *2:* 377–387 (1990).

175 Vojtiskova, M.; Lengerova, A.: On the possibility that thymus-mediated alloantigenic stimulation results in tolerance response. Experientia *21:* 661 (1965).

176 Staples, P.J.; Gery, I.; Waksman, B.H.: Role of thymus in tolerance. III. Tolerance to bovine gamma globulin after direct injections of antigen into the shielded thymus of irradiated rats. J. Exp. Med. *124:* 127 (1966).

177 Gery, I.; Waksman, B.H.: Role of the thymus in tolerance. V. Suppressive effect of treatment with nonaggregated and aggregated bovine gamma globulin in specific immune responses in normal adult rats. J. Immunol. *98:* 446 (1967).

178 Posselt, A.M.; Barker, C.F.; Tomaszewski, J.E.; Markmann, J.F.; Choti, M.A.; Naji, A.: Induction of donor-specific unresponsiveness by intrathymic islet transplantation. Science *249:* 1293 (1990).

179 Herold, K.C.; Montag, A.G.; Buckingham, F.: Induction of tolerance to autoimmune diabetes with islet antigens. J. Exp. Med. *176:* 1107–1114 (1992).

180 Ohzato, H.; Monaco, A.P.: Induction of specific unresponsiveness (tolerance) to skin allografts by intrathymic donor-specific splenocyte injection in antilymphocyte serum-treated mice. Transplantation *54:* 1090–1095 (1992).

181 Starzl, T.E.; Demetris, A.J.; Trucco, M.; Ramos, H.; Zeevi, A.; Rudert, W.A.; Kocova, M.; Ricordi, C.; Ildstad, S.; Murase, N.: Systemic chimerism in human female recipients of male livers. Lancet *340:* 876–877 (1992).

182 Starzl, T.E.; Demetris, A.J.; Murase, N.; Ildstad, S.; Ricordi, C.; Trucco, M.: Cell migration, chimerism, and graft acceptance. Lancet *339:* 1579–1582 (1992).

183 Opelz, G.; Mickey, M.R.; Terasaki, P.I.: Blood transfusions and kidney transplants: Remaining controversies. Transplant. Proc. *13:* 136 (1981).

184 Singal, DP.; Ludwin, D.; Blajchman, M.A.: Blood transfusion and renal transplantation. Br. J. Haematol. *61:* 595 (1985).
185 Miller, R.G.; Phillips, R.A.: Reduction of the in vitro cytotoxic response produced by in vivo exposure to semiallogeneic cells: Recruitment of active suppression? J. Immunol. *117:* 1913 (1976).
186 Sheng-Tanner, X.; Miller, R.G.: Correlation between lymphocyte-induced donor-specific tolerance and donor cell recirculation. J. Exp. Med. *176:* 407–413 (1992).
187 Lagaaij, E.L.; Hennemann, I.P.H.; Ruigrok, M.; de Haan, M.W.; Persihn, G.G.; Termijtelen, A.; Hendriks, G.F.J.; Weimer, W.; Claas, F.H.J.; van Rood, J.J.: Effect of one HLA-DR antigen-matched and completely HLA-DR-mismatched blood transfusion on survival of heart and kidney allografts. N. Engl. J. Med. *321:* 701 (1989).
188 Lafferty, K.J.; Bootes, A.; Dart, G.; Talmage, D.W.: Effect of organ culture on the survival of thyroid allografts in mice. Transplantation *22:* 138–149 (1976).
189 Lafferty, K.; Prowse, S.; Simeonovic, C.; Warren, H.S.: Immunobiology of tissue transplantation: A return to the passenger leukocyte concept; in Paul, W.E.; Fathman, C.G.; Metzgar, H. (eds): Annual Review of Immunology, pp. 143–173 (Annual Reviews, Inc., Palo Alto 1983).
190 Sollinger, H.W.; Burkholder, P.M.; Rasmus, W.R.; Bach, F.H.: Prolonged survival of xenografts after organ culture. Surgery *81:* 74 (1977).
191 Lechler, R.; Batchelor, J.: Restoration of immunogenicity to passenger cell depleted kidney allografts by the addition of donor strain dendritic cells. J. Exp. Med. *155:* 31–41 (1982).
192 Faustman, D.; Hauptfield, V.; Lacy, P.; Davie, J.: Prolongation of murine islet allograft survival by pretreatment of islets with antibody directed to Ia determinants. Proc. Natl Acad. Sci. USA *78:* 5156–5159 (1981).
193 Rosengard, B.R.; Kortz, E.O.; Guzzetta, P.C.; Sundt, T.M., III; Ojikutu, C.A.; Alexander, R.B.; Sachs, D.H.: Transplantation in miniature swine: Analysis of graft-infiltrating lymphocytes provides evidence for local suppression. Hum. Immunol. *28:* 153–158 (1990).
194 Hardy, M.A.; Lau, H.; Weber, C.; Reemtsma, K.: Pancreatic islet transplantation: Induction of graft acceptance by ultraviolet irradiation of donor tissue. Ann. Surg. *200:* 441 (1984).
195 Deeg, H.J.: Ultraviolet irradiation in transplantation biology: Manipulation of immunity and immunogenicity. Transplantation *45:* 845–851 (1988).
196 Qin, S.X.; Wise, M.; Cobbold, S.P.; Leong, L.; Kong, Y.C.; Parnes, J.R.; Waldmann, H.: Induction of tolerance in peripheral T cells with monoclonal antibodies. Eur. J. Immunol. *20:* 2737–2745 (1990).
197 Cobbold, S.P.; Martin, G.; Waldmann, H.: The induction of skin graft tolerance in major histocompatibility complex-mismatched or primed recipients: Primed T cells can be tolerized in the periphery with anti-CD4 and anti-CD8 antibodies. Eur. J. Immunol. *20:* 2747–2755 (1990).
198 Chen, Z.; Cobbold, S.; Metcalfe, S.; Waldmann, H.: Tolerance in the mouse to major histocompatibility complex-mismatched heart allografts, and to rat heart xenografts, using monoclonal antibodies to CD4 and CD8. Eur. J. Immunol. *23:* 805–810 (1992).
199 Shizuru, J.; Gregory, A.K.; Chao, C.-B.; Fathman, C.G.: Islet allograft survival after a single course of treatment of recipient with antibody to L3T4. Science *237:* 278 (1987).
200 Shizuru, J.; Seydel, K.B.; Flavin, T.F.; Wu, A.; Kon, C.; Hoyt, E.; Fujimoto, N.; Billingham, M.; Starnes, V.; Fathman, C.G.: Induction of donor-specific unresponsive-

ness to cardiac allografts in rats by pretransplant anti-CD4 monoclonal antibody therapy. Transplantation *50:* 366 (1990).

201 Alters, S.E.; Shizuru, J.A.; Ackerman, J.; Grossman, D.; Seydel, K.B.; Fathman, C.G.: Anti-CD4 mediates clonal anergy during transplantation tolerance induction. J. Exp. Med. *173:* 491–494 (1991).
202 Isobe, M.; Yagita, H.; Okumura, K.; Ihara, A.: Specific acceptance of cardiac allograft after treatment with antibodies to ICAM-1 and LFA-1. Science *255:* 1125–1127 (1992).

Richard S. Lee, Harvard Medical School and Transplantation Unit,
Department of Surgery, Massachusetts General Hospital, Boston, MA 02114 (USA)

Granstein RD (ed): Mechanisms of Immune Regulation.
Chem Immunol. Basel, Karger, 1994, vol 58, pp 259–290

Oral Tolerance: A Biologically Relevant Pathway to Generate Peripheral Tolerance against External and Self Antigens

Aharon Friedman, Ahmad Al-Sabbagh, Leonilda M.B. Santos,
Jacqueline Fishman-Lobell, Malu Polanski, Mercy Prabhu Das,
Samia J. Khoury, Howard L. Weiner

Center for Neurological Diseases, Brigham and Women's Hospital and Harvard Medical School, Boston, Mass., USA

Introduction and Historical Perspectives

During the life span of an organism, cells of the acquired (clonotypic) immune system undergo a series of processes in order to prevent reactivity to self antigens. The outcome of these processes is a state of immune tolerance (i.e. immunological unresponsiveness) to self [1–3]. Development of tolerance to self antigens has been explained by several mechanisms which include clonal deletion [4–7], clonal anergy [7–10], and clonal suppression [11–13]. Tolerance of T lymphocytes is achieved in thymic and peripheral (postthymic or extrathymic) environments, and tolerance of B lymphocytes is achieved in bone marrow as well as in the periphery [3, 14–16]. Peripheral tolerance is of major importance for self integrity, since many self antigens are not encountered in thymus or bone marrow [3, 14]. Self integrity is further ensured by the induction of tolerance against foreign (or external) 'beneficial' molecules, such as dietary antigens, in order to prevent 'unwanted' immune responses, such as food allergy [17–19]. Tolerance of lymphocytes to external antigens is induced in the periphery, and is probably explained by the same mechanisms leading to peripheral tolerance of self [20].

Immunological tolerance to external antigens may be experimentally demonstrated by feeding a protein antigen in solution, and by testing immune responsiveness directed against the same antigen after parenteral immunization [20–23]. Similar results may be obtained by the inhalation of arerosolized antigen or antigen 'dust' [24–26]. The unresponsive state obtained by feeding or inhaling antigen is antigen-specific, systemic, and mediated by T lymphocytes [20, 27–30]. Other factors such as age [31], genetic background [32–34], and nutritional status [35] were also shown to be critical for induction of tolerance. Achievement of tolerance by feeding, designated oral tolerance (OT), has been known for over a 100 years [20, 36], and is considered to be a biologically relevant pathway for inducing tolerance against dietary antigens [37]. Early studies, performed between the turn of the century and 1945, were aimed at exploiting OT as a means of treating food and skin allergies [36]. More recent studies were aimed at defining the immunological and intestinal structural basis for OT [20–23, 37]. These studies emphasized the complex structure of gut-associated lymphoid tissue (GALT), a tissue unique in its capacity to generate both an immune response and immunological tolerance [36].

Recently, interest in OT has been increased due to the provocative possibility of using OT as a therapeutic tool for reinstating tolerance to self antigens involved in autoimmune diseases [38–41]. Thus, we have investigated OT as a means to suppress experimental and spontaneous autoimmune diseases [38], and as a means to increase graft survival [42, 43]. Others have suggested the use of OT as a possible means for reducing allergic reactions [18–20, 37]. In addition, OT has become an important factor to consider in the design of enterally administered vaccines [36]. Hence, a better understanding of OT will reveal the mechanisms underlying the distinction between response and tolerance and could ultimately lead to promising clinical applications. The objective of the present review is to outline the structural and physiological basis of OT, to evaluate proposed mechanisms for OT, and finally, to summarize clinical applications for OT. For detailed accounts of early studies on OT, the reader is referred to several recent reviews [20–23].

The Intestinal Immune System

Ontogeny
Cells of the intestinal immune system originate in the bone marrow. The migration and colonization of the gut by these cells is a continuous process that begins in the fetus and continues well after parturition. The kinetics of migration and colonization differ between species, but have essentially the same result. In the human, development of the gastrointestinal tract is well advanced by 10–11 weeks of gestation [44]. At this time lymphocytes, macrophages and dendritic cells appear in the lamina propria (LP), intraepithelial lymphocytes (IEL) between enterocytes (1.5/100 epithelial cells), as well as primary in Peyer's patches (PP), containing class II positive and CD4+ cells, in the mucosa [45] (see fig. 1). Most mucosal T

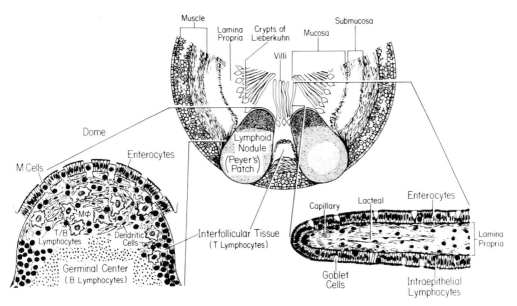

Fig. 1. Histological structure of GALT. Cross-section of murine small intestine illustrates the main areas of gut lymphoid tissue: LP, IEL and lymphoid nodules (PP). The LP is continous throughout submucosa, mucosa and villi. Immune cells are diffusely distributed in LP and are supported by a fine mesh of reticular fibers produced by local fibroblasts. Antigen is absorbed by enterocytes and diffuses via the basement membrane to capillaries or lacteals. PP include a germinal center, populated mainly by proliferating and mature B lymphocytes, and a dome area. The dome forms a unique structure capable of uptaking and presenting particulate antigen; the cells in this structure consist of M cells, interspersed between enterocytes, responsible for antigen uptake and transfer, and underlying macrophages and dome dendritic cells. The distal surface of M cells has a dendritic morphology which might function in direct transfer of antigen to underlying macrophages, dendritic cells and migrating lymphocytes (B and T).

lymphocytes are derived from the thymus [45]. The IEL compartment contains a subpopulation of T lymphocytes that develops extrathymically, probably in the intestinal epithelium itself [14, 46–49]. These cells express $\gamma\delta$ TcR and have been described in several species [50], and appear in mucosal tissue after the appearance of thymus-derived T lymphocytes [45]. By 14–16 weeks of gestation there is an increase in both cell volume and type: IEL become more abundant (12.5/100 epithelial cells), T lymphocytes appear in the LP, and B lymphocytes appear within PP – without cellular compartmentalization [45]. This trend increases by 20 weeks of gestation, which is characterized by the appearance of additional cells of the innate immune system in the LP (neutrophils, eosinophils, basophils and mast cells), and by the cellular compartmentalization of PP [45]. The epithelium above PP becomes specialized and membrane or microfold (M) cells become apparent [51]. Since prenatal mucosal lymphocytes respond to stimuli by proliferation and cytokine secretion, it seems that the mucosal immune system in the human is functionally mature prior to birth [45]. In the rodent, however, most intestinal lymphoid tissue is virtually absent by birth; in this species colonization occurs later, and becomes mature by 1–3 weeks of age, depending upon the type of rodent [51, 52].

Cells of the immune system enter mucosal tissue by extravasation through high endothelial venules (HEV) [18, 53]. Homing to HEV is achieved by binding of homing receptors on lymphocytes to vascular addressins on HEVs [53–55]. Homing patterns of lymphocytes are central to the understanding of lymphocyte function in the gut. If homing is random then lymphocyte populations in an organism should be functionally similar [56]. If, however, homing is organ-specific then lymphoid populations could be functionally distinct [54]. B lymphocytes appear to migrate in a random manner [53]. This, however, does not seem to be the case for T lymphocytes which appear to migrate in an organ-specific manner [57–59]. IEL express a lymphocyte-homing receptor for PP that does not bind to lymph node HEV addressins (integrin β_7 [60]). Interestingly, it was found that expression of addressins on vascular endothelium is regulated by cytokines (TGF-β down-regulates whereas IFN-γ and IL-4 up-regulate) [61], which indicates that, in addition to organ specificity, the rate of lymphocyte homing and extravasation might also be regulated.

Structure: Anatomy, Histology and Lymphocyte Populations

The anatomy and histology of GALT is only briefly summarized here, with emphasis on the small intestine; for more detailed descriptions the reader is referred to recent reviews by Brandtzaeg et al. [18, 19], Mayhofer [51] and Owen and Ermak [62].

In a typical cross section of the small intestine, GALT is distinctively present in 4 foci (fig.1): the mucosa (mucosa-associated lymphoid tissue – MALT), LP, PP and between epithelial cells as IEL [62]. Scaffolding of GALT is provided by the intestinal stroma – mainly fibroblasts. Lymphoid cells are scarce in epithelial glands (i.e. crypts of Lieberkühn, Brunner glands and others), and are virtually absent from the muscularis mucosa and from the tunica muscularis [62]. Lymphoid cells are diffusely present in LP and MALT without any apparent internal organization. Migration of cells (mainly lymphocytes) occurs freely between these areas [63]. In contrast, PP have profound internal organization and cell compartmentalization [51]. The PP consists of lymphoid follicles that occupy the full thicknesss of the intestinal mucosa. The basic structure of the PP is the lymphoid follicle. It consists of a subepithelial dome of comparatively low cell density, a mantle (corona) of closely packed small lymphocytes, and an eccentric inner area consisting of lymphoblasts – the germinal center [50]. Between each follicle lies an area of heterogeneous lymphoid tissue – the interfollicular, or parafollicular tissue – consisting of HEV, fibroblasts and a mixture of small and large lymphocytes [63]. Selective immunochemical staining reveals that the dome, corona and germinal center are predominantly populated by B lymphocytes, whereas T lymphocytes are abundant in the parafollicular area, but are only scattered in follicular corona and dome (all appear to be thymus-derived) [64–66]. The T lymphocytes present in the dome and corona are predominantly CD4+ cells of both Th1 and Th2 phenotypes, whereas T lymphocytes of the parafollicular area contain both CD4+ and CD8+ phenotypes [64, 66].

The follicular dome and overlaying epithelial layer have a unique structure that enables antigen sampling [62]. An important physical barrier throughout the gastrointestinal tract is the coating of mucus produced by goblet cells. This layer physically holds microorganisms and particles away from the mucosal surface and traps particles in a gel which is carried down the gastrointestinal tract by peristalsis. In the dome epithelium however, goblet cells are often absent or diminished in number, thus reducing the thickness of the mucus layer [62]. The epithelial barrier is further reduced by M cells. These are epithelial cells adapted for vesicular transport of a wide range of particles from the lumen to lymphoid cells within the dome [67–70]. They have short widely spaced microvilli and lack a rigid internal cytoskeleton which makes them easily displaced and deformed by migrating lymphoid cells in the dome [70]. Cellular processes invaginate the basolateral intercellular space (microfolds), and form pockets through which lymphoid cells enter and exit without disrupting M cell membranes [62]. M cells are renewed within the same crypts that renew enterocytes and goblet cells, indicating that all might originate from a common stem cell [64]; the mechanism controlling M cell differentiation and frequency is not known.

The number of follicles within PP, their size, and distribution throughout the small intestine vary from species to species and with age [71]. In humans, PP in the duodenum are small, consisting of a few follicles [71]. They increase in size more distally in the ileum, and are largest in the terminal ileum [71]. The size and distribution of PP roughly correlates with the distribution of endogenous enteric microflora. In mice and rats, PP contain from 2 to 11 follicles and are more or less uniform in size throughout the small intestine [72]. In ruminants, PP vary in size and in character [72]. PP in the proximal intestine resemble those of man and rodents, with M cells forming only a portion of the follicle-associated epithelium. In the distal ileum, however, PP are bigger and contain large numbers of M cells. The increase of PP number and reduction of the epithelial barrier in the distal ileum is explained by the increase of bacterial density towards the colon [62].

IEL reside adjacent to the columnar epithelial layer of small intestine villi, and constitute 15–20% of cell number. Virtually all IEL lie on the basal membrane between enterocytes; their movement between enterocytes is marginally possible, and is limited by enterocyte desmosomal structures [50, 73]. They are uniformly distributed throughout the villus with a lower density present near the crypts [73]. Morphologically, most IEL appear larger than resting lymphocytes, though they are not blasts [74, 75]. The IEL population possesses a number of unique features that are distinct from lymphocytes residing in other lymphoid tissues (TcR usage, surface molecules and biological functions) [50]. About 90% of mononuclear cells isolated from murine intestinal epithelium express CD3 and an associated TcR (>90% in the human [50]) [76, 77]; the number of IEL that are sIg+ (B lymphocytes) is virtually zero [50]. This indicates that the majority of IEL are of T lymphocyte lineage. Seventy-five percent of CD3+ murine IEL are CD8+, CD4− (80% in the human [50]) [76, 77]. Other subpopulations and frequencies are: CD8−, CD4+ – 5–10%, CD8+, CD4− – 5%. Studies have shown that 45–65% of murine CD8+, CD4− IEL use the $\gamma\delta$ TcR [78], and it is this population that has been shown to be of extrathymic origin [46–49] (the frequency of this population is apparently lower in the human [50]; see previous section too). Other CD8+, CD4− cells (35–55%) use the $\alpha\beta$ TcR [78]. Another distinctive feature of the $\gamma\delta$ TcR, CD8+, CD4− IEL is their usage of the homodimeric α/α form of CD8 [49]. $\alpha\beta$ TcR, CD8+, CD4− IEL express the heterodimeric α/β form of CD8, are of thymic origin, and arrive from nearby PP [49]. Murine $\gamma\delta$ IEL express $V\gamma7$ and $V\delta1$ or $V\delta4$ genes which are distinct from the $\gamma\delta$ TcR cells in other tissues [76–78]. Collectively, these data indicate the unique structure of intestinal IEL. The immunological significance of IEL is unclear. Murine $\alpha\beta$ TcR IEL have been shown to provide help [79], whereas murine $\gamma\delta$ IEL, which account for more than 50% of IEL, are associated with cytotoxicity [80], NK function [81], and have been shown to abrogate OT [79] (this will be discussed in more detail in the following section).

Function: Interpretation of Immunogenic and Tolerogenic Signals

The most intriguing feature of GALT is its capacity to generate potent immune responses on one hand, and to induce peripheral tolerance (OT) to external antigens on the other [19, 36, 82] (fig. 2). Interestingly, both

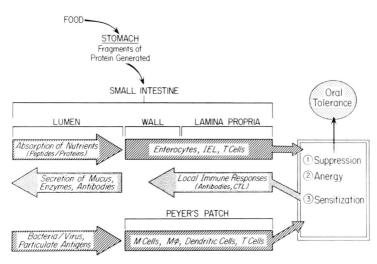

Fig. 2. Immune functions of GALT. Intestinal immune functions include the induction of local responses against pathogens in the gut lumen, and the generation of systemic tolerance. The generation of oral tolerance occurs in the PP and/or as a result of antigen absorption via enterocytes. The induction of local immune responses also occurs in PP.

processes require antigen stimulation [82], involve cytokine production [19], and might occur at the same time – the first leading to potent local and systemic immune responses, while the latter leads to systemic antigen-specific nonresponsiveness [83].

The generation of acquired (specific) immune responses in the small intestine is believed to occur in PP [18, 19] (fig. 2). It is suggested that antigen is taken up by M cells and is transferred to the dome area. Here, either directly or via class II-positive macrophages and/or dendritic cells, antigen is presented to T lymphocytes, which undergo stimulation and proliferation in the dome and parafollicular areas. At the same time, B lymphocytes are stimulated in the dome, proliferate and thus form follicular germinal centers. The terminal differentiation of B lymphocytes is determined by cytokines secreted by differentiating T lymphocytes in the parafollicular areas. Following differentiation, effector cells migrate to the mucosa and LP, where antibody is secreted and/or cell-mediated lysis is performed. Antibodies secreted in mucosa and LP are IgM, IgA and IgE. IgA is pinocytosed by means of receptor-mediated endocytosis into enterocytes and then secreted into the intestinal lumen; IgM might be secreted in a similar manner or via bile epithelium [84]. IgE remains in the mucosa bound to local mast cells [84]. IgG is predominantly released into the bloodstream [84]. Thus, B lymphocyte-directed responses are mainly local and indicate the presence of Th2 CD4+ lymphocytes [19, 85, 86].

The site of tolerance generation is not known. PP have been suggested as a possible site for suppressing excessive immune responses and for generation of OT [18, 19, 38, 87]. This is a logical possibility, since PP are a predominant site of lymphocyte differentiation, and inhibition of differenitiation would lead to diminished immune responses. Suppression is proposed to be induced by CD4+ T lymphocyte suppressor inducers and mediated by CD8+ suppressor T lymphocytes [19, 20, 38]. Suppression might be achieved in PP, mucosa, LP, or by IEL. Local effects of suppression are suggested to be mediated by plieotropic cytokines that affect lymphocytes as well as other cells of the immediate environment, i.e. IL-6, TGF-β, TNF-α, IFN-γ [38, 61, 88].

The above scenario assumes that immune responses occur prior to initiation of suppression. Is this the case for OT too? Most data indicate the opposite: a single, or repeated feeding of a soluble protein does not lead to any detectable immune response prior to parenteral immunization (IgA, IgE, IgG, DTH or T lymphocyte proliferation) [20]. Furthermore, OT has to be abrogated to enable local (and systemic) immune responses [89, 90]. There is evidence, however, to show that repeated feedings, especially those of particulate antigens, can lead to weak, transient local immune responses (mainly antibody production) prior to the domination of tolerance [20, 91–94]. Thus, in contrast to suppression of excessive immune responses, in order to generate OT a tolerogenic signal is delivered prior to, or concurrently with, the induction of a detectacle immune response. What determines the dominance, and thereby the immunological outcome, of concurrent tolerogenic and immunogenic signals? Are both signals generated at the same site and by the same cells? These questions remain unanswered by available data, however, some possible resolutions are suggested in the following.

In general, the signal generated might reflect previous contact with the same antigen [95]. Thus, antigen-derived material absorbed from the gut has a dual immunogenic and tolerogenic potential, such that the net effect of feeding is strongly influenced by responses to past and concurrent contacts with specific antigen [83]. Though the gut may either respond or become tolerant to a given antigen, evidence is available to show that both do not occur at the same time [96]. Other studies indicate that in young immature animals the two signals might be generated concurrently with the ensuing dominance of OT [97, 98]. We assume, therefore, that antigen-specific tolerogenic and immunogenic signals are *not* generated simultaneously; a given antigen might induce either one of the two, but not both together.

Whether or not a tolerogenic/immunogenic signal is deliverd in response to a new antigen depends upon the type and structure of antigen and the route of entry [20]. Though dietary antigens are likely to be degraded by the time they reach the small intestine, several studies have indicated that degradation is partial, and that in fact intact antigen is absorbed too [32, 99, 100]. Absorbed antigen – undegraded or partially degraded – might have a key role in the generation of OT [101], since serum containing antigen, absorbed from the gut 1 h after feeding, transfers OT [35, 102, 103]. In addition, we have found that passively transferred anti-OVA antibodies prevented the generation of OVA-specific OT [Melamed and Friedman, in preparation]. Soluble antigen (intact or degraded) is probably not absorbed by M cells which are specialized to take up particulate antigen such as bacteria [20, 62]. Thus, a particulate bacterial antigen absorbed by M cells would deliver an immunogenic signal via PP. Soluble antigen (degraded or intact) is more likely to be absorbed by enterocytes, passed through the cell body, secreted into the basal lamina, and from there diffuses into the LP capillaries and lacteals. This route of entry passes by IEL and LP lymphoid tissue and is most likely to induce tolerogenic signals. In addition. the presence of antigen in the absence of inflammatory costimulatory signals could generate anergy [104].

Tolerogenic signals might be generated by CD4+, TcR $\alpha\beta$ suppressor-inducer T lymphocytes, or by CD8+, TcR $\alpha\beta$ suppressor T lymphocytes capable of producing inhibitory cytokines such as TGF-β [38]. What, if any, is the role of TcR $\gamma\delta$ T lymphocytes in the generation of OT? These cells have been proposed to form a first line of defense at epithelial surfaces by destroying cells altered by transformation, infection or other forms of damage [105]. This hypothesis is supported by several IEL functions such as cytotoxicity [106, 107], recognition of antigen associated with nonclassical MHC proteins expressed by enterocytes [108], and recognition of mycobacteria-derived superantigens and heat-shock proteins [108–110]. Peripheral tolerance, induced by nasal administration of antigen, has been transferred by TcR $\gamma\delta$ IEL from airway epithelium [111], and by TcR $\gamma\delta$ T lymphocytes from spleens [112]. However, intestinal TcR $\gamma\delta$ IEL were reported to abrogate OT when transferred to orally tolerized recipients [79, 90, 113, 114]. These findings were interpreted to indicate that intestinal IEL protect local IgA responses under conditions of OT [90]. A possible regulatory role for TcR $\gamma\delta$ IEL (whether they generate or abrogate OT) is supported by the panel of cytokines they have been shown to release: IL-2, IL-3, IL-6, IFN-γ, TGF-β [88], and IL-5 [115]. These cytokines regulate B and T lymphocyte physiology as well as that of epithelium [88], and HEV endothelium [61], and thereby indicate possible integrative regulation of the immediate environment. Hence, monitoring the state of tolerance (by either generating or abrogating) might be considered as a 'first-line' function too.

Hence, the distinction between tolerance and immune response is dependent upon antigen structure, route of antigen internalization, site of exposure to antigen within the gut, and previous contact with antigen.

Another possible pathway for generation of OT, which is not mutually exclusive, assumes that peripheral tolerance is not induced locally, but rather is induced systemically upon transfer of intact antigen, or its peptides, into the circulation [32, 99, 100, 116]. Thus, OT could be the result of an overflowing of degraded-soluble dietary antigens into the bloodstream after being absorbed by enterocytes. These antigens, distributed via blood and lymph vessels, might induce tolerance in peripheral lymph nodes, including PP. Hence, peripheral or oral tolerance would be generated in designated lymphoid organs and not only in the diffuse lymphoid tissue of the gut, and is dependent upon the route of antigen entry. Consistent with this notion are studies showing that enterally administered aqueous antigen induces tolerance [116, 117] that orally fed protein antigens are found in the blood within 1 h of feeding [100], and that transfer of serum containing anti-OVA antibodies inhibits the establishment of OVA-specific OT [Melamed and Friedman, in preparation]. Immunogenic signals, as described above, are generated in the gut by uptake of particulate antigen by M cells and concomitant PP stimulation.

Mechanisms for Oral Tolerance

General Considerations
Peripheral tolerance is the process by which cells of the acquired immune system are tolerized by antigens not present in thymus or bone marrow at the time of differentiation. Peripheral tolerance to self antigens has been shown to occur in both T and B lymphocytes, with clonal deletion and anergy being the major mechanisms [3]; immune suppression has been considered to be a major mechanism for regulating excessive immune responses, whether or not directed against self antigens [3]. Peripheral tolerance to external antigens, as represented by OT, is different in several important aspects: (1) the effector cells are believed to be T lymphocytes, with B lymphocyte responsiveness in OT, or peripheral tolerance for that matter, being controlled by tolerized T lymphocytes [2, 20]; (2) clonal deletion does not occur following oral tolerization [20], and (3) both suppression and anergy have been found to be major mechanisms for maintaining OT [20, 38, 118–120].

Induction of OT to an antigen is measured by the absence, or reduction, of specific immune responses that should be evoked by parenteral immunization with the same antigen following feeding [20, 82, 121–126]. Parenteral immunization is given in hind footpads or in multiple subcutaneous sites, and involves emulsification of antigen in complete Freund's adjuvant (CFA), aluminum hydroxide or use of antigen-pulsed peritoneal exudate cells [120, 126–128]. The assays commonly used for measuring OT are detection of antibody production [129], delayed-type hypersensitivity (DTH) [28], contact sensitivity [130] (all these are a measure of OT in vivo), hemolytic plaque-forming cells (PFC [29]), T lymphocyte proliferation [29] and IL-2 production [118, 119] (the last three assays are a measure of OT in vitro). Reduction of disease severity is used as an assay to evaluate OT against self antigens in both experimental and spontaneous autoimmune diseases [38]. We have recently shown altered cytokine patterns in brains of rats orally tolerized against guinea pig myelin basic protein (GP-MBP) [131], and have consequently begun evaluating OT by measuring cytokine release in PP of orally tolerized mice [132] as well as mRNA expression of these cytokines [133].

OT may be evoked by feeding both soluble and particulate antigens (i.e. red blood cells [cf. 35, 134]) as well as by means of gastric intubation [20]. The feeding regimens used to induce OT differ [128]. Thus, OT may be induced by a single feeding of a protein antigen [35, 121, 126, 127, 135, 136], or by several intermittent feedings [38, 93, 137, 138]. When a single feeding regimen is used, larger doses are required to induce tolerance (approximately 10 mg depending on antigen and species [82]). An intermittent regimen usually requires smaller dosages (i.e. 0.25 mg per feeding for 5 intermittent feedings [107, 139]). A single feeding is sufficient to induce complete T lymphocyte unresponsiveness [126], but several feedings are required to block off antibody production [20] [Melamed and Friedman, unpubl. data]. Feeding regimens might have profound effects on mechanisms of unresponsiveness expressed by tolerized T lymphocytes [116, 140–142]. OT has been successfully induced in the mouse, rat, human, pig, dog [for review, see 82], and chicken [Friedman et al., unpubl. data], and less successfully so in the rabbit and ruminants [82]. OT is easily induced in immunologically mature subjects, but not in immunologically immature, suckling pups [23, 82, 143].

The induction of OT is antigen-specific and systemic [20]. Thus, immunization of mice orally tolerized against OVA with a mixture of antigens in CFA revealed unresponsiveness only to apply to OVA and not to other antigens present in the immunizing mixture (HSA and PPD) [126]; this pattern of unresponsiveness was observed in all lymphoid organs tested (popliteal, inguinal, mesenteric and cervical lymph nodes, PP, thoracic duct lymphocytes and spleen [120]). The antigenic epitopes responsible for generating tolerance might differ from those responsible for generating immune responses [141]. The state of unresponsiveness following OT has been reported to be transient [20, 126, 136]. Thus, if feeding of antigen was terminated, immune responses against the tolerizing antigen reappeared within 21–65 days, depending on the feeding regimen, mouse strain and antigen used [20, 126 136]. If, however, antigen persistence was maintained, unresponsiveness did not subside for the entire experimental period reported [136]. Other factors that have been reported to affect OT are strain differences and maternal influences [for review, see 83].

As discussed above, the mechanism by which orally administered antigen induces tolerance most probably relates to the structure of protein antigens, and route of entrance via GALT. Several mechanisms have been reported to be responsible for mediating OT: anti-idiotypic antibodies, immune complexes, biologically filtered antigen and regulatory T lymphocytes [reviewed in 20]. The following discussion assumes that the primary mechanism responsible for mediating OT involves regulatory T lymphocytes; the validity of other mechanisms is discussed in Mowat [20].

Clonal Deletion

Of the three mechanisms for immune tolerance mentioned above, clonal deletion seems the least likely to explain OT, since OT may be specifically abrogated [20]. Thus, treatment of orally tolerized animals with cyclophosphamide, 2′-deoxyguanosine, or by transfer of TcR $\gamma\delta$ IEL results in complete recovery of responsiveness [79, 144, 145].

Anergy

Anergy has only recently been demonstrated as a possible mechanism for OT [118–120]. Anergy is defined as a state of T lymphocyte unresponsiveness characterized by absence of proliferation, IL-2 production and diminished expression of IL-2R [146, 147]. Anergy may be experimentally differentiated from clonal deletion by demonstrating the presence of antigen-specific TcR clontypes [15], or by release from the anergic state which is accomplished by preculture of cells in IL-2 [104]. Under these conditions we have shown that a single feeding of 20 mg OVA induced a state of anergy in OVA-specific T lymphocytes: cells did not respond to OVA by proliferation, OVA stimulation did not induce IL-2 production or IL-2R expression, and the nonresponsive state was reversed by preculture of tolerized cells in IL-2 [119, 120]. One other study has indirectly demonstrated anergy as a mechanism for OT: Whitacre et al. [118] reported diminished IL-2 and IFN-γ production in rats orally fed myelin basic protein (MBP) in the presence of the soybean protease inhibitor; however, anergy was not confirmed in this study by TcR analysis or by IL-2-driven release. We have recently shown that the induction of anergy depends upon antigen dosage and frequency of feeding [142].

Suppression

Most available data support suppression as a mechanism for OT [20, 38]. Immunological suppression is classically demonstrated by the suppression of antigen-specific immune responses by T lymphocytes [148, 149]. Specific suppression means that both suppressor and responding cells are specific for the same antigen [148], though cells might be responding to different epotopes on the same antigen [149]. Bystander suppression means that antigen (or epitope)-specific suppressor cells suppress immune responses of cells specific for *other* antigens (or epitopes) [38]. In either case, the suppressor cells are specifically activated and driven by antigen [38]. Both specific and bystander suppression are active forms of suppression and are probably mediated by inhibitory cytokines (see next section) [38], though direct or cognitive suppression cannot be formally ruled out at this time [148]. The phenotype of a 'professional' suppressor cell is a CD8+ TcR $\alpha\beta$ T lymphocyte [149], the activation of which is dependent upon the presence of a CD4+ suppressor inducer T lymphocyte [149, 150].

OT was primarily shown to be transferred by splenic T lymphocytes into syngeneic, irradiated, naive recipients [151]; this indicated that OT was mediated by T lymphocytes. The splenic T lymphocytes transferring OT were implicated as active suppressor cells because they were capable of transferring OT to normal recipients [28]. At the same time, it was also shown that orally tolerized recipients suppressed responses of normal transferred spleen cells [128]. PP and mesenteric lymph nodes were also shown to harbor antigen-specific suppressor cells following feeding [130, 152]. These cells suppressed anti-sheep erythrocyte responses of normal cells cotransferred into sublethally irradiated recipients, and were lost consequently to anti-Thy 1.2 treatment [152]. Further support for a suppressor cell-mediated mechanism for OT was attained by abrogating suppression by means of cyclophosphamide [153–155] and by in vitro cell mixing experiments [20, 92, 135, 156]. Suppression has also been abrogated by treatment of tolerized cells with an anti "suppressor cell" antiserum [89, 157, 158]. In most cases, the cells responsible for these functional observations were phenotypically characterized as a CD8+ TcR $\alpha\beta$ T lymphocytes [20]. Several reports indicated, however, that suppression could be mediated by CD8+ TcR $\gamma\delta$ T lymphocytes and by migrating CD4+ TcR $\alpha\beta$, cyclophosphamide-resistant, suppressor-inducer T lymphocytes [87, 111, 112]. CD8+ TcR $\gamma\delta$ suppressor T lymphocytes were demonstrated in airway epithelium, but not in the gut [79, 111–114].

Not all attempts to abrogate, or transfer T lymphocyte-mediated suppression were successful; these studies implicated other T lymphocyte-transferable mechanisms in the generation of OT [140, 159]. The involvement of additional mechanisms in OT was indicated by successful transfer of OT by serum from fed animals, providing the serum transferred was collected 1 h after feeding [100]. The effect of serum might, however, be explained by activation of specific suppressor cells by factors (antigenic fragments) present in serum 1 h after feeding [154].

More recent studies on OT have led to the description of a different form of suppression-bystander suppression (see above) [38]. Bystander suppression was shown to be mediated by CD8+ TcR $\alpha\beta$ T lymphocytes [139, 160]. These cells, after being specifically activated and driven by antigen, were capable of suppressing, via cytokine release, responses of other T lymphocyte populations to their respective antigens [139, 160]. This form of suppression is most intriguing, since the oral administration of a nondisease-inducing portion of an autoantigen represents an antigen-specific method by which an autoimmune disease can be immune-regulated [161]. The remainder of this review addresses this possibility.

Clinical Application of Oral Tolerance

Autoimmune Diseases and Allergies

One of the primary goals in developing effective therapy for auto-immune diseases is to specifically suppress autoreactive immune processes without affecting the rest of the immune system. We and others have been investigating antigen-driven peripheral tolerance as a means to suppress autoimmune processes using OT because of its inherent clinical applicability [38–41] (see fig. 3).

In order to test whether feeding an autoantigen could suppress an experimental autoimmune disease, the Lewis rat model of experimental autoimmune encephalomyelitis (EAE) was studied [161]. With increasing dosages of GP-MBP, the incidence and severity of disease was suppressed, as well as proliferative responses of lymph node cells to MBP. Antibody responses to MBP were decreased but not as dramatically as proliferative responses. Thus, it appears that oral tolerance to MBP, as to other non-self antigens [20], preferentially suppresses cellular immune responses. EAE is associated with inflammatory cells that accumulate in the central nervous system (CNS). In animals fed MBP, there was a marked decrease in the number of cells infiltrating the nervous system. In order to determine the duration of protection following feeding, animals were fed 3 times prior to immunization and then selected groups were immunized at weekly inter-

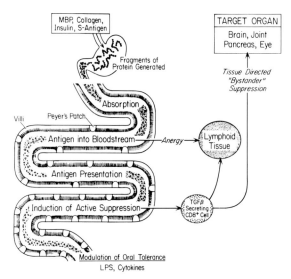

Fig. 3. Suppression of autoimmunity by oral administration of autoantigen. An autoantigen is fed and leads to the generation of anergy and/or active suppression. Anergy is postulated to occur by absorption of antigenic fragments into the bloodstream, and active suppression by the generation of regulatory cells in PP. Both mechanisms lead to a state of systemic unresponsiveness. Active suppression is mediated by a TGF-β-secreting CD8+ T lymphocyte, or other regulatory cells that secrete inhibitory cytokines such as IL-4 or IL-10. The regulatory cells migrate to the target organ where they encounter the oral tolerogen, and suppress inflammation via the release of inhibitory cytokines (bystander suppression).

vals. Animals were protected for approximately 2–3 months after this feeding regimen.

As previously stated, in bystander suppression the oral administration of a nondisease-inducing portion of an autoantigen represents an antigen-specific method by which an autoimmune disease can be immune-regulated [161]. With this in mind we investigated whether nonencephalitogenic portions of MBP could suppress EAE; to do so, fragments of MBP and synthetic peptides were orally administered [161]. Suppression of disease occurred by feeding fragments or synthetic peptides prior to immunization with MBP in CFA. There was some suggestion that nonencephalitogenic fragments were more potent in generating suppression than encephalitogenic fragments. This possibility has now been confirmed by further studies [162]: bystander suppression has been demonstrated in the EAE model in which orally administered MBP suppresses proteolipid protein (PLP)-induced disease in the SJL mouse.

We also noted that feeding bovine MBP suppressed EAE in the Lewis rat and in the strain 13 guinea pig, showing cross-species tolerization [163].

Nonetheless, in a recent series of experiments, it appears that homologous MBP is a more potent oral tolerogen for EAE than heterologous MBP [164].

The majority of studies related to mechanisms of OT suggest that active suppression is generated following exposure of antigen via the gut [20]. To test this mechanism in the EAE model, mesenteric lymph nodes and spleen cells were adoptively transferred from animals fed MBP into naive animals that were then immunized with MBP in CFA. We found that protection was adoptively transferred and that such protection was dependent on CD8+ T lymphocytes [165]. Splenic and mesenteric T lymphocytes from fed animals also suppressed in vitro proliferative responses and antibody production in vivo. In vitro suppression was also mediated by CD8+ T lymphocytes. The suppression was antigen-specific in that T lymphocytes from animals fed MBP suppressed MBP responses, but not responses to mycobacteria.

To further study the mechanism of OT in the EAE model, we studied the ability of cells from MBP-fed animals to suppress proliferative responses of an MBP or OVA T lymphocyte line using a transwell system. We found that cells from MBP-tolerized animals could suppress either an MBP or an OVA line across the transwell provided that the cells from the fed animals were triggered in vitro with the oral tolerogen [139]. The factor responsible for the suppression was identified as TGF-β. CD8+ cells from animals fed MBP released TGF-β in vitro when stimulated with the fed antigen. Furthermore, in vivo administration of anti-TGF-β antibody abrogated OT. Detailed immunohistology was performed in animals orally tolerized with MBP and in animals naturally recovering from EAE [131]. Brains from OVA-fed animals at the peak of disease showed perivascular infiltration with activated mononuclear cells which secreted the inflammatory cytokines IL-1, IL-2, TNF-α, IFN-γ, IL-6 and IL-8. Inhibitory cytokines TGF-β and IL-4 and prostaglandin E$_2$ (PGE$_2$) were absent. In MBP orally tolerized animals there was a marked reduction of the perivascular infiltrate and down-regulation of the inflammatory cytokines. In addition, there was up-regulation of the inhibitory cytokine TGF-β. When bacterial lipopolysaccharide (LPS) was fed in addition to MBP, protection against EAE was enhanced and was associated with elevated IL-4 and PGE$_2$ in the brain [131, 166]. In control recovering animals (day 18) staining for inflammatory cytokines was diminished and there was up-regulation of TGF-β and IL-4. These results suggest that the suppression of EAE, whether induced by oral tolerization or occurring due to natural recovery, is related to the secretion of inhibitory cytokines or factors that actively suppress the inflammatory process in the target organ.

EAE in the Lewis rat is usually an acute monophasic illness. For oral administration of autoantigens to have clinical applicability, it must be effective in disease processes where activated autoreactive cells already exist [39]. To study the ability of OT to suppress ongoing autoimmune processes, experiments were performed in relapsing models of EAE [163]. A relapsing model of EAE occurs in the Lewis rat following injection of spinal cord homogenate plus adjuvant. Oral administration of MBP to animals recovering from their first attack significantly suppressed the second attack and decreased histological manifestations of the disease. In addition, cell-mediated immunity as measured by DTH and antimyelin antibody responses, was also suppressed. A more chronic model of EAE occurs in the strain 13 guinea pig. In a series of experiments, guinea pigs were injected with white matter homogenate in CFA, and upon recovering from the first attack, were fed 10 mg of bovine myelin or BSA, 3 times weekly over a 3-month period. In animals fed the bovine myelin preparation there was a diminution in frequency of attacks, and a decrease in demyelination in the spinal cord and certain portions of the white matter. These results demonstrate that orally administered myelin antigens can suppress chronic relapsing EAE and that induction of tolerance via the oral route is relevant to the therapy of human demyelinating disorders such as multiple sclerosis.

In addition to oral exposure to antigen, the body is constantly exposed to inhaled antigen which affects the bronchial associated lymphoid tissue in a fashion similar to GALT [24, 25]. To determine the effect of this route of antigen exposure on EAE, MBP was aerosolized to Lewis rats [167]. Either a single or intermittent dosage of aerosolized MBP completely abrogated clinical EAE. CNS inflammation, DTH reactions and antibody responses to MBP were also significantly reduced in aerosol-treated animals. Aerosolization of histone, a basic protein of similar charge as MBP, had no effect. Aerosolization was more effective than oral administration of MBP over a wide dose range (0.005–5 mg) suggesting that protection via inhalation was not merely secondary to gastric absorption of aerosolized antigen. Splenic T cells isolated from animals that had inhaled aerosolized MBP adoptively transferred protection to naive animals immunized with MBP. Aerosolization of MBP to animals with relapsing EAE after recovery from the first attack decreased subsequent attack severity, as well as antibody and DTH responses to MBP. Thus, aerosolization of an autoantigen is a highly potent method to down-regulate an experimental T cell-mediated autoimmune disease and suggests that exposure of antigen to lung mucosal surfaces preferentially generates immunologic tolerance [24, 25, 122, 167].

To further assess oral tolerization as a method to treat autoimmune diseases, studies were performed in experimental autoimmune uveitis

(EAU). Oral administration of S-antigen (S-Ag), a retinal autoantigen that induces EAU, prevented or markedly diminished the clinical appearance of S-Ag-induced disease as measured by ocular inflammation [123, 168]. Furthermore, oral administration of S-Ag also markedly diminished uveitis induced by the uveitogenic M and N fragments of the S-Ag. Oral administration of S-Ag did not prevent MBP-induced EAE. In vitro studies demonstrated a significant decrease in proliferative responses to the S-Ag in lymph node cells draining the site of immunization from fed versus nonfed animals. Furthermore, the addition of splenocytes from S-Ag-fed animals to cultures of a CD4+ S-Ag-specific lymphocyte line profoundly suppressed the line's response to the S-Ag, whereas these splenocytes had no effect on a PPD-specific lymphocyte line. The antigen-specific in vitro suppression was blocked by anti-CD8 antibody, demonstrating that suppression was dependent on CD8+ T lymphocytes. As in EAE, EAU was severely depressed by feeding S-Ag-related peptides that were either uveitogenic, cross-reactive or synthetic [169–171].

Oral tolerance was also tested in the adjuvant arthritis model. Previous investigators have demonstrated suppression of collagen-induced arthritis by feeding collagen type II [172–174]. We studied adjuvant arthritis (AA) in the rat, another well-characterized and more fulminant form of experimental arthritis [175]. Adjuvant arthritis is induced by injection of *Mycobacterium tuberculosis* (MT) into the base of the tail. Attempts to suppress adjuvant arthritis by oral administration of MT were not successful. Nonetheless, we found that oral administration of chicken collagen type II (CII), given at a dose of 3 μg per feeding, consistently suppressed the development of AA. A decrease in DTH responses to CII was observed that correlated with suppression to AA. Oral administration of collagen type I also suppressed AA; only minimal effects were seen with collagen type III. Suppression was antigen-specific in that feeding CII did not suppress EAE, and feeding MBP did not suppress AA. Suppression of AA could be adoptively transferred by T cells from CII-fed animals and could be obtained when CII was fed after disease onset. These results suggest that autoimmunity to CII may have a pathogenic role in AA. Alternatively, suppression of AA by CII may be related to the phenomenon of antigen-driven bystander suppression [139].

NOD diabetic mice spontaneously develop an autoimmune form of diabetes associated with insulitis. This is a naturally occurring disease and the autoimmune nature of the disease is suggested by lymphocytic infiltration of the islets of Langerhans which precedes the destruction of insulin-producing β cells. A variety of immunomodulatory treatments have been studied in the NOD mouse and immunsuppressive therapy that affects T lymphocyte function has been successful. To test OT as a mode of therapy, we administered porcine insulin at a dose of 1 mg orally twice a week for 5 weeks and then weekly until 1 year of age [176]. The severity of lymphocytic infiltration of pancreatic islets was reduced by oral administration of insulin and there was a delay in the onset of diabetes. A decreased incidence of diabetes was seen in animals followed for 1 year. As expected, orally administered insulin had no metabolic affect on blood glucose levels. Furthermore, splenic T cells from animals orally treated with insulin adoptively transferred protection against diabetes, demonstrating that oral insulin generates active cellular mechanisms that suppress disease. Additional studies have demonstrated the ability to suppress insulitis by administering insulin peptides or the A or B chain of insulin. Given the mechanism of antigen-driven bystander suppression and the role of TGF-β and IL-4 in OT, our results do not definitively implicate autoreactivity to insulin as a pathogenic mechanism in the NOD mouse.

The clinical relevance of OT in the autoimmune state has been studied in several autoimmune-prone mouse strains [177, 178]. These strains, B × SB, MRL-lpr/lpr and NZB, differed in their capacity to become orally tolerized after feeding of several protein antigens [178]. The presence of selective enteric tolerance suggested aberrations in the cellular control of peripheral tolerance [178]. Oral tolerance against bovine γ-globulin and casein failed in the NZB/W mouse strain, which serves as a murine model for systemic lupus erythematosus (SLE) in the human; failure of OT was antigen-specific and age-dependent, and appeared to parallel antibody patterns seen in human SLE [177]. Thus, spontaneous nonorgan-specific autoimmunity might reflect a defect of peripheral tolerance both to self and external antigens. These studies indicate that the use of OT as a therapeutic approach to autoimmunity might only be successful for organ-specific autoimmune diseases, whereas nonorgan-specific diseases, chronic inflammatory diseases, and allergic diseases might not be inhibited by OT.

In contrast to this prediction, OT has been shown to protect against chronic immune complex-mediated nephritis and immune complex disease [179–181]. Furthermore, experimental blunting of OT and promotion of systemic IgG and IgM led to an increase in nephritis severity [180]. Thus, it appears that in spite of the above mentioned failures, OT might be an effective approach for treating both organ-specific autoimmune diseases as well as systemic immune response-mediated diseases [20, 37, 38, 179–182].

Transplantation

The success of using OT to induce peripheral tolerance against self antigens in the autoimmune state promoted experiments designed to test OT as a means for prolonging allograft survival [42, 43, 183]. In an initial series of experiments, we fed splenocytes from Wistar-Furth rats to Lewis rats and studied the accelerated allograft rejection model. We found that oral tolerance to spleen cells prevents sensitization by skin grafts, and transforms accelerated rejection of vascularized cardiac allografts to an acute form typical of unsensitized recipients. In addition, the mixed lymphocyte response in vitro and DTH responses were suppressed following oral administration of antigen [43]. In a second series of experiments we have found induction of immunity and oral tolerance with polymorphic class II major histocompatibility complex allopeptides in the rat. Inbred Lewis rats were immunized or fed class II synthetic MHC allopeptides. In vivo these animals developed DTH responses. Furthermore, oral administration of the allopeptide mixture daily for 5 days before immunization reduced DTH responses both to the allopeptide mixture and to the allogenic splenocytes. This reduction was antigen-specific [183], and was accompanied by elevation of intragraft IL-4, but was not associated with increased IL-4 [184]. The prevention of accelerated graft rejection was also observed if alloantigens were administered intravenously [184]. Similar findings were reported in the pig [42]: islet allograft survival was prolonged by means of preoral tolerization of islet tissue. Thus, oral tolerance may be of benefit in down-regulating alloreactivity associated with transplantation.

Vaccine Design

The development of vaccines to be administered via the oral route is an important approach for disease prevention in both humans and animals; however, OT may be a real impediment in design of orally administered synthetic vaccines [19]. For example, if successful, a rabies oral vaccine could be distributed via bait in the wild as part of a worldwide strategy to prevent the spread of rabies [185–187]. Since the oral route might generate an immune response as well as tolerance, vaccine efficacy will depend upon its capacity to override tolerogenic stimuli [19]. The problem of OT is expected to be significant in development of synthetic nonparticulate vaccines, which might be tolerogenic as conventional protein antigens. This problem might be overcome by protecting vaccines during their delivery to the lower gut by formulating particulate vaccines resistant to digestion [188, 189], by delivering antigen in recombinant bacteria [190], and by constructs expressing inflammatory molecules, such as LPS or cholera toxin, both shown to be effective stimulators of oral immunization [95, 121, 140, 153, 154, 191, 192]; the use of LPS as a modulator of intestinal immunity is of special interest since it has been shown to increase OT as well as increasing immunity [166].

Summary and Conclusions

OT is a relevant biological pathway for generating peripheral tolerance against both self and external antigens with minimal side effects (fig. 3). This route might, therefore, contain promising potential for the treatment of autoimmune and allergic diseases in the human (fig. 3). Thus, oral administration of autoantigens suppresses experimental autoimmune diseases (EAE, EAU, AA, collagen-induced arthritis, NOD diabetes) in a disease- and antigen-specific manner, and oral administration of alloantigens has led to increase of allograft survival. OT might be important in treatment of immune complex diseases and food allergies. OT is mediated by T lymphocytes using at least two nonmutually exclusive mechanisms: suppression and anergy. Suppression can be adoptively transferred by CD8+ T lymphocytes which act by releasing TGF-β and IL-4 following antigen-specific triggering. Antigen-driven tissue-directed suppression occurs following oral administration of an antigen from the target organ, even if it is *not* the disease-inducing antigen (bystander suppression). Thus, synthetic peptides can induce OT, and tolerogenic epitopes of antigen may be different from the autoreactive epitope. Due to the promising results in animal models, OT is being tested in clinical trials in multiple sclerosis, rheumatoid arthritis and uveitis [193, 194].

References

1   Janeway, C.A.: The immune response evolved to discriminate infectious nonself from noninfectious self. Immunol. Today *13:* 11–16 (1992).
2   Mitchison, N.A.: Specialization, tolerance, memory, competition, latency, and strife among T cells. Annu. Rev. Immunol. *10:* 1–12 (1992).
3   Kroemer, G.: Martinez-A.C.: Mechanisms of self tolerance. Immunol. Today *13:* 401–404 (1992).
4   Kappler, J.W.; Staerz, U.D.; White, J.; Marrack, P.: Self tolerance eliminates T cells specific for Mls-modified products of the major histocompatibility complex. Nature *332:* 35–40 (1988).
5   Jones, L.A.; Chin, T.; Longo, D.L.; Kruisbeek, A.M.: Peripheral clonal elimination of functional T cells. Science *250:* 1726–1729 (1990).
6   Webb, S.: Morris, C.; Sprent, J.: Extrathymic tolerance of mature T cells: Clonal elimination as a consequence of immunity. Cell *63:* 1249–1256 (1990).
7   Ramsdell, F.; Fowlkes, B.J.: Clonal deletion versus clonal anergy: The role of the thymus in inducing self tolerance. Science *248:* 1342–1348 (1990).
8   Rammensee, H.G.; Kroschewski, R.; Frangoulis, B.: Clonal anergy in mature Vβ6$^+$ T lymphocytes on immunizing Mls-1^a mice with Mls-1^a expressing cells. Nature *339:* 541–544 (1989).
9   Burkly, L.C.; Kanagawa, O.; Brinster, R.L.; Flavel, R.A.: T cell tolerance by clonal anergy in transgenic mice with nonlymphoid expression of MHC class II I-E. Nature *342:* 564–566 (1989).

10   Ramsdell, F.; Lantz, T.; Fowlkes, B.J.: A nondeletetional mechanism of thymic self
     tolerance. Science 246: 1038–1041 (1989).

11   Gershon, R.K.; Kondo, K.: Infectious immunological tolerance. Immunology 21: 903–
     914 (1971).

12   Germain, R.N.; Benacerraf, B.: A single major pathway of T lymphocyte interactions
     and in antigen-specific immune suppression. Scand. J. Immunol. 13: 1–10 (1981).

13   Tada, T.; Asano, Y.; Sano, K.: Present understanding of suppressor T cells. Res.
     Immunol. 140: 291–294 (1989).

14   Abo, T.: Extrathymic differentiation of T lymphocytes and its biological function.
     Biomed. Res. 13: 1–25 (1992).

15   Miller, J.F.A.P.; Moraham, G.: Peripheral T cell tolerance. Annu. Rev. Immunol. 10:
     51–70 (1992).

16   Goodnow, C.C.: Transgenic mice and analysis of B-cell tolerance. Annu. Rev. Immunol.
     10: 489–518 (1992).

17   Strobel, S.: Mechanisms of gastrointestinal immunoregulation and food induced injury
     to the gut. Eur. J. Clin. Nutr. 45: suppl.1, pp. 1–9 (1990).

18   Brandtzaeg, P.; Baklien, K.; Bjerke, K.; Rognum, T.O.; Scott, H.; Valnes, K.: Nature
     and properties of the human gastrointestinal immune system; in Miller, K.; Nicklin, S.
     (eds): Immunology of the Gastrointestinal Tract. Boca Raton, CRC Press, 1987, pp.
     1–88.

19   Brandtzaeg, P.: Overview of the mucosal immune system. Curr. Top. Microbiol.
     Immunol. 146: 13–28 (1989).

20   Mowat, A.M.: The regulation of immune responses to dietary protein antigens. Im-
     munol. Today 8: 93–98 (1987).

21   Chiller, J.M.; Glasebrook, A.L.: Oral tolerance and the induction of T cell unresponsive-
     ness. Monogr. Allergy. Basel, Karger, 1988, vol. 24, pp. 256–265.

22   Challacombe, S.J.: Cellular factors in the induction of mucosal immunity by oral
     immunization. Adv. Exp. Med. Biol. 216B: 887–899 (1987).

23   Strobel, S.; Fergusen, A.: Oral tolerance: Induction and modulation. Klin. Pädiatr. 197:
     297–301 (1985).

24   Holt, P.G.; Leivers, S.: Tolerance induction via antigen inhalation: Isotype specificity,
     stability, and involvement of suppressor T cells. Int. Arch. Allergy Appl. Immunol. 67:
     155–160 (1982).

25   Holt, P.G.; Batty, J.E.; Turner, K.J.: Inhibition of specific IgE response in mice by
     pre-exposure to inhaled antigen. Immunology 42: 409–417 (1981).

26   Gilmour, M.I.; Wathes, C.M.; Taylor, F.G.: Serum antibody responses in mice to
     intermittent inhalation of ovalbumin dust. Int. Arch. Allergy Appl. Immunol. 95:
     285–288 (1991).

27   Challacombe, S.J.; Tomasi, T.B.: Systemic tolerance and secretory immunity after oral
     immunization. J. Exp. Med. 152: 1459–1472 (1980).

28   Miller, S.D.; Hanson, D.G.: Inhibition of specific immune responses by feeding protein
     antigens. IV. Evidence for tolerance and specific active suppression of cell-mediated
     immune responses to ovalbumin. J. Immunol. 123: 2344–2350 (1979).

29   Titus, R.; Chiller, J.: Orally-induced tolerance. Definition at the cellular level. Int. Arch.
     Allergy Appl. Immunol. 65: 323–338 (1981).

30   Thomas, H.C.; Parrott, D.M.V.: The induction of tolerance to a soluble protein antigen
     by oral administration. Immunology 27: 631–639 (1974).

31   Hanson, D.G.: Ontogeny of orally induced tolerance to soluble proteins in mice. I.
     Priming and tolerance in newborns. J. Immunol. 127: 1518–1524 (1981).

32 Stokes, C.R.; Swarbrick, E.T.; Soothill, J.F.: Genetic differences in immune exclusion and partial tolerance to ingested antigens. Clin. Exp. Immunol. *52:* 678–684 (1983).

33 Lamont, A.G.; Mowat, A.M.; Browning, M.J.; Parrott, D.M.: Genetic control of oval tolerance to ovalbumin in mice. Immunology *63:* 737–739 (1988).

34 Vaz, N.M.; Rios, M.J.; Lopes, L.M.; Gontijo, C.M.; Castanheira, E.B.; Jacquemart, F.; Andrade, L.A.: Genetics of susceptibility to oral tolerance to ovalbumin. Braz. J. Med. Biol. Res. *20:* 785–790 (1987).

35 Lamont, A.G.; Gordon, M.; Ferguson, A.: Oral tolerance in protein deprived mice. II. Evidence of normal 'gut processing' of ovalbumin, but suppressor cell deficiency, in deprived mice. Immunology *61:* 339–343 (1987).

36 Mestecky, J.; McGhee, J.R.: Oral immunization: Past and present. Curr. Top. Microbiol. Immunol. *146:* 3–12 (1989).

37 Emancipator, S.N.; Lamm, M.E.: Oral tolerance as a protective mechanism against hypersensitivity disease. Monogr. Allergy. Basel, Karger, 1988, vol. 24, pp. 244–250.

38 Weiner, H.L.; Zhang, Z.J.; Khoury, S.J.; Miller, A.; Al-Sabbagh, A.; Brod, S.A.; Lider, O.; Higgins, P.; Sobel, R.; Nussenblatt, R.B.; Hafler, D.A.: Antigen-driven peripheral immune tolerance. Suppression of organ-specific autoimmune diseases by oral administration of autoantigens. Ann. N.Y. Acad. Sci. *636:* 227–232 (1991).

39 Thompson, H.S.; Staines, N.A.: Could specific oral tolerance be a therapy for autoimmune disease? Immunol. Today *11:* 396–399 (1990).

40 Staines, N.A.: Oral tolerance and collagen arthritis. Br. J. Rheumatol. *30:* suppl. 2, pp. 40–43 (1991).

41 Staines, N.A.: Oral tolerance and collagen arthritis. Br. J. Rheumatol. *31:* 283–284 (1992).

42 Hrstka, J.; Hesse, U.J.; Hrstka, V.; Danis, J.; Schmitz-Rode, M.; Tunggal, B.; Hordegen, P.; Peters, S.: Prolongation of islet allograft survival following immunologic conditioning by antig feeding in the pig. Transplant. Proc. *24:* 663–664 (1992).

43 Sayegh, M.H.; Zhang, Z.J.; Hancock, W.W.; Kwok, C.A.; Carpenter, C.B.; Weiner, H.L.: Down-regulation of the immune response to histocompatibility antigen and prevention of sensitization by skin allografts by orally administered alloantigen. Transplantation *53:* 163–166 (1992).

44 Moxey, P.C.; Trier, J.S.: Specialised cell types in the human fetal intestine. Anat. Rec. *191:* 269–285 (1979).

45 MacDonald, T.T.; Spencer, J.: Ontogeny of the mucosal immune response. Springer Semin. Immunopathol. *12:* 129–137 (1990).

46 Rocha, B.; Vassalli, P.; Guy-Brand, D.: The extrathymic T-cell development pathway. Immunol. Today *13:* 449–454 (1992).

47 Mosley, R.L.; Styre, D.; Klein, J.R.: Differentiation and functional maturation of bone marrow-derived intestinal epithelial T cells expressing membrane T cell receptor in athymic radiation chimeras. J. Immunol. *145:* 1369–1375 (1990).

48 Bandeira, A.; Itohara, S.; Bonneville, M.; Burlen-Defranoux, O.; Mota-Santos, T.; Coutinho, A.; Tonegawa, S.: Extrathymic origin of intestinal intraepithelial lymphocytes bearing T-cell antigen receptor $\gamma\delta$. Proc. Natl Acad. Sci. USA *88:* 43–47 (1991).

49 Guy-Grand, D.; Cerf-Bensussan, N.; Malissen, B.; Malassis-Seris, M.; Briottet, C.; Vassalli, P.: Two gut intraepithelial CD8+ lymphocyte populations with different T cell receptors: A role for the gut epithelium in T cell differentiation. J. Exp. Med. *173:* 471–481 (1991).

50 Mowat, A.M.: Human intaepithelial lymphocytes. Springer Semin. Immunopathol. *12:* 165–190 (1990).

51 Mayrhofer, G.: Physiology of the intestinal immune system; in Newby, T.J.; Stokes, C.R. (eds): Local Immune Responses of the Gut. Boca Raton, CRC Press, 1984, pp. 1–98.

52 Challacombe, S.J.: The investigation of secretory and systemic immune responses to ingested material in animal models; in Miller, K.; Nicklin, S. (eds): Immunology of the Gastrointestinal Tract. Boca Raton, CRC Press, 1987, pp. 99–126.

53 Jalkenen, S.: Lymphocyte homing to the gut. Springer Semin. Immunopathal. *12:* 153–164 (1990).

54 Stamper, H.B.; Woodruff, J.J.: Lymphocyte homing into lymph nodes: In vitro demonstration of the selective affinity of circulating lymphocytes for high-endothelial venules. J. Exp. Med. *144:* 828–833 (1976).

55 Butcher, E.C.: The regulation of lymphocyte traffic. Curr. Top. Microbiol. Immunol. *128:* 85–122 (1986).

56 Westerman, J.; Blaschke, V.; Zimmermann, G.; Hirschfeld, U.; Papst, R.: Random entry of circulating lymphocyte subsets into peripheral lymph nodes and Peyer's patches: No evidence in vivo of a tissue-specific migration of B and T lymphocytes at the level of high endothelial venules. Eur. J. Immunol. *22:* 2219–2223 (1992).

57 Gallatin, W.M.; Weissman, I.L.; Butcher, E.C.: A cell surface molecule involved in organ-specific homing of lymphocytes. Nature *303:* 30–34 (1983).

58 Rasmussen, R.A.; Chin, Y.H.; Woodruff, J.J.; Easton, T.G.: Lymphocyte recognition of lymph node high endothelium. VII. Cell surface proteins involved in adhesion defined by monoclonal anti-$HEBF_{LN}$ (A.11). J. Immunol. *135:* 19–24 (1985).

59 Chin, Y.H.; Rasmussen, R.A.; Woodruff, J.J.; Easton, T.G.: A monoclon anti-$HEBF_{PP}$ antibody with specificity for lymphocyte surface molecules mediating adhesion to Peyer's patch high endothelium of the rat. J. Immunol. *136:* 2556–2561 (1986).

60 Hu, M.C.T.; Crowe, D.T.; Weissman, I.L.; Holzmann, B.: Cloning and expression of mouse integrin $\beta_p (\beta_7)$: A functional role in Peyer's patch-specific lymphocyte homing. Proc. Natl Acad. Sci. USA *89:* 8254–8258 (1992).

61 Chin, Y.H.; Cai, J.P.; Xu, X.M.: Transforming growth factor-$\beta 1$ and IL-4 regulate the adhesiveness of Peyer's patch high endothelial venule cells for lymphocytes. J. Immunol. *148:* 1106–1112 (1992).

62 Owen, R.L; Ermak, T.H.: Structural specializations for antigen uptake and processing in the digestive tract. Springer Semin. Immunopathol. *12:* 139–152 (1990).

63 Bhalla, D.K.; Owen, R.L.: Cell renewal and migration In lymphoid follicles of Peyer's patch and cecum – An autoradiographic study in mice. Gastroenterology *82:* 232–242 (1982).

64 Ermak, T.H.; Owen, R.L.: Differential distribution of lymphocytes and accessory cells in mouse Peyer's patches. Anat. Rec. *215:* 144–152 (1986).

65 Sobhon, P.: The light and electron microscopic studies of Peyer's patches in non-germ-free adult mice. J. Morphol. *135:* 457–481 (1971).

66 Witmer, M.D.; Steinman, R.M.: The anatomy of peripheral lymphoid organs with emphasis on accessory cells: Light-microscopic immunocytochemical studies of mouse spleen, lymph node, and Peyer's patch. Am. J. Anat. *170:* 465–481 (1984).

67 Egberts, H.J.A.; Brinkhoff, M.G.M.; Mouwen, J.M.V.M.: Biology and pathology of the intestinal M-cell. A review. Anat. Q. *7:* 333–336 (1985).

68 Bhalla, D.K.; Owen, R.L.: Migration of B and T lymphocytes to M cells in Peyer's patch follicle epithelium: An autoradiographic and immunocytochemical study in mice. Cell. Immunol. *81:* 105–117 (1983).

69 Von Rosen, L.; Podjaski, B.; Bettman, I.; Otto, H.F.: Observations on the ultrastructure and function of the so-called 'microfold' or 'membraneous' cells (M cells) by means of

peroxidase as a tracer. An experimental study with special attention to the physiological parameters of resorption. Virchows Arch [A] *390:* 289–312 (1981).

70 Wolf, J.L.; Bye, W.A.: The membranous epithelial (M) cell and the mucosal immune system. Annu. Rev. Med. *35:* 95–112 (1984).

71 Cornes, J.S.: Number, size and distribution of Peyer's patches in the human small intestine. I. The development of Peyer's patches. Gut *6:* 225–235 (1965).

72 Smith, M.W.; Peacock, M.A.: M-cell distribution in follicle-associated epithelium of mouse Peyer's patches. Am. J. Anat. *519:* 167–175 (1980).

73 Meader, R.D.; Landers, D.F: Electron and light microscopic observations on relationships between lymphocytes and intestinal epithelium. Am. J. Anat. *121:* 763–773 (1967).

74 Marsh, M.N.: Studies of intestinal lymphoid tissue. I. Electron microscopic evidence of 'blast transformation' in epithelial lymphocytes of mouse small intestine mucosa. Gut *16:* 665–674 (1975).

75 Greenwood, J.H.; Austin, L.L.; Dobbins, W.O.: In vivo characterisation of human intestinal intraepithelial lymphocytes. Gastroenterology *85:* 1023–1035 (1983).

76 Goodman, T.; Lefrancois, L.: Expression of the γ-δ T cell receptor on intestinal CD8$^+$ intraepithelial lymphocytes. Nature *333:* 855–858 (1988).

77 Bonneville, M.; Janeway, C.A.; Ito, K.; Haser, W.; Isida, I.; Nakanishi, N.; Tonegawa, S.: Intestinal intraepithelial lymphocytes are a distinct set of T cells. Nature *336:* 479–481 (1990).

78 Bonneville, M.; Itohara, S.; Krecko, E.G.; Mombaerts, P.; Isida, I.; Kastsuki, M.; Berns, A.; Farr, A.G.; Janeway, C.A.; Tonegawa, S.: Transgenic mice demonstrate that epithelial homing of γ/δ T cells is determined by cell lineages independent of T cell receptor specificity. J. Exp. Med. *171:* 1015–1026 (1990).

79 Fujihashi, K.; Taguchi, T.; Aicher, W.K.; McGhee, J.R.; Bluestone, J.A.; Eldridge, J.H.; Kiyono, H.: Immunoregulatory functions for murine intraepithelial lymphocytes: γ/δ T cell receptor-positive (TCR$^+$) cells abrogate oral tolerance while α/β TCR$^+$ T cells provide B cell help. J. Exp. Med. *175:* 695–707 (1992).

80 Klein, J.R.; Kagnoff, M.F.: Nonspecific recruitment of cytotoxic effector cells in the intestinal mucosa of antigen-primed mice. J. Exp. Med. *160:* 1931–1936 (1984).

81 Carman, P.S.; Ernst, P.B.; Rosenthal, K.L.; Clark, D.A.; Befus, A.D.; Bienenstock, J.: Intraepithelial leukocytes contain a unique subpopulation of NK-like cytotoxic cells active in the defence of gut epithelium to enteric murine coronavirus. J. Immunol. *136:* 1548–1553 (1986).

82 Stokes, C.R.: Induction and control of intestinal immune responses; in Newby, T.J.; Stokes, C.R. (eds): Local Immune Responses of the Gut. Boca Raton, CRC Press, 1984, pp. 97–142.

83 Hanson, D.G.; Vaz, N.M.; Rawlings, L.A.; Lynch, J.M.: Inhibition of specific immune responses by feeding protein antigens. II. Effects of prior passive and active immunization. J. Immunol. *122:* 2261–2266 (1979).

84 Newby, T.J.: Protective immune responses in the intestinal tract; in Newby, T.J.; Stokes, C.R. (eds): Local Immune Responses of the Gut. Boca Raton, CRC Press, 1984, pp. 143–198.

85 Xu-Amano, J.; Aicher, W.K.; Taguchi, T.; Kiyono, H.; McGhee, J.R.: Selective induction of Th$_2$ cells in murine Peyer's patches by oral immunization. Int. Immunol. *4:* 433–445 (1992).

86 Wilson, A.D.; Baily, M.; Williams, N.A.; Stokes, C.R.: The in vivo production of cytokines by mucosal lymphocytes immunized by oral administration of keyhole limpet hemocyanin using cholera toxin as an adjuvant. Eur. J. Immunol. *21:* 2333–2339 (1991).

87   Mattingly, J.A.: Immunological suppression after oral administration of antigen. III. Activation of suppressor-inducer cells in the Peyer's patches. Cell. Immunol. *86:* 46–52 (1984).

88   Barrett, T.A.; Gajewski, T.F.; Danielpour, D.; Chang, E.B.; Beagley, K.W.; Bluestone, J.A.: Differential function of intraepithelial lymphocyte subsets. J. Immunol. *149:* 1124–1130 (1992).

89   Mowat, A.M.; Lamont, A.G.; Strobel, S.; Mackenzie, S.: The role of antigen processing and suppressor T cells in immune responses to dietary proteins in mice. Adv. Exp. Med. Biol. *216A:* 709–720 (1987).

90   Fujihashi, K.; Kiyono, H.; Beagley, K.W.; Elridge, J.H.; McGhee, J.R.: Role of the GALT contrasuppressor T cell circuit in isotype-specific immunoregulation. Adv. Exp. Med. Biol. *237:* 649–653 (1988).

91   Sedgwick, J.D.; Holt, P.G.: Induction of IgE-secreting cells and IgE isotype-specific suppressor cells in the respiratory lymph nodes of rats in response to antigen inhalation. Cell. Immunol. *94:* 182–194 (1985).

92   MacDonald, T.T.: Immunosuppression caused by antigen feeding. I. Evidence for the activation of a feedback suppressor pathway in the spleens of antigen-fed mice. Eur. J. Immunol. *12:* 767–773 (1982).

93   Matthews, J.B.; Fivaz, B.H.; Sewell, H.F.: Serum and salivary antibody responses and the development of oral tolerance after oral and intragastric antigen administration. Int. Arch. Allergy Appl. Immunol. *65:* 107–113 (1981).

94   Guatam, S.C.; Chikkala, N.F.; Battisto, J.R.: Oral administration of the contact sensitizer trinitrochlorobenzene: Initial sensitization and subsequent appearance of a suppressor population. Cell. Immunol. *125:* 437–448 (1990).

95   Troncone, R.; Ferguson, A.: Gliadin presented the gut induces oral tolerance in mice. Clin. Exp. Immunol. *72:* 284–287 (1988).

96   Elson, C.O.; Ealding, W.: Cholera toxin feeding did not induce oral tolerance in mice and abrogated oral tolerance to an unrelated protein antigen. J. Immunol. *133:* 2892–2897 (1984).

97   Telemo, E,; Bailey, M.; Miller, B.G.; Stokes, C.R.; Bourne, F.J.: Dietary antigen handling by mother and offspring. Scand. J. Immunol. *34:* 689–696 (1991).

98   Nicklin, S.; Miller, K.: Naturally acquired tolerance to dietary antigen: Effect of in utero and perinatal exposure on subsequent immune competence in the rat. J. Reprod. Immunol. *10:* 167–176 (1987).

99   Swarbrick, E.T.; Stokes, C.R.; Soothill, J.F.: Absorption of antigens after oral immunization and the simultaneous induction of specific systemic tolerance. Gut *20:* 121–125 (1979).

100 Peng, H.J.; Turner, M.W.; Strobel, S.: The generation of a 'tolerogen' after the ingestion of ovalbumin is time-dependent and unrelated to serum levels of immunoreactive antigen. Clin. Exp. Immunol. *81:* 510–515 (1990).

101 Bruce, M.G.; Ferguson, A.: Oral tolerance produced by gut-processed antigen. Adv. Exp. Med. Biol. *216A:* 721–731 (1987).

102 Kafnoff, M.F.: Effects of antigen-feeding on intestinal and systemic immune responses. IV. Similarity between the suppressor factor in mice after erythrocyte-lysate injection and erythrocyte feeding. Gastroenterology *79:* 54–61 (1980).

103 Bruce, M.G.; Ferguson, A.: The influence of intestinal processing on the immunogenicity and molecular size of absorbed, circulating ovalbumin in mice. Immunology *59:* 295–300 (1986).

104 DeSilva, D.R.; Urdahl, K.B.; Jenkins, M.K.: Clonal anergy is induced in vitro by T cell receptor occupancy in the absence of proliferation. J. Immunol. *147:* 3261–3267 (1991).

105 Janeway, C.A.; Jones, B.; Hayday, A.: Specificity and function of T cells bearing $\gamma\delta$ receptors. Immunol. Today 9: 73–76 (1988).

106 Offit, P.A.; Dudzik, K.I.: Rotavirus specific cytoxic T lymphocytes appear at the intestinal mucosal surface after rotavirus infection. J. Virol. 63: 3507–3512 (1989).

107 Lefrancois, L.; Goodman, T.: In vivo modulation of cytolytic activity and Thy-1 expression of TCR$\gamma\delta^+$ intraepithelial lymphocytes. Science 243: 1716–1718 (1989).

108 Houlden, B.A.; Matis, L.A.; Cron, R.Q.; Widacki, S.M.; Brown, G.D.; Pampeno, C.; Meruelo, D.; Bluestone, J.A.: A TCR$\gamma\delta$ cell recognizing a novel TL-encoded gene product. Cold Spring Harb. Symp. Quant. Biol. 54: 45–55 (1989).

109 Haregewoin, A.; Soman, G.; Hom, R.C.; Finberg, R.W.: Human $\gamma\delta^+$ T cells response to mycobacterial heat-shock protein. Nature 340: 309–312 (1989).

110 Born, W.; Hall, L.; Dallas, A.; Boymel, J.; Shinnick, T.; Young, D.; Brennan, P.; O'Brien, R.: Recognition of a peptide antigen by heat shock-reactive $\gamma\delta$ T lymphocytes. Science 249: 67–79 (1990).

111 McMenamin, C.; Schon-Hegrad, M.; Oliver, J.; Girn, B.; Holt, P.G.: Regulation of IgE responses to inhaled antigens: Cellular mechanisms underlying allergic sensitization versus tolerance induction. Int. Arch. Allergy Appl. Immunol 94: 78–82 (1991).

112 McMenamin, C.; Oliver, J.; Girn, B.; Holt, B.J.; Kees, U.R.; Thomas, W.R.; Holt, P.G.: Regulation of T-cell sensitization at epithelial surfaces in the respiratory tract: Suppression of IgE responses to inhaled antigens by CD3$^+$ TCR$\alpha^-\beta^-$ lymphocytes (putative γ/δ T cells). Immunology 74: 234–239 (1991).

113 Fujihashi, K.; Kiyono, H.; Aicher, W.K.; Green, D.R.; Singh, B.; Elridge, J.H.; McGhee, J.R.: Immunoregulatory function of CD3$^+$, CD4$^-$, and CD8$^-$ T cells. $\gamma\delta$ T cell receptor-positive T cells from nude mice abrogate oral tolerance. J. Immunol. 143: 3415–3422 (1989).

114 Fujihashi, K.; Taguchi, T.; McGhee, J.R.; Elridge, J.H.; Bruce, M.G.; Green, D.R.; Singh, B.; Kiyono, H.: Regulatory function for murine intraepithelial lymphocytes. Two subsets of CD3$^+$, T cell receptor-1$^+$ intraepithelial lymphocyte T cells abrogate oral tolerance. J. Immunol. 145: 2010–2019 (1990).

115 Taguchi, T.; Aicher, W.K.; Fujihashi, K.; Yamamoto, M.; McGhee, J.R.; Bluestone, J.A.; Kiyono, H.: Novel function for intraepithelial lymphocytes: Murine CD3$^+$, $\gamma\delta$ TCR T cells produce IFN-γ and IL-5. J. Immunol. 147: 3736–3744 (1991).

116 Miller, A.; Zhang, A.J.; Prabdu-Das, M.; Sobel, A.; Weiner, H.L.: Active suppression vs. clonal anergy following oral or IV administration of MBP in actively and passively induced EAE. Neurology 42: suppl. 3, p. 301 (1992).

117 Burstein, H.J.; Shea, C.M.; Abbas, A.K.: Aqueous antigens induce in vivo tolerance selectivity in IL-2 and IFN-γ-producing (Th$_1$) cells. J. Immunol. 148: 3687–3691 (1992).

118 Whitacre, C.C.; Gienapp, I.E.; Orosz, C.G.; Bitar, D.: Oral tolerance in experimental autoimmune encephalomyelits. III. Evidence for clonal anergy. J. Immunol. 147: 2155–2163 (1991).

119 Friedman, A.; Melamed, D.: T cell anergy is a mechanism for oral tolerance (abstract). 8th International Congress of Immunology, Budapest 1992.

120 Melamed, D.; Friedman, A.: Direct evidence for anergy in T lymphocytes tolerized by oral administration of ovalbumin. Eur. J. Immunol. 23: 935–942 (1993).

121 Hanson, D.G.; Vaz, N.M.; Maia, L.C.; Hornbrook, M.M.; Lynch, J.M.; Roy, C.A.: Inhibition of specific immune responses by feeding protein antigens. Int. Arch. Allergy Appl. Immunol. 55: 526–532 (1977).

122 Sedgwick, J.D.; Holt, P.G.: Down-regulation of immune responses to inhaled antigen: Studies on the mechanism of induced suppression. Immunology 56: 635–642 (1985).

123 Thurau, S.R.; Caspi, R.R.; Chan, C.C.; Weiner, H.L.; Nussenblatt, R.B.: Immunological suppression of experimental autoimmune uveitis. Fortschr. Ophthalmol. *88:* 404–407 (1991).

124 Kitamura, K.; Kiyono, H.; Fujihashi, K.; Elridge, J.H.; Beagley, K.W.: Isotype-specific immunoregulation. Systemic antigen induces splenic T contrasuppressor cells which support IgM and IgG subclass but not IgA responses. J. Immunol. *140:* 1385–2392 (1988).

125 Carvalho, C.R.; Vaz, N.M.: Specific responses to two unrelated antigens in mice made orally tolerant to one of them. Braz. J. Med. Biol. Res. *23:* 861–864 (1990).

126 Melamed, D.; Friedman, A.: Modification of the immune response by oral tolerance: Antigen requirements and interaction with immunogenic stimuli. Cell. Immunol. *146:* 412–420 (1993).

127 Cowdery, J.S.; Curtin, M.F.; Steinberg, A.D.: Effect of prior intragastric antigen administration on primary and secondary anti-ovalbumin responses of C57Bl/6 and NZB mice. J. Exp. Med. *156:* 1256–1261 (1982).

128 Hanson, D.G.; Vaz, N.M.; Maia, L.C.; Lynch, J.M.: Inhibition of specific immune responses by feeding protein antigens. III. Evidence against maintenance of tolerance to ovalbumin by orally induced antibodies. J. Immunol. *123:* 2337–2343 (1979).

129 Silverman, G.A.; Peri, B.A.; Rothberg, R.M.: Systemic antibody responses of different species following ingestion of soluble protein antigens. Dev. Comp. Immunol. *6:* 737–746 (1982).

130 Gautam, S.C.; Battisto, J.R.: Orally induced tolerance generates an efferently acting suppressor T cell and an acceptor T cell that together down-regulate contact sensitivity. J. Immunol. *135:* 2975–2983 (1985).

131 Khoury, S.J.; Hancock, W.W.; Weiner, H.L.: Oral tolerance to myelin basic protein and natural recovery from experimental autoimmune encephalomyelitis are associated with down-regulation of inflammatory cytokines and differential up-regulation of transforming growth factor-β, interleukin-4, and prostaglandin E expression in the brain. J. Exp. Med. *176:* 1355–1364 (1992).

132 Santos, L.M.B.; Al-Sabbagh, A.; Londono, A.; Weiner, H.L.: Oral tolerance to myelin basic protein induces TGF-β secreting T cells in Peyer's patches. J. Immunol. *150:* (8-pt II): 115A (1993).

133 Fishman-Lobell, J.; Friedman, A.; Weiner, H.L.: Differential in vivo cytokine gene expression in lymph nodes of orally tolerized mice. J. Immunol. *150:* (8-pt II): 114A (1993).

134 Mattingly, J.; Waxman, B.: Immunologic suppression after oral administration of antigen. II. Antigen-specific helper and suppressor factors produced by spleen cells of rats fed sheep erythrocytes. J. Immunol. *125:* 1044 (1980).

135 Cowdery, J.S.; Johlin, B.J.: Regulation of the primary in vitro response to TNP-polymerized ovalbumin by T suppressor cells induced by ovalbumin feeding. J. Immunol. *132:* 2783–2789 (1984).

136 Strobel, S.; Ferguson, A.: Persistence of oral tolerance in mice fed ovalbumin is different for humoral and cell-mediated immune responses. Immunology *60:* 317–318 (1987).

137 Kiyono, H.; McGhee, J.R.; Wannemuehler, M.J.; Michalek, S.M.: Lack of oral tolerance in C3H/HeJ mice. J. Exp. Med. *155:* 605–610 (1982).

138 Carr, R.I.; Hardtke, M.A.; Katilus, J.; Sadi, D.: Orally induced tolerance to casein in mice on normal mouse chow. Clin. Immunol. Immunopathol. *40:* 497–504 (1986).

139 Miller, A.; Lider, O.; Weiner, H.L.: Antigen-driven bystander suppression following oral administration of antigens. J. Exp. Med. *174:* 791–798 (1991).

140 Mowat, A.M.; Thomas, M.J.; MacKenzie, S.; Parrott, D.M.: Divergent effects of bacterial lipopolysaccharide on immunity to orally administered protein and particulate antigens in mice. Immunology *58:* 677–683 (1986).

141 Miller, A.; Prabhu-Das, M.; Weiner, H.L.: Epitopes of myelin basic protein (MBP) that trigger TGF-β release following oral tolerization to MBP are different from encephalitogenic epitopes. FASEB J. 6: 1686 (1992).

142 Friedman, A.; Weiner, H.L.: Induction of anergy and/or active suppression in oral tolerance is determined by frequency of feeding and antigen dosage. J. Immunol. 150: (8-pt II): 4A (1993).

143 Strobel, S.; Ferguson, A.: Immune responses to fed protein antigens in mice. II. Systemic tolerance or priming is related to age at which antigen is first encountered. Pediatr. Res. 18: 588–594 (1984).

144 Mowat, A.M.; Strobel, S.; Drummond, H.E.; Ferguson, A.: Immunological response to fed protein antigens in mice. I. Reversal of tolerance to ovalbumin by cyclophosphamide. Immunology 45: 105–113 (1982).

145 Mowat, A.M.: Depletion of suppressor T cells by 2′-deoxyguanosine abrogates tolerance in mice fed ovalbumin and permits the induction of intestinal delayed-type hypersensitivity. Immunology 58: 179–184 (1986).

146 Schwartz, R.H.: A cell culture model for T lymphocyte clonal anergy. Science 248: 1349–1356 (1990).

147 Jenkins, M.K.; Schwartz, R.H.: Antigen presentation by chemically modified splenocytes induces antigen-specific T cell unresponsiveness in vitro and in vivo. J. Exp. Med. 165: 302–319 (1987).

148 Sercarz, E.; Krzych, U.: The distinctive specificity of antigen-specific suppressor T cells. Immunol. Today 12: 111–118 (1991).

149 Green, D.R.; Flood, P.M.; Gershon, R.K.: Immunoregulatory T-cell pathways. Annu. Rev. Immunol. 1: 439–463 (1983).

150 Morimoto, C.; Reinherz, E.L.; Borel, Y.; Schlossman, S.F.: Direct demonstration of the human suppressor inducer subset by anti-T cell antibodies. J. Immunol. 130: 157–161 (1983).

151 Richman, L.K.; Chiller, J.M.; Brown, W.R.; Hanson, D.G.; Vaz, N.M.: Enterically induced immunological tolerance. I. Induction of suppressor T lymphocytes by intragastric administration of soluble proteins. J. Immunol. 12: 2429–2433 (1978).

152 MacDonald, T.T.: Immunosuppression caused by antigen feeding. II. Suppressor T cells mask Peyer's patches B cell priming to orally administered antigen. Eur. J. Immunol. 13: 138–142 (1983).

153 Kay, R.A.; Ferguson, A.: The immunological consequences of feeding cholera toxin. I. Feeding cholera toxin suppresses the induction of systemic delayed-type hypersensitivity but not humoral immunity. Immunology 66: 410–415 (1989).

154 Kay, R.A.; Ferguson, A.: The immunological consequences of feeding cholera toxin. II. Mechanisms responsible for the induction of oral tolerance for DTH. Immunology 66: 416–421 (1989).

155 Strobel, S.; Mowat, A.M.; Drummond, H.E.; Pickering, M.G.; Ferguson, A.: Immunological responses to fed protein antigens in mice. II. Oral tolerance for CMI is due to activation of cyclophosphamide-sensitive cells by gut-processed antigen. Immunology 49: 451–456 (1983).

156 Green, D.R.; Gold, J.; St Martin, S.; Gershon, R.; Gershon, R.K.: Microenvironmental immunoregulation: Possible role of contrasuppressor cells in maintaining immune responses in gut-associated lymphoid tissues. Proc. Natl Acad. Sci. USA 79: 889–892 (1982).

157 Mowat, A.M.; Lamont, A.G.; Parrott, D.M.: Suppressor T cells, antigen-presenting cells and the role of I-J restriction in oral tolerance to ovalbumin. Immunology 64: 141–145 (1988).

158 Ferguson, A.; Mowat, A.M.; Strobel, S.: Abrogation of tolerance to fed antigen and induction of cell-mediated immunity in the gut-associated lymphoreticular tissues. Ann. N.Y. Acad. Sci. *409:* 486–497 (1983).

159 Gautam, S.C.; Battisto, J.R.: Suppression of contact sensitivity and cell-mediated lympholysis by oral administration of hapten is caused by different mechanisms. Cell. Immunol. *78:* 295–304 (1983).

160 Miller, A.; Lider, O.; Roberts, A.B.; Sporn, M.; Weiner, H.l.: Suppressor T cells generated by oral tolerization to myelin basic protein suppress both in vitro and in vivo immune responses by the release of TGF-β following antigen specific triggering. Proc. Natl Acad. Sci. USA *89:* 421–425 (1992).

161 Higgins, P.; Weiner, H.L.: Suppression of experimental autoimmune encephalomyelitis by oral administration of myelin basic protein and its fragments. J. Immunol. *140:* 440–445 (1988).

162 Al-Sabbagh, A.; Miller, A.; Sobel, R.A.; Weiner, H.L.: Suppression of PLP induced EAE in the SJL mouse by oral administration of MBP. Neurology *42:* suppl. 3, p. 346 (1992).

163 Brod, S.A.; Al-Sabbagh, A.; Sobel, R.A.; Hafler, D.A.; Weiner, H.L.: Suppression of experimental autoimmune encephalomyelitis by oral administration of myelin antigens. IV. Suppression of chronic relapsing disease in the Lewis rat and strain 13 guinea pig. Ann. Neurol. *29:* 615–622 (1991).

164 Miller, A.; Lider, O.; Al-Sabbagh, A.; Weiner, H.L.: Suppression of experimental autoimmune encephalomyelitis by oral administration of myelin basic protein. V. Hierarchy of suppression by myelin basic protein from different species. J. Neuroimmunol. *39:* 243–250 (1992).

165 Lider, O.; Santos, L.M.B.; Lee, C.S.Y.; Higgins, P.J.; Weiner, H.L.: Suppression of experimental autoimmune encephalomyelitis by oral administration of myelin basic protein. II. Suppression of disease and in vitro immune responses is mediated by antigen-specific CD8$^+$ T lymphocytes. J. Immunol. *142:* 748–752 (1989).

166 Khoury, S.J.; Lider, O.; Al-Sabbagh, A.; Weiner, H.L.: Suppression of experimental autoimmune encephalomyelitis by oral administration of myelin basic protein. III. Synergistic effect of lipopolysaccharide. Cell. Immunol. *131:* 302–310 (1990).

167 Weiner, H.L.; Al-Sabbagh, A.; Sobel, R.: Antigen driven peripheral immune tolerance: Suppression of experimental autoimmune encephalomyelitis by aerosol administration of myelin basic protein (abstract). FASEB J. *4:* 2102 (1990).

168 Nussenblatt, R.B.; Caspi, R.R.; Mahdi, R.; Chan C.C.; Roberge, F.; Lider, O.; Weiner, H.L.: Inhibition of S-antigen induced experimental autoimmune uveoretinitis by oral induction of tolerance with S-antigen. J. Immunol. *144:* 1689–1695 (1990).

169 Vrabec, T.R.; Gregerson, D.S.; Dua, H.S.; Donoso, L.A.: Inhibition of Experimental autoimmune uveoretinitis by oral administration of S-antigen and synthetic peptides. Autoimmunity *12:* 175–184 (1992).

170 Singh, V.K.; Kalra, H.K.; Yamaki, K.; Shinohara, T.: Suppression of experimental autoimmune uveitis in rats by the oral administration of the uveitopathogenic S-antigen fragment as a cross-reactive homologous peptide. Cell. Immunol. *139:* 81–90 (1992).

171 Thurau, S.R.; Chan, C.C.; Suh, E.; Nussenblatt, R.B.: Induction of oral tolerance to S-antigen induced experimental autoimmune uveitis by a uveitogenic 20 mer peptide. J. Autoimmun. *4:* 507–516 (1991).

172 Nagler-Anderson, C.; Bober, L.A.; Robinson, M.E.; Siskind, G.W.; Thorbecke, F.J.: Suppression of type II collagen-induced arthritis by intragastric administration of soluble type II collagen. Proc. Natl Acad. Sci. USA *83:* 7443–7446 (1986).

173 Thompson, H.S.G.; Staines, N.A.: Gastric administration of type II collagen delays the onset and severity of collagen-induced arthritis in rats. Clin. Exp. Immunol. *64:* 581–586 (1986).

174 Thompson, H.S.; Staines, N.A.: Suppression of collagen-induced arthritis with pregastrically or intravenously administered type II collagen. Agents Actions *19:* 318–319 (1986).

175 Zhang, J.Z.; Lee, C.S.Y.; Lider, O.; Weiner, H.L.: Suppression of adjuvant arthritis in Lewis rats by oral administration of type II collagen. J. Immunol. *145:* 2489–2493 (1990).

176 Zhang, J.A.; Davidson, L.; Eisenbarth, G.; Weiner, H.L.: Suppression of diabetes in NOD mice by oral administration of porcine insulin. Proc. Natl Acad. Sci. USA *88:* 10252–10256 (1991).

177 Carr, R.I.; Tilley, D.; Forsyth, S.; Etheridge, P.; Sadi, D.: Failure of oral tolerance in $(NZB \times NZW)F_1$ mice is antigen specific and appears to parallel antibody patterns in human systemic lupus erythematosus. Clin. Immunol. Immunopathol. *42:* 298–310 (1987).

178 Miller, M.L.; Cowdery, J.S.; Laskin, C.A.; Curtin, M.F.; Steinberg, A,D.: Heterogeneity of oral tolerance in autoimmune mice. Clin. Immunol. Immunopathol. *31:* 231–240 (1984).

179 Browning, M.J.; Parrott, D.M.: Protection from chronic immune complex nephritis by a single dose of antigen administered by the intragastric route. Adv. Exp. Med. Biol. *216B:* 1619–1625 (1987).

180 Gesualdo, L.; Lamm, M.E.; Emancipator, S.N.: Defective oral tolerance promotes nephritogenesis in experimental IgA nephropathy induced by oral immunization. J. Immunol. *145:* 3684–3691 (1990).

181 Devey, M.E.; Bleasdale, K.: Antigen feeding modifies the course of antigen-induced immune complex disease. Clin. Exp. Immunol. *56:* 637–644 (1984).

182 Mowat, A.M.: The role of antigen recognition and suppressor cells in mice with oral tolerance to ovalbumin. Immunology *56:* 253–260 (1985).

183 Sayegh, M.H.; Khoury, S.J.; Hancock, W.H.; Weiner, H.L.; Carpenter, C.B.: Induction of immunity and oral tolerance with polymorphic class II major histocompatibility complex allopeptides in the rat. Proc. Natl Acad. Sci. USA *89:* 7762–7766 (1992).

184 Hancock, W.W.; Sayegh, M.H.; Kwok, C.A.; Weiner, H.L.; Carpenter, C.B.: Oral but not intravenous alloantigen prevents accelerated allograft rejection by selective intragraft Th2 cell activation. Transplantation; in press (1993).

185 Koprowski, H.: Rabies oral immunization. Curr. Top. Microbiol. Immunol. *146:* 137–151 (1989).

186 Rupprechet, C.E.; Wiktor, T.J.; Hamir, A.N.; Dietzschold, B.; Wunner, B.; Glickman, L.T.; Koprowski, H.: Oral immunization and protection of racoons (*Procyon lotor*) with a vaccinia rabies glycoprotein recombinant virus vaccine. Proc. Natl Acad. Sci. USA *83:* 7947–7950 (1986).

187 Steck, F.A.; Wandeler, A.; Bischel, B.; Capt, S.; Schneider, L.: Oral immunization of foxes against rabies. Abl. Vet. Med. *29:* 372–376 (1982).

188 McGhee, J.R.; Mestecky, J.: Oral immunization: A summary. Curr. Top. Microbiol. Immunol. *146:* 233–237 (1989).

189 Elridge, J.H.; Gilley, R.M.; Staas, J.K.; Moldoveanu, Z.; Mrulbroek, J.A.; Tice, T.R.: Biodegradable microspheres: Vaccine delivery system for oral immunization. Curr. Top. Microbiol. Immunol. *146:* 59–68 (1989).

190 Curtiss, R.; Kelly, S.M.; Gulig, P.A.; Nakayama, K.: Selective delivery of antigens by recombinant bacteria. Curr. Top. Microbiol. Immunol. *146:* 35–50 (1989).

191 Umesaki, Y.; Setoyama, H.: Immune response of mice to orally administered asialo-GM1-specific rabbit IgG in the presence or absence of cholera toxin. Immunology *75:* 386–388 (1992).

192 Elson, C.O.: Cholera toxin and its subunits as potential oral adjuvants. Curr. Top. Microbiol. Immunol. *146:* 29–34 (1989).

193 Marx, J: Testing of autoimmune therapy begins. Science *252:* 27–28 (1991).

194 Weiner, H.L.; Mackin, G.A.; Matsui, M.; Orav, E.J.; Khoury, S.J.; Dawson, D.M.; Hafler, D.A.: Double-blind pilot trial of oral tolerization with myelin antigens in multiple sclerosis. Science *259:* 1321–1324 (1993).

Aharon Friedman, PhD, Center for Neurological Diseases, Brigham and Women's Hospital and Harvard Medical School, 75 Francis Street, Room 104, Boston, MA 02115 (USA)

Granstein RD (ed): Mechanisms of Immune Regulation.
Chem Immunol. Basel, Karger, 1994, vol 58, pp 291–313

Mechanisms of Ultraviolet Radiation Carcinogenesis[1]

Stephan Grabbe, Richard D. Granstein

Department of Dermatology, University of Münster, FRG; Cutaneous Biology
Research Center, Massachusetts General Hospital, Harvard Medical School,
Boston, Mass., USA

Ultraviolet radiation (UVR) has pleomorphic effects on the cellular
constituents of the skin and the whole organism, affecting a variety of
subcellular structures and different cell systems. In general, they can be
separated into acute or short-term effects and long-term effects. *Acute*
responses of cutaneous tissue to UVR, which are measurable within hours
or days, include the induction of keratinocyte and melanocyte proliferation, migration of epidermal Langerhans cells (LC) out of the epidermis,
production and release of cytokines and other mediators, production of
biologically active vitamin D_3, as well as dose-dependent cellular cytotoxicity for LC and keratinocytes (KC). These alterations lead to secondary
activation of vascular cells, leukocytes, and interstitial tissue cells. These
acute UV effects are measurable even after a single exposure of mammalian
skin to UVR and are portrayed in schematic form in figure 1.

The most prominent effects of *chronic* UV irradiation are an alteration of
connective tissue structure, the modulation of T-cell-mediated immune
responses and the induction of various cutaneous malignancies. This chapter will focus on the carcinogenic effects of UVR with special emphasis on
the role of UV-induced immunosuppression as a pathogenetic factor for
UV carcinogenesis.

Carcinogenic Effects of UVR

The carcinogenic effect of sunlight was first recognized almost a
century ago [1]. Later, the action spectrum of the carcinogenic effects of

[1] Supported in part by NIH Grant AR40667 (R.D.G.).

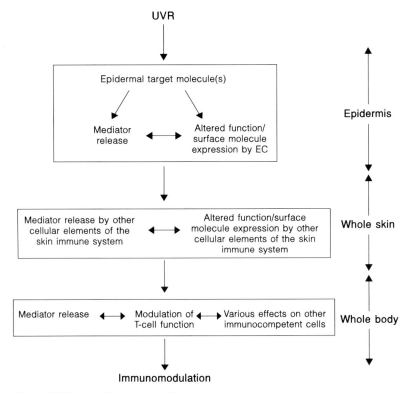

Fig. 1. UVR-induced immunomodulation.

sunlight was found to be within the UV range [2, 3], and was further narrowed down to wavelength areas between 260 and 300 nm [45]. The work of Urbach et al. [5] established that UVB radiation (290–320 nm) was far more effective in inducing skin tumors in experimental animals than UVA (320–400 nm) or UVC (200–290 nm). Since UVB, as opposed to UVC (which is completely absorbed by atmospheric ozone and thus not present at the terrestrial surface), reaches human skin in significant doses, UVB radiation is the most relevant wavelength for the induction of skin cancer in man. In addition to natural UVR, therapeutic application of UVA in conjunction with the photosensitizer 8-methoxypsoralen (PUVA) has also been shown to be carcinogenic in man as well as in experimental animals [6, 7].

UV-Induced Neoplastic Transformation

In vitro, the mutagenicity of UVR was demonstrated by irradiation of keratinocyte or fibroblast cultures, which showed altered morphology and growth characteristics after UVR exposure and gave rise to immortalized cell clones, which became tumorigenic after re-injection into nude mice [8–10]. In vivo, the carcinogenic effect of UVB radiation can be demonstrated by prolonged exposure of mice (typically 20–40 kJ/m^2 3× weekly for 10 weeks), which develop squamous cell carcinomas and spindle cell tumors (most of which are undifferentiated squamous cell carcinomas) [11]. In this regard, UVR has been recognized as a complete carcinogen, acting as a tumor initiator as well as a tumor promotor.

In humans, a significant correlation between sunlight exposure and the incidence of basal cell carcinomas, squamous cell carcinomas and lentigo maligna melanomas has been observed [12–14]. Since these tumors are usually located on light-exposed areas, occur more frequently in outdoor than in indoor workers, are more frequent in fair skinned than in deeply pigmented people, and are also more frequent in Caucasians living in equatorial areas, epidemiological data strongly suggest UVR as a significant risk factor for non-melanoma skin cancer. While in these tumors the cumulative UV dose appears to be the relevant risk factor, acute high-dose UV exposures, especially in early childhood, appear to be an important risk factor for the pathogenesis of other human skin cancers (superficial spreading melanoma) [13–15]. It remains unclear if the acute exposure itself causes the increased risk of skin cancer or if it is a marker for individuals likely to have high lifetime exposures to UVR.

Subcellular Targets for Carcinogenic Effects of UVR

The induction of malignant transformation by UVR requires absorption of the photon energy within the skin, eventually leading to an alteration of growth-controlling genes within the cellular DNA. One of the most prominent effects of UVR on DNA is the dimerization of pyrimidine bases, leading to the formation of cyclobutane rings as well as 6,4-hydroxy-dihydropyrimidines and other photoproducts [16]. These cyclobutane dimers are characterized by a covalent link between neighboring nucleotide bases and generally prevent transcription of the gene, in which the photoadduct is located. If the affected gene is a growth-regulating gene (e.g. oncogene or tumor suppressor gene), malignant transformation of the cell may result, especially when more than one of these genes is affected simultaneously. To counteract these deleterious effects of UVR on cellular DNA, evolution has established a number of repair mechanisms, including excision repair ('dark repair') as well as photoreactivation [for review, see 17, 18]. Evidence to support this concept includes observations of UVR-induced cyclobutane dimer formation, UV-induced DNA changes detectable by specific antibodies, and increased tumorigenesis in patients with xeroderma pigmentosum, a disease characterized by a deficient ability to repair the UV-induced DNA damage [19–21]. In some organisms such as man and the opossum, *Monodelphis domestica*), UVR-induced cyclobutane dimers can be repaired by a photoreactivating enzyme, which itself is activated by long-wave UVA and visible irradiation. Ley et al. [22] recently presented evidence that photoreactivation effectively reduces the number of UVR-induced tumors in *M. domestica*, indicating that UV-induced cyclobutane dimer formation is directly involved in the pathogenesis of UV carcinogenesis. Furthermore, a direct pathogenetic role of UV-induced cyclobutane dimer formation within the DNA in UV carcinogenesis is implicated by studies performed in a fish model [20]. Alternatively, formation of radicals leading to secondary DNA alteration, or the alteration of gene regulatory elements (e.g. transcription factors) may also be involved in subcellular mechanisms of a UV-induced neoplastic transformation. In particular, recent data indicate that exposure of HeLa cells to UVR leads to activation of Src tyrosine kinases, as well as induction of Ha-Ras, Raf-1 and MAP-2 gene transcripts. This signal then increases phosphorylation of c-Jun, potentiating its activity [23, 24]. In addition to c-Jun, c-Fos was also found to be induced after UVR [25], strengthening the evidence that AP-1 is a critical target molecule for UVR.

UV-Induced Immunomodulation

In addition to direct carcinogenesis, UV irradiation also induces a specific modulation of the immune response towards haptens and proteins, as well as neoplastically transformed cells. The UVR-induced alteration of T-cell-mediated immune responses (contact hypersensitivity (CHS) and delayed-type hypersensitivity (DTH) responses) have been reviewed elsewhere in this issue [M. Perez et al., pp. 314–330]. Many of the phenomena observed in the CHS and/or DTH system can also be observed in the immune responses towards cutaneous malignancies induced by UVR and will only be mentioned briefly. However, in addition to a global alteration of immune responses by UVR, some immunologic phenomena of UV carcinogenesis are associated specifically with UVR-induced neoplasms and do not apply for other tumor types or antigens. The wavelengths responsible for the immunomodulatory effects of UVR range between 250 and 320 nm, and irradiation doses required for immunosuppression in experimental animals as well as in man are readily achievable by natural sunlight exposure [26]. Typical dose regimens for experimental UVB irradiation are listed in table 1.

Table 1. Experimental UVB irradiation regimen (approximate figures, may vary significantly with cell/tissue type and irradiation source)

Effect	Cumulative dose of UVB
In vitro irradiation of cell suspensions	
Modulation of cell function	ca. $10-500$ J/m^2
Cytotoxicity	$>$ca. 200 J/m^2
In vivo UVB irradiation	
Local immunosuppression	$0.5-5$ kJ/m^2, 4 daily doses of 1 kJ/m^2
Systemic immunosuppression	>20 kJ/m^2, variable dose regimens
Carcinogenesis	>500 kJ/m^2, $20-50$ kJ/m^2 3$\times$ weekly for 10 or more weeks

UV-Absorbing Molecules Mediating the Immunomodulatory Effects of UVR

Since the effective wavelengths are almost entirely absorbed within the epidermis, target structures for transmission of the immunomodulatory effects of UVR need to be located epidermally. The target molecule responsible for the immunomodulatory effects of UVR may also be DNA, since studies in the marsupial *M. domestica* suggest that repair of UV-induced DNA damage by virtue of photoreactivation abrogates the immunosuppressive effects of UV exposure [27, 28]. Moreover, studies performed by Kripke et al. [29] demonstrate that topical application of liposome-encapsulated T$_4$ endonuclease abrogated pyrimidine dimer formation and immunomodulatory effects of UVR. Finally, the absorption spectrum of DNA corresponds to the action spectrum of the UV-induced immunosuppression, with a peak around 300 nm [17, 30].

However, recent work suggests that, besides DNA, epidermal urocanic acid (UCA) may also have a role as a photoreceptor for UVR and mediator of its immunosuppressive effects [31, 32]. UV irradiation of UCA, which is a major component of the stratum corneum, leads to a trans $\rightarrow$ cis isomerization of UCA, and cis-UCA has been found to inhibit CHS and DTH responses in a similar fashion to UVR itself. Moreover, removal of the stratum corneum before UV irradiation prevented subsequent immunosuppression, systemic administration of UCA was found to be immunosuppressive and topical application of cis-UCA led to an increased incidence of malignancies in UV-carcinogenesis experiments [33]. It is not entirely clear, however, whether UCA exerts these effects directly or via induction of secondary mediators (e.g. cytokines, prostaglandins). Evidence exists suggesting that UCA can induce both TNFα and prostaglandin production (see below).

Additional mechanisms, such as the generation of (oxygen) radicals or absorption by cellular or nuclear proteins, may also be involved in UV-induced immunomodulation [for review, see 17, 18, 23].

Relationship between UV-Induced Immunomodulation and Carcinogenesis

Epidemiological Data

Clinical and experimental evidence strongly suggest that the immunomodulatory effect of UVR also leads to a deficient immune response against UV-induced cutaneous malignancies. The fact that immunosuppressed patients (e.g. transplant patients) show a greatly increased incidence of squamous cell carcinomas, basal cell carcinomas [34–36] and lentigo maligna melanomas [12, 37] clearly demonstrates a correlation between the degree of immunocompetence and skin cancer formation. Moreover, in these patients skin cancers usually develop within UV-exposed areas [12, 38]. With the exception of lymphomas [39], other neoplasmas are not significantly increased in these patients. Similar results were obtained using animal studies [40, 41]. Clinical studies have demonstrated that UVR can indeed suppress CHS responses in humans [42]. Moreover, recent results suggest that skin cancer patients have an increased sensitivity to UVR-induced suppression of CHS, compared to age-matched controls without skin cancer [43, 44]. These observations led to the concept that an immunosurveillance mechanism exists, which constantly screens the skin for malignantly transformed cells and eliminates or represses them immunologically [45–47]. Although this concept is not generally accepted due to a lack of direct experimental evidence, the epidemiological studies mentioned above strongly suggest the existence of tumor immune surveillance mechanisms, at least against UV-induced neoplasms within the skin.

Cellular and Soluble Mediators Involved in UV-Induced Modulation of Tumor Immunity

In mice, the immune response against UV-induced cutaneous neoplasms and its alteration by UVR has been studied in some detail. Studies performed by Kripke et al. [40, 48, 49] demonstrated that UV-induced skin tumors were generally poorly transplantable to normal syngeneic hosts, but grew progressively if this host had been UV-irradiated prior to transplantation of the tumor. Subcarcinogenic doses of UVR were sufficient to inhibit tumor rejection [40, 50–53]. The UV-induced alterations of tumor growth were not restricted to UV-exposed areas but could also be observed at nonirradiated sites, provided the irradiation was performed with UVB doses above approximately 5 kJ [26, 51, 53]. Moreover, normal mice, surgically anastomosed to UVR-treated mice, were found to lose their ability to reject UVR-induced tumors [51]. Transfer of splenic T cells from either tumor-bearing or from nontumor-bearing, UV-irradiated mice also led to a suppression of the ability of normal mice to reject UVR-induced tumors [54, 55]. Thus, the induction of suppressor T lymphocytes directed against epitopes on UV-induced tumors was believed to be responsible for the suppressed immune response by UV-irradiated hosts to these tumors [56–59]. These suppressor T cells (T_s) generally recognize UV-induced tumors as a group and do not permit progressive growth of non-UV-induced neoplasms (e.g. chemically induced tumors) [58, 60–63]. Their phenotype is generally Thy-1$^+$, Lyt-1$^+$, Lyt-2$^-$, L3T4$^+$, I-A$^-$, but varies between different tumor systems and wavelengths used for irradiation [55, 64–66]. Antigen presentation appears to play a major role in the generation of these T_s, since the suppression is MHC-restricted [67–72]. Functionally, I-J restriction seems to be involved in the induction of T_s [73–76]. It appears that T_s act in an early step of the generation of immunity, since they specifically block the generation of tumor-specific T-helper cells but are unable to suppress tumor immunity once it has been induced [56, 58, 59, 65, 71, 77–80]. Cloned UV T_s cell lines have been established in vitro, which have some characteristics of T-helper-2 (Th2) cells [64]. Thus, suppressor cells formed during UV carcinogenesis resemble those observed in the UV-mediated suppression of CHS or DTH responses and may be generated via identical mechanisms. However, since significant variability between T_s phenotypes and functions have been observed depending on the experimental conditions used, it is still unclear whether T_s are generated via a common (and as yet undefined) mechanism or are regulated individually for various types of immune responses.

Besides the generation of tumor-specific T_s cells, other mechanisms may also be involved in ths inhibition of tumor immune responses after UV irradiation. Studies showing that transplantation of UV-irradiated skin also results in immunosuppression indicate that skin-derived factors and/or cells may play a role in this regard [81–89]. Most importantly, soluble factors have been demonstrated within the skin and also within the systemic circulation of UV-irradiated mice that appear capable of modulating tumor immune responses [90, 91]. Among those that have been characterized so far, IL-1, IL-1ra, IL-6, IL-10 and TNFα may be the most important. Exciting new evidence has recently been presented which shows that administration of IL-10 is able to mimic some of the immunomodulatory effects of UVR [92, 93]. Furthermore, information generated by Streilein and co-workers [94–96] suggests that suppression of CHS responses within the skin by UVR may be mediated at least in a part by UV-induced secretion of TNFα. Not all strains of mice are equally susceptible to the immunomodulatory effects of UVR, indicating that the degree of UV-induced immunosuppression may be genetically determined [97]. In mice, a pleomorphism within the TNFα gene locus may be responsible for the differential susceptibility of various strains to the immunomodulatory effects of UVR [94–97]. Preliminary evidence suggests that this may also be true in humans, where UV-responsive and UV-resistant individuals could be distinguished by the ability or inability to sensitize through UV-irradiated skin [42, 98]. Interestingly, UV-susceptible individuals were found to have a higher incidence of UV-induced cutaneous malignancies [43, 44]. The exact mechanisms of action of these cytokines for the induction of T_s and UV-induced immunomodulation are, however, still unclear.

Functional alterations of other cellular constituents of the immune system have also been observed after UVR, some of which may also be in part cytokine-mediated. For example, an inhibition of natural killer cell activity has been observed in UV-irradiated mice, which could at least partially be attributed to UV-induced secretion of IL-6 [99, 100]. Remarkably, however, several types of immune reactions are not affected by UVR, such as antibody production, macrophage function, allograft rejection or the inflammatory responses to irritants, underlining the selectivity of UV-induced immunomodulation [40, 52, 54, 101]. These immunomodulatory effects of UVR are summarized in table 2.

Prostaglandins and urocanic acid (as discussed above) have also been suggested to play a role in UV-induced immunosuppression, since some of its features could be either blocked by administration of neutralizing antibodies against or inhibitors of these factors or mimicked by administration of these factors themselves [33, 85, 102–106]. Interestingly, KC exposed to UVR have been shown to release factors that suppress the induction of CHS and DTH as well as inhibiting IL-1 activity [106–108]. Indeed, UVB radiation and UVA radiation each induce KC to release a distinct immunomodulatory factor [106]. UVB radiation results in the release of a factor that inhibits DTH but not CHS [106], and some of this activity, at least, appears to relate to the release of IL-10 [93]. Exposure of KC to UVA radiation results in the release of a factor that inhibits CHS but not DTH [106]. The molecular nature of contra-IL-1 activity and the inhibitor of CHS remains unknown. Hypothetically, it is possible that release of these factors in vivo after UVR exposure may mediate, at least in part, the immunosuppression observed. This is best demonstrated in vivo for IL-10. It is not clear, however, whether these immunomodulatory mediators act independently or are stimulated by a common underlying mechanism. For example, urocanic acid has been found to stimulate TNF production within the skin and both have been demonstrated to increase the concentration of prostaglandin E_2 [103, 109–116]. Taken together, at least some of the nonantigen-specific phenomena observed after UV irradiation appear to be mediated by cytokines, some of which may finally act via prostaglandin-dependent mechanisms.

Modulation of Cutaneous Antigen Presentation by UVR

Since UV-induced immunomodulation is genetically restricted, antigen-specific and appears to occur early during the generation of immunity, antigen-presenting cells (APC) may be crucially involved in the immunomodulatory effects of UVR. Various experimental systems have been established to examine UV effects on APC function. However, to correctly assess the effects of UVR on LC function, it is important to distinguish between sensitization (i.e. antigen presentation to unprimed T cells) and elicitation (i.e. antigen presentation to primed, antigen-specific T cells) of immune responses. Moreover, different assay systems assessing APC function may test different components of APC function and thus require different costimulatory conditions. Therefore, the experimental systems used for data generation should be carefully considered for an evaluation of the effects of UVR on APC function.

UV-induced alterations of function and morphology of epidermal LC, the most relevant APC within the epidermis, include a reduction of LC dendricity, decreased MHC class II expression, and emigration of LC out of the epidermis into the afferent lymphatics [67, 68, 71, 117–121]. Upon morphologic evaluation, an altered number and morphology of APC has been observed in various types of human and murine tumors, including UV-induced skin tumors [122–126]. Functionally, $I\text{-}A^+$ macrophages were found to be necessary for the induction of cytotoxic T-lymphocyte responses against UV-induced tumor, and treatment of mice with silica or trypan blue (agents which are cytotoxic for phagocytosing cells) shortened the latency period of skin cancer formation during UV irradiation [122, 127]. Moreover, an impaired APC function was observed during the latency period of UV carcinogenesis [122]. UV-irradiated LC were found to be less capable of generating immunity and tolerogenic signals are transmitted to the immune system after sensitization through UV-irradiated skin in primary and, in some systems, secondary immune responses [67, 122, 123, 128–131]. However, UVB radiation exposure of the sites of elicitation of CHS in mice augments the response [132]. This effect may be partially mediated by UVB-induced TNF release [95]. Other studies demonstrated that in vitro UV-irradiated LC can present antigen to Th2 clones but are unable to present antigen to Th1 clones and, indeed, induce anergy in Th1 clones [119, 133–135]. Additionally, LC surface molecule expression is also

Table 2. UV effects on immune function

A. Molecular target of UVR

---

1. DNA
Nuclear DNA
Cell membrane-bound DNA (?)

2. Urocanic acid (UCA)
trans → cis isomerization of UCA

3. Others
Transcription factors (AP-1, NF-κB (?))
Src tyrosine kinase
Proto-oncogenes (c-Fos, c-Jun, Ha-ras)
Membrane and/or cellular proteins (?)

B. Direct effects of UVR within the skin

---

1. Langerhans cells (LC)
LC death (high doses of UVR only)
Emigration of LC out of the epidermis, alteration of LC trafficking
Alteration of Ag processing
Alteration of LC surface molecules (adhesion molecules, cytokine receptors)
Alteration of LC cytokine production (IL-1β, IL-6, TNFα, MIP-1α)

2. Other cutaneous cells
Dose-dependent activation/increased proliferation (keratinocytes (KC)) or cell death
  (KC: 'sunburn cells')
Production and release of cytokines (KC)
  IL-1α
  IL-1ra
  Contra-IL-1
  IL-6
  IL-10
  TNFα
Production of prostaglandins
Production of immunomodulatory hormones and neurotransmitters (KC, MC), e.g.
  α-MSH
Photochemical transformation of 7-dehydrocholesterin to cholecalciferol
  (provitamin D)
Altered expression of adhesion molecules
Emigration out of the epidermis (Thy-1$^+$ DETC (mouse))
Immigration into the epidermis (OKM-5$^+$ dermal macrophages (human), T and B
  lymphocytes)

C. Secondary effects of UVR on immune function

---

1. Systemic alteration of immunity and inflammation
Cytokine production/mediator release by dermal cells (mast cells, dermal macrophages,
  fibroblasts, etc.) and/or extracutaneous cells
Elevation of serum levels of IL-1, TNFα, IL-6, acute phase proteins (high doses of UVR,
  >1–2 minimal erythemal doses (MED))

Table 2. (cont.)

Vascular changes (cytokine production and adhesion molecule expression by endothelial cells, vascular permeability)

Induction of UV-specific antigenic determinants on malignantly transformed as well as on normal cells

No inhibition of inflammatory response to irritants, in vitro proliferative response to mitogens, antibody responses, allogeneic mixed lymphocyte reaction or allograft rejection

2. Effects on antigen-presenting cells (APC)

Inhibition of APC function in vivo

  Low dose UVB (<1,000 J/m^2) in vivo: number of LC decreases, local suppression of cutaneous antigen presentation function (presentation of hapten in unprimed, but not in most primed systems), number of LC recovers with time even during chronic irradiation.

  High dose UVB (20–40 kJ/m^2) in vivo: systemic suppression of APC function

Inhibition of the sensitization phase of CHS and DTH responses to haptens and protein antigens, augmentation of the elicitation phase of CHS responses; inhibition of CHS and DTH by UVR involves different mechanisms

Augmentation of the elicitation phase of CHS in mice

Inhibition of tumor antigen presentation by epidermal APC after irradiation in vitro for both induction and elicitation of tumor immunity

No inhibition of macrophage function in UV-irradiated mice

3. Effects on T cells

Generation of transferable T-suppressor (T$_s$) cell activity (high-dose UVB)

  Heterogeneous phenotype of UVB-induced UV-T$_s$ (most are Thy-1$^+$, Lyt-1$^-$, L3T4$^-$, I-A$^-$, and I-J$^+$, others are Thy-1$^+$, Lyt-1$^+$, Lyt-2$^-$, L3T4$^+$, I-A$^-$)

  UV-T$_s$ are present in the spleen, lymph nodes and thymus of UV-irradiated mice

  UV-induced immunosuppression varies from model to model with regard to the appropriate conditions of dose and regimen of radiation and timing of antigen administration, and the different antigens examined (contact sensitizers, UV tumor antigens, bacterial antigens); once induced, however, the T$_s$ are antigen-specific

  UV-T$_s$ do not appear to affect already established T$_h$ cell activity

  Generation of T$_s$ is not prevented by thymectomy or splenectomy prior to UV irradiation

  UV-induced immunosuppression is MHC-restricted

UV carcinogenesis

  UV-T$_s$ appear before clinically visible tumors develop

  Cloned UV-T$_s$ lines have been established in vitro

  UV-T$_s$ activity can be transferred to unirradiated or x-irradiated mice, rendering them incapable of rejecting otherwise immunogenic UV-induced tumors, but leaving immunity towards other neoplasms (e.g. chemically induced) unaltered

  UV-T$_s$ recognize most UV-induced tumors as a group; they do not affect the incidence of extracutaneous malignancies (e.g. mammary carcinomas)

  UV-T$_s$ inhibit the generation of antitumor CTL and/or DTH responses (depending on the tumor cell line)

  UV-T$_s$ specifically block the generation of tumor-specific T$_h$ cells

---

altered after UVR, including the expression of MHC molecules, adhesion molecules and other molecules potentially involved in antigen presentation [117, 118, 136–137]. These changes of LC function are observed after in vivo irradiation (as above) or in vitro irradiation of LC with low doses of UVR ($20–200 \text{ J/m}^2$). Alterations of LC morphology and surface marker expression after low-dose UVR may in part be mediated via cytokines and some have recently been attributed to the effects of locally secreted TNFα [138] and IL-1β [139]. In addition to direct effects on APC, higher doses of UVR lead to a systematic alteration of APC function [51]. Again, cytokines may have a significant role in mediating these effects. As described above, exposure of keratinocytes to UVA radiation results in release of an inhibitor of CHS while IL-10 is produced by epidermal cells after UVB exposure and may be responsible for the UVR effects on DTH responses and on Th1 cell stimulation by APC [92, 93, 140–142].

Tumor Antigen Presentation by Epidermal APC

Interestingly, many cutaneous carcinogens were also found to alter LC morphology and function in a similar fashion as UVR, indicating that most, if not all, agents with a known carcinogenic effect on cutaneous tissue also affect LC function. APC have been found to bind tumor-associated antigens (TAA) in tumor-bearing mice [143, 144]. A series of investigations performed by Halliday et al. [126, 145, 146] indicates that LC migrate into cutaneous neoplasms, possibly due to chemotactic effects of a tumor-derived cytokine. Moreover, human studies suggest a correlation between the number of tumor-infiltrating APC and clinical prognosis [147]. Functional investigation of APC function in tumor-bearing animals revealed both impaired and increased APC activity [148, 149], and splenic adherent cells were shown to present TAA in vivo and in vitro [150, 151]. Tumor rejection induced by local application of cytokines may also be mediated via APC [152, 153]. Thus, APC may play a significant role in the induction and modulation of tumor-specific immune responses. However, no correlation between LC number within cutaneous tumors and type or degree of tumor immunity was found in one specific tumor system [126], and no difference was seen in the average reduction of LC density after UVR in patients with basal cell carcinoma compared to normal healthy volunteers [154]. Likewise, skin cancer patients were not found to have lower minimal erythema dose (MED) than cancer-free patients, and UV-induced reduction of LC density was unrelated to the MED [154]. Thus, erythematogenic, carcinogenic and immunosuppressive effects of UVR, although related, may be mediated via different mechanisms.

These observations demonstrate that evidence supporting a role for APC in tumor immune responses is largely indirect and partially conflicting

[71, 121–123, 126, 143, 144, 148–152, 155–162]. Therefore, a series of studies has recently been initiated to directly assess the presentation of TAA by epidermal APC [163–165]. In these studies, LC-enriched epidermal cell suspensions were exposed to TAA and injected into naive syngeneic mice to assess the ability of LCs to induce tumor immunity, as well as into tumor-immune mice, to assess the elicitation of tumor-immune responses by TAA-exposed LC. These studies were performed using a chemically induced spindle cell tumor, S1509a, which grows progressively in syngeneic hosts and is mildly immunogenic, as indicated by the appearance of tumor-specific T-helper cells and concomitant tumor immunity [156, 159]. It was found that freshly isolated epidermal APC were poorly capable of inducing tumor-specific immunity, but were potently able to elicit tumor-specific DTH responses in primed, tumor-immune mice [163]. Further studies indicated that cytokines are crucially involved in the regulation of tumor antigen presentation by epidermal APC, and that primary and secondary immune responses are differentially modulated by these cytokines. For example, incubation of epidermal APC in GM-CSF potently enhanced LC tumor antigen presentation in unprimed systems [163]. Thus, naive mice could be successfully immunized against a subsequent tumor challenge by prior injection of GM-CSF-incubated, tumor-antigen-exposed epidermal APC. For some other cytokines, a down-regulatory effect on tumor antigen presentation was observed, since short-term exposure of GM-CSF-incubated epidermal APC to TNFα abrogated their ability to induce tumor immunity [165]. However, the elicitation of secondary tumor immune responses (DTH response after injection of tumor-antigen-exposed epidermal APC in tumor-immune mice) was potently enhanced after incubation of epidermal APC in TNFα [165]. IL-1α was also found to inhibit tumor-antigen presentation by epidermal APC for the induction of tumor-specific immunity, apparently via an induction of TNFα production [164]. TGFβ was not found to affect tumor-antigen presentation by epidermal APC in this system and also did not inhibit the elicitation of tumor-specific DTH in primed, tumor-immune mice. In vitro exposure of epidermal APC to UVR dose dependently inhibited both their ability to induce tumor-specific immune responses as well as to elicit DTH responses against this tumor after injection into tumor-immune mice [165].

Taken together, these data indicate that epidermal APC are able to present TAA both for the induction as well as for the elicitation of tumor-specific immune responses. Their effectiveness in presenting TAA, however, is potently modulated by cytokines. Thus, the local cytokine environment within the skin appears to be crucial for the generation and elicitation of tumor-immune responses. Since UV irradiation modulates cutaneous cytokine production, a cytokine-mediated alteration of cutaneous tumor-antigen presentation may contribute to the suppressed immune response towards UV-induced neoplasms in chronically UV-exposed skin. Thus, the increased incidence of skin cancers in chronically UV-exposed skin may, at least partially, be due to an altered presentation of tumor-associated antigens on malignantly transformed cells by cutaneous APC. Studies investigating the role of APC in tumor immunity therefore not only reveal further insights into antigen-specific tumor-host immune responses, but may also lead to a novel approach towards the development of antigen-specific tumor immunotherapy [162].

References

1   Unna, P.G.: Die Histopathologie der Hautkrankheiten (Hirschwald, Berlin 1894).
2   Findlay, G.M.: Ultraviolet light and skin cancer. Lancet ii: 1070–1073 (1928).
3   Rausch, H.P.: Baumann, C.A.: Tumor production in mice with ultraviolet irradiation. Am J Cancer 35: 55–58 (1939).
4   Blum, H.: Carcinogenesis by Ultraviolet Light (Princeton University Press, Princeton 1959).
5   Urbach, F.: Epstein, J.H.; Forbes, P.D.: Ultraviolet carcinogenesis: Experimental, global and genetic aspects; in Fitzpatrick, Pathak, Harber, Harker, Seiji, Kukita (eds): Sunlight and Man, pp. 258–283 (University of Tokyo Press, Tokyo 1974).
6   Alcalay, J.; Bucana, C.; Kripke, M.L.: Cutaneous pigmented melanocytic tumor in a mouse treated with psoralen plus ultraviolet A radiation. Photodermatol Photoimmunol Photomed 7: 28–31 (1990).
7   Stern, R.S.; Laird, N.; Melski, J.; Parrish, J.A.; Fitzpatrick, T.B.; Bleich, H.L.: Cutaneous squamous-cell carcinoma in patients treated with PUVA. N Engl J Med 310: 1156–1161 (1984).
8   Ananthaswamy, H.N.; Kripke, M.L.: In vivo transformation of primary cultures of neonatal BALB/c mouse epidermal cells with ultraviolet B radiation. Cancer Res 41: 2882–2890 (1981).
9   Sutherland, B.M.; Delihas, N.C.; Oliver, R.P.; Sutherland, J.C.: Action spectra for ultraviolet light-induced transformation of human cells to anchorage independent growth. Cancer Res 4: 2211–2214 (1981).
10 Pawlowski, A.; Herberman, A.: Heterotransplantation of human basal cell carcinomas in nude mice. J Invest Dermatol 72: 310–313 (1979).
11 Van der Leun, J.C.: UV carcinogenesis. Photochem Photobiol 39: 861–868 (1984).
12 Newell, G.R.; Sider, J.G.; Bergfelt, L.; Kripke, M.L.: Incidence of cutaneous melanoma in the United States by histology with special reference to the face. Cancer Res 48: 5036–5041 (1988).
13 Jung, E.: Photokarzinogenese der Haut; in Macher, Kolde, Bröcker (eds): Jahrbuch der Dermatologie 1992/93. Licht und Haut, pp. 179–186 (Biermann, Zülpich 1993).
14 Sober, A.J.; Rhodes, A.R.; Mihm, M.C.; Fitzpatrick, T.B.: Neoplasms: malignant melanoma, in Fitzpatrick, Eisen, Wolff, Freedberg (eds): Dermatology in General Medicine, vol. 1, pp. 947–966 (McGraw-Hill, New York 1987).
15 Donawho, C.K.; Kripke, M.L.: Evidence that the local effect of ultraviolet radiation on the growth of murine melanomas is immunologically mediated. Cancer Res 51: 4176–4181 (1991).

16 Sutherland, B.M.; Harber, L.C.; Kochevar, I.E.: Pyrimidine dimer formation and repair in human skin. Cancer Res *40:* 3181–3185 (1980).

17 Kochevar, I.E.; Pathak, M.A.; Parrish, J.A.: Photophysics, photochemistry, and photobiology; in Fitzpatrick, Eisen, Wolff, Freedberg, Austen (eds): Dermatology in General Medicine, 3rd ed., vol. 1, pp. 1441–1451 (McGraw-Hill, New York 1987).

18 Schauder, S.: Lichtschutz; in Macher, Kolde, Bröcker (eds): Jahrbuch der Dermatologie 1992/93. Licht und Haut, pp. 187–217 (Bierman, Zülpich 1993).

19 Epstein, J.H.; Fukuyama, K.; Read, W.B.; Epstein, W.L.: Defect of DNA synthesis in skin of patients with xeroderma pigmentosum demonstrated in vivo. Science *168:* 1477–1479 (1970).

20 Hart, R.W.; Setlow, R.B.; Woodhead, A.D.: Evidence that pyrimidine dimers in DNA can give rise to tumors. Proc Natl Acad Sci USA *74:* 5574–5579 (1977).

21 Tan, E.M.; Stoughton, R.B.: Ultraviolet light-induced damage to deoxyribonucleic acid in human skin. J Invest Dermatol *52:* 537–543 (1969).

22 Ley, R.D.; Applegate, L.A.; Fry, R.J.M.; Sanchez, A.B.: Photoreactivation of ultraviolet radiation-induced skin and eye tumors of *Monodelphis domestica.* Cancer Res *51:* 6539–6542 (1991).

23 Devary, Y.; Gottlieb, R.A.; Smeak, T.; Karin, M.: The mammalian ultraviolet response is triggered by activation of Src tyrosine kinases. Cell *71:* 1081–1091 (1993).

24 Radler-Pohl, A.; Sachsenmaier, C.; Gebel, S.; Auer, H.P.; Bruder, J.T.; Rapp, U.; Angel, P.; Rahmsdorf, H.J.; Herrlich, P.: UV-induced activation of AP-1 involves obligatory extranuclear steps including Raf-1 kinase. EMBO J *12:* 1005–1012 (1993).

25 Shah, G.; Ghoh, R.; Amstad, P.A.; Cerrutti, P.A.: Mechanism of induction of c-fos by ultraviolet B (290–320 nm) in mouse JB6 epidermal cells. Cancer Res *53:* 38–45 (1993).

26 DeFabo, E.C.; Kripke, M.L.: Wavelength dependence and dose-rate independence of UV radiation-induced immunologic unresponsiveness of mice to a UV-induced fibrosarcoma. Photochem Photobiol *32:* 183–188 (1980).

27 Applegate, L.A.; Ley, R.D.; Kripke, M.L.: Photoreaction of UVR-induced immune suppression in *Monodelphis domestica.* Photochem Photobiol *49:* 93s (1989).

28 Applegate, L.A.; Ley, R.D.; Alcalay, J.; Kripke, M.L.: Identification of the molecular target for the suppression of contact hypersensitivity by ultraviolet radiation. J Exp Med *170:* 1117–1131 (1989).

29 Kripke, M.L.; Cox, P.A.; Alas, L.G.; Yarosh, D.B.: Pyrimidine dimers in DNA initiate systemic immunosuppression in UV-irradiated mice. Proc Natl Acad Sci USA *89:* 7516–7520 (1992).

30 Hillenkamp, F.; Grabbe, S.: Die physikalischen Grundlagen der Wirkung optischer Strahlung auf die Haut, in Macher, Kolde, Bröcker (eds): Licht und Haut, pp. 1–30 (Biermann, Zülpich 1992).

31 Noonan, F.P.; Bucana, C.; Sauder, D.N.; DeFabo, E.C.: Mechanism of systemic immune suppression by UV irradiation in vivo. II. The UV effects on number and morphology of murine Langerhans' cells and the UV-induced suppression of contact hypersensitivity have different wavelength dependencies. J Immunol *132:* 2408–2416 (1984).

32 Noonan, F.P.; DeFabo, E.C.; Morrison, H.: Cis-urocanic acid, a product formed by ultraviolet B irradiation of the skin, initiates an antigen presentation defect in splenic dendritic cells in vivo. J Invest Dermatol *90:* 92–99 (1988).

33 Noonan, F.P.; DeFabo, E.C.: Immunosuppression by ultraviolet B radiation: Initiation by urocanic acid. Immunol Today *13:* 250–254 (1992).

34 Kinlen, L.J.; Sheil, A.G.; Peto, J.; Doll, R.: Collaborative United Kingdom-Australia study of cancer in patients treated with immunosuppressive drugs. Br J Med *ii:* 1461–1466 (1979).

35 Gupta, A.K.; Cardella, C.J.; Haberman, H.F.: Cutaneous malignant neoplasms in
    patients with renal transplants. Arch Dermatol *112:* 1288–1293 (1986).
36 Hoxtell, E.O.; Mandel, J.S.; Murray, S.S.; Schuman, L.M.; Goltz, R.W.: Incidence of
    skin cancer after renal transplantation. Arch Dermatol *113:* 436–438 (1977).
37 Greene, M.H.; Young, T.I.; Clark, W.H., Jr.: Malignant melanoma in renal transplant
    patients. Lancet *i:* 1196–1198 (1981).
38 Schmieder, G.J.; Yoshikawa, T.; Mata, S.M.; Streilein, J.W.; Taylor, J.R.: Cumulative
    sunlight exposure and the risk of developing skin cancer in Florida. J Dermatol Surg
    Oncol *18:* 517–522 (1992).
39 Penn, I.; Hammond, W.; Brettschneider, L.; Starzl, T.E.: Malignant lymphomas in
    transplantation patients. Transplant Proc *1:* 106–112 (1969).
40 Kripke, M.L.: Immunologic parameters of ultraviolet carcinogenesis. J Natl Cancer Inst
    57: 211–215 (1976).
41 Ebbesen, P.: Enhanced lymphoma incidence in BALB/c mice after ultraviolet light
    treatment. J Natl Cancer Inst *67:* 1077–1078 (1981).
42 Cooper, K.D.; Oberhelman, L.; Hamilton, T.A.; Baadsgaard, O.; Terhune, M.; LeVee,
    G.; Anderson, T.; Koren, H.: UV exposure reduces immunization rates and promotes
    tolerance to epicutaneous antigens in humans: Relationship to dose, CD1a-DR+
    epidermal macrophage induction, and Langerhans cell depletion. Proc Natl Acad Sci
    USA *89:* 8497–8501 (1992).
43 Streilein, J.W.: Immunogenetic factors in skin cancer. N Engl J Med *325:* 884–886 (1992).
44 Yoshikawa, T.; Rae, V.: Bruins-Slot, W.; van den Berg, J.W.; Taylor, R.; Streilein, J.W.:
    Susceptibility to effects of UVB radiation induction of contact hypersensitivity as a risk
    factor for skin cancer in humans. J Invest Dermatol *95:* 530 (1990).
45 Burnet, F.M.: Immunological surveillance in neoplasia. Transplant Rev *7:* 3–51 (1971).
46 Burnet, F.M.: Immunologic Surveillance (Pergamon Press, Oxford 1970).
47 Romerdahl, C.A.; Okamoto, H.; Kripke, M.L.: Immune surveillance against cutaneous
    malignancies in experimental animals. Immunol Ser *46:* 749–767 (1989).
48 Kripke, M.L.: Antigenicity of murine skin tumors induced by ultraviolet light. J Natl
    Cancer Inst *53:* 1333–1336 (1974).
49 Kripke, M.L.: Latency, histology and antigenicity of tumors induced by ultraviolet light
    in three inbred mouse strains. Cancer Res *37:* 1395–1400 (1977).
50 Hong, S. R.; Roberts, L.K.: Cross-reactive tumor antigens in the skin of mice exposed
    to subcarcinogenic doses of ultraviolet radiation. J Invest Dermatol *88:* 154–160 (1987).
51 Fisher, M.S.; Kripke, M.L.: Systemic alteration induced in mice by ultraviolet light
    irradiation and its relationship to ultraviolet carcinogenesis. Proc. Natl Acad Sci USA
    74: 1688–1692 (1977).
52 Kripke, M.L.; Lofgreen, J.S.; Beard, J.; Jessup, J.M.; Fisher, M.S.: In vivo immune
    responses of mice during carcinogenesis by ultraviolet irradiation. J Natl Cancer Inst *59:*
    1227–1230 (1977).
53 DeFabo, E.C.; Kripke, M.L.: Dose-response characteristics of immunologic unrespon-
    siveness to UV-induced tumors produced by UV irradiation of mice. Photochem
    Photobiol *30:* 385–390 (1979).
54 Spellman, C.W.; Woodward, J.G.; Daynes, R.A.: Modification of immunological poten-
    tial by ultraviolet radiation. I. Immune status of short-term UV-irradiated mice.
    Transplantation *24:* 112–119 (1977).
55 Spellman, C.W.; Daynes, R.A.: Modification of immunological potential by ultraviolet
    radiation. II. Generation of suppressor cells in short-term UV-irradiated mice. Trans-
    plantation *24:* 120–126 (1977).

56   Fortner, G.W.; Kripke, M.L.: In vivo reactivity of splenic lymphocytes from normal and
     UV-irradiated mice against syngeneic UV-induced tumors. J. Immunol *118:* 1483–1487
     (1977).
57   Fisher, M.S.; Kripke, M.L.: Further studies on the tumor-specific suppressor cells
     induced by ultraviolet radiation. J Immunol *121:* 1139–1144 (1978).
58   Kripke, M.L.; Thorn, R.M.; Lill, P.H.; Civin, P.I.; Pazmino, N.H.; Fisher, M.S.:
     Further characterization of immunological unresponsiveness induced in mice by ultravi-
     olet radiation. Growth and induction of nonultraviolet-induced tumors in ultraviolet-
     irradiated mice. Transplantation *28:* 212–217 (1979).
59   Fisher, M.S.; Kripke, M.L.: Suppressor T lymphocytes control the development of
     primary skin cancer in ultraviolet-irradiated mice. Science *216:* 1133–1134 (1982).
60   Spellman, C.W.; Daynes, R.A.: Cross-reactive transplantation antigens between UV-
     irradiated skin and UV-induced tumors. Photodermatology *1:* 164–169 (1984).
61   Roberts, L.K.; Lynch, D.H.; Daynes, R.A.: Evidence for two functionally distinct
     cross-reactive tumor antigens associated with ultraviolet light and chemically induced
     tumors. Transplantation *33:* 352–360 (1982).
62   Hostetler, L.W.; Romerdahl, C.A.: Kripke, M.L.: Specificity of antigens on UV radia-
     tion-induced antigenic tumor cell variants measured in vitro and in vivo. Cancer Res. *49:*
     1207–1213 (1989).
63   Hostetler, L.W.; Kripke, M.L.: Origin and significance of transplantation antigens
     induced on cells transformed by UV radiation, in Greene, Hamaoka (eds): Development
     and Recognition of the Transformed Cell, pp. 307–329 (Plenum Publishing, New York
     1987).
64   Roberts, L.K.; Spellman, C.W.; Warner, N.L.: Establishment of a continuous T-cell line
     capable of suppressing anti-tumor immune responses in vivo. J Immunol *131:* 514–519
     (1983).
65   Roberts, L.K.; Samlowski, W.E.; Daynes, R.A.: Immunological consequences of ultravi-
     olet radiation exposure. Photodermatology *3:* 284–298 (1986).
66   Ullrich, S.E.; Kripke, M.L.: Mechanism in the suppression of tumor rejection produced
     in mice by repeated UV irradiation. J Immunol *133:* 2786–2790 (1984).
67   Kripke, M.L.; McClendon, E.: Studies on the role of antigen-presenting cells in the
     systemic suppression of contact hypersensitivity by UVB radiation. J Immunol *137:*
     443–447 (1986).
68   Gurish, M.F.; Lynch, D.H.; Daynes, R.A.: Changes in antigen-presenting cell function
     in the spleen and lymph nodes of ultraviolet-irradiated mice. Transplantation *33:*
     280–284 (1982).
69   Urban, J.L.; Burton, R.C.; Holland, J.M.; Kripke, M.L.; Schreiber, H.: Mechanisms of
     syngeneic tumor rejection. Susceptibility of host-selected progressor variants. J Exp Med
     155: 557–573 (1982).
70   Burnham, D.K.; Mak, C.K.; Webster, R.J.: Daynes, R.A.: Relationship between in-
     ducible H-2 expression and the immunogenicity of murine skin neoplasms. I. Evidence
     that the immunogenicity of ultraviolet radiation- and chemically-induced tumors is
     associated with their susceptibility to gamma-interferon-mediated enhancement of H-
     2^{Kk} expression. Transplantation *47:* 533–542 (1989).
71   Daynes, R.A.; Burnham, D.K.; Dewitt, C.W.; Roberts, L.K.; Drueger, G.G.:
     The immunobiology of ultraviolet-radiation carcinogenesis. Cancer Surv *4:* 51–99
     (1985).
72   Perry, L.; Dorf, M.; Benacerraf, F.; Greene, M.: Regulation of immune response to
     tumor antigen: Interference with syngeneic tumor immunity by anti-IA alloantisera.
     Proc Natl Acad Sci USA *76:* 918–924 (1979).

73 Noma, T.; Usui, M.; Dorf, M.E.: Characterization of the accessory cells involved in suppressor T-cell induction. J Immunol *134:* 1374–1380 (1985).

74 Halliday, G.M.; Wood, R.C.; Muller, H.K.: Presentation of antigen to suppressor cells by a dimethylbenz(a)anthracene-resistant, Ia-positive, Thy-1-negative, I-J-restricted epidermal cell. Immunology *69:* 97–103 (1990).

75 Lowy, A.; Tominaga, A.; Drebin, J.A.; Takaoki, M.; Benacerraf, B.: Greene, M.I.: Identification of an I-J$^+$ antigen-presenting cell required for third order suppressor cell activation. J Exp Med *157:* 353–358 (1983).

76 Granstein, R.D.; Askari, M.; Whitaker, D.; Murphy, G.F.: Epidermal cells in activation of suppressor lymphocytes: Further characterization. J Immunol *138:* 4055–4062 (1987).

77 Kripke, M.L.: Immunologic mechanisms in UV-radiation carcinogenesis. Adv Cancer Res *34:* 69–106 (1981).

78 Woodward, J.G.; Daynes, R. A.: Cell-mediated immune response to syngeneic UV-induced tumors. I. The presence of tumor-associated macrophages and their possible role in the in vitro generation of cytotoxic lymphocytes. Cell Immunol *41:* 304–319 (1978).

79 Kripke, M.L.: Effects of UV radiation on tumor immunity. J Natl Cancer Inst *82:* 1392–1396 (1990).

80 Ullrich, S.E.; Yee, G.K.; Kripke, M.L.: Suppressor lymphocytes induced by epicutaneous sensitization of UV-irradiated mice control multiple immunological pathways. Immunology *58:* 185–190 (1986).

81 Palaszynski, E.W.; Kripke, M.L.: Transfer of immunological tolerance to ultraviolet-radiation-induced skin tumors with grafts of ultraviolet-irradiated skin. Transplantation *36:* 465–466 (1983).

82 Baadsgaard, O.; Fox, P.; Cooper, K.D.: Human epidermal cells from ultraviolet light-exposed skin preferentially activate autoreactive CD4$^+$2H4$^+$ suppressor-inducer lymphocytes in CD8$^+$ suppressor/cytotoxic lymphocytes. J Immunol *140:* 1738–1744 (1988).

83 Baadsgaard, O.; Salvo, B.; Mannie, A.; Dass, B.; Fox, D.A.; Cooper, K.D.: In vivo ultraviolet exposed human epidermal cells activate T suppressor cell pathways that involve CD4$^+$CD45RA$^+$ suppressor-inducer T cells. J Immunol *145:* 2854–2861 (1990).

84 Yee, G.K.; Ullrich, S.E.; Kripke, M.L.: The role of suppressor factors in the regulation of immune responses by ultraviolet radiation-induced suppressor T lymphocytes. II. Activity of suppressor cell culture sonicates. Cell Immunol *121:* 88–98 (1989).

85 Yee, G.K.; Levy, J.G.; Kripke, M.L.; Ullrich, S.E.: The role of suppressor factors in the regulation of immune responses by ultraviolet radiation-induced suppressor T lymphocytes. III. Isolation of a suppressor factor with the B16G monoclonal antibody. Cell Immunol *126:* 255–267 (1990).

86 Pope, B.L.: Activation of suppressor T cells by low-molecular weight factors secreted by spleen cells from tumor-bearing mice. Cell Immunol *93:* 364–374 (1985).

87 Satoh, T.; Tokura, Y.; Satoh, Y.; Takigawa, M.: Ultraviolet-induced suppressor T cells and factor(s) in murine contact photosensitivity. III. Mode of action of T-cell suppressor factor(s) and interaction with cytokines. Cell Immunol *131:* 120–131 (1990).

88 Schwarz, T.; Urbanska, A.; Gschnait, F.; Luger, T.A.: Inhibition of the induction of contact hypersensitivity by a UV-mediated epidermal cytokine. J Invest Dermatol *87:* 289–291 (1986).

89 Harriott-Smith, T.G.; Halliday, W.J.: Circulating suppressor factors in mice subjected to ultraviolet irradiation and contact sensitization. Immunology *57:* 207–211 (1985).

90   Luger, T.A.; Schwarz, T.: Epidermal cell-derived cytokines; in Bos (ed.): Skin Immune
      System, pp. 257–283 (CRC Press, Boca Raton 1989).
91   Schwarz, T.: The significance of epidermal cytokines in UV-induced immune suppres-
      sion. Hautarzt 39: 642–646 (1988).
92   Bogdan, C.; Vodovotz, Y.; Nathan, C.: Macrophage deactivation by interleukin-10. J
      Exp Med 174: 1549–1555 (1991).
93   Rivas, J.M.; Ullrich, S.E.: Systemic suppression of delayed-type hypersensitivity by
      supernatants from UV-irradiated keratinocytes. An essential role for keratinocyte-
      derived IL-10. J Immunol 149: 3865–3871 (1992).
94   Yoshikawa, T.; Streilein, J.W.: Tumor necrosis factor-alpha and ultraviolet B light have
      similar effects on contact hypersensitivity in mice. Reg Immunol 3: 139–144 (1991).
95   Yoshikawa, T.; Kurimoto, I.; Streilein, J.W.: Tumor necrosis factor-alpha mediates
      ultraviolet light B-enhanced expression of contact hypersensitivity. Immunology 76:
      264–271 (1992).
96   Yoshikawa, T.: Streilein, J.W.: Genetic basis of the effects of ultraviolet light B on
      cutaneous immunity. Evidence that polymorphism at the Tnfa and Lps loci governs
      susceptibility. Immunogenetics 32: 398–405 (1990).
97   Streilein, J.W.; Bergstresser, P.R.: Genetic basis of ultraviolet-B effects on contact
      hypersensitivity. Immunogenetics 27: 252–258 (1988).
98   Vermeer, M.; Schmieder, G.J.; Yoshikawa, T.; van den Berg, J.W.; Metzman, M.S.;
      Taylor, J.R.; Streilein, J.W.: Effects of ultraviolet B light on cutaneous immune
      responses of humans with deeply pigmented skin. J Invest Dermatol 97: 729–734 (1991).
99   Luger, T.; Krutmann, J.; Kirnbauer, R.; Urbanski, A.; Schwarz, T.; Klappacher, G.;
      Kock, A.; Micksche, M.; Malejczyk, J.; Schauer, E.: IFN-beta-2/IL-6 augments the
      activity of human natural killer cells. J Immunol 143: 1206–1209 (1989).
100   Urbanski, A.: Schwarz, T.; Neuner, P.; Krutmann, J.; Kirnbauer, R.; Kock, A.; Luger,
      T.A.: Ultraviolet light induces increased circulating interleukin-6 in humans. J Invest
      Dermatol 94: 808–811 (1990).
101   Norbury, K.C.; Kripke, M.L.; Budmen, M.B.: In vitro reactivity of macrophages and
      lymphocytes from ultraviolet-irradiated mice. J Natl Cancer Inst 59: 1231–1235 (1977).
102   Zimmer, T.: Jones, P.P.: Combined effects of tumor necrosis factor-alpha, prostaglandin
      E_2, and corticosterone on induced Ia expression on murine macrophages. J Immunol
      145: 1167 (1990).
103   Chung, H.-T.; Burnham, D.K.; Robertson, B.; Roberts, L.K.; Daynes, R.A.: Involve-
      ment of prostaglandins in the immune alterations caused by the exposure of mice to
      ultraviolet radiation. J Immunol 137: 2478–2484 (1986).
104   Betz, M.; Fox, B.S.: Prostaglandin E_2 inhibits production of Th1 lymphokines but not
      of Th1 lymphokines. J Immunol 146: 108–113 (1991).
105   Greaves, M.W.; Camp, R.D.R.; Prostaglandins, leukotrienes, phospholipase, platelet
      activation factor, and cytokines: An integrated approach to inflammation of human
      skin. Arch Dermatol Res 280: S33–S41 (1988).
106   Kim, T.Y.; Kripke, M.L.; Ullrich, S.E.: Immunosuppression by factors released from
      UV-irradiated epidermal cells: Selective effects on the generation of contact and delayed
      hypersensitivity after exposure to UVA or UVB radiation. J Invest Dermatol 94: 26–32
      (1990).
107   Schwartz, T., et al.: Inhibition of the induction of contact hypersensitivity by a
      UV-mediated epidermal cytokine. J Invest Dermatol 87: 289 (1986).
108   Schwartz, T., et al.: UV-irradiated epidermal cells produce a specific inhibitor of
      interleukin-1 activity. J Immunol 138: 1457 (1987).

109   Gordon, C.; Wofsy, D.: Effects of recombinant murine tumor necrosis factor-alpha on immune function. J Immunol *144:* 1753 (1990).

110   Robertson, B.; Gahring, L.; Newton, R.; Daynes, R.: In vivo administration of interleukin-1 to normal mice depresses their capacity to elicit contact hypersensitivity responses: Prostaglandins are involved in this modification of immune function. J Invest Dermatol *88:* 380 (1987).

111   Fujii, T.; Igarashi, T.; Kishimoto, S.: Significance of suppressor macrophages for immunosurveillance of tumor-bearing mice. J Natl Cancer Inst *78:* 509–517 (1987).

112   Jun, B.-D.; Roberts, L.K.; Cho, B.-K.; Robertson, B.; Daynes, R.A.: Parallel recovery of epidermal antigen-presenting cell activity and contact hypersensitivity responses in mice exposed to ultraviolet irradiation: The role of a prostaglandin-dependent mechanism. J Invest Dermatol *90:* 311–316 (1988).

113   Rheins, L.A.; Barnes, L.; Amornsiripanitch, S.; Collins, C.E.; Nordlund, J.J.: Suppression of the cutaneous immune response following topical application of the prostaglandin PGE$_2$. Cell Immunol *106:* 33–42 (1986).

114   Roberts, L.K.; Jun, B.-D.; Law, M.Y.L.: Cis-urocanic acid-induced immunosuppression is associated with a prostaglandin-dependent mechanism (abstract). J Invest Dermatol *94:* 572 (1990).

115   Demidem, A.; Taylor, J.R.; Grammer, S.F.; Streilein, J.W.: Influence of microenvironmental factors on human Langerhans cell function in vitro. Reg Immunol *3:* 297–304 (1990).

116   Kurimoto, I.; Streilein, J.W.: Cis-urocanic acid suppression of contact hypersensitivity induction is mediated via tumor necrosis factor-alpha. J Immunol *148:* 3072–3078 (1992).

117   Stingl, G.; Gazze-Stingl, L.A.; Aberer, W.; Wolff, K.: Antigen presentation by murine epidermal cells and its alteration by UVB light. J Immunol *127:* 1707–1714 (1981).

118   Stingl, L.A.; Sauder, D.N.; Iljima, M.; Wolff, K.; Pehamberger, H., Stingl, G.: Mechanism of UV-B induced impairment of the antigen-presenting capacity of murine epidermal cells. J Immunol *130:* 1586–1591 (1983).

119   Simon, J.C.; Krutmann, J.; Elmets, C.A.; Bergstresser, P.R.; Cruz, P.D.: Ultraviolet B-irradiated antigen-presenting cells display altered accessory signaling for T cell activation: Relevance to immune responses initiated in skin. J Invest Dermatol *98:* 66s–69s (1992).

120   Kripke, M.L.; Munn, C.G.; Jeevan, A.; Tang, J.-M.; Bucana, C.: Evidence that cutaneous antigen-presenting cells migrate to regional lymph nodes during contact sensitization. J Immunol *145:* 2833–2838 (1990).

121   Alcalay, J.; Kripke, M.L.: Antigen-presenting activity of draining lymph node cells from mice painted with a contact allergen during ultraviolet carcinogenesis. J Immunol *146:* 1717–1721 (1991).

122   Alcalay, J.; Craig, J.; Kripke, M.: Alterations in Langerhans cells and Thy-1$^+$ dendritic epidermal cells in murine epidermis during the evolution of ultraviolet radiation-induced skin cancers. Cancer Res *49:* 4591–4596 (1989).

123   Bergfelt, L.; Bucana, C.; Kripke, M.: Alterations in Langerhans cells during growth of transplantable murine tumors. J Invest Dermatol *91:* 129–135 (1988).

124   Alcalay, J.; Goldberg, L.; Wolf, J.; Kripke, M.: Variations in the number and morphology of Langerhans cells in the epidermal component of squamous cell carcinomas. Arch Dermatol *125:* 917–920 (1989).

125   Yamaji, K.; Matsui, M.; Saida, T.: Increased densities of Langerhans cells in the epidermis of skin tumors. J Dermatol *14:* 20–24 (1987).

126 Halliday, G.M.; Reeve, V.E.; Barnetson, R.S.C.: Langerhans cell migration into ultraviolet light-induced squamous skin tumors is unrelated to anti-tumor immunity. J Invest Dermatol *97:* 830–834 (1991).

127 Norbury, K.C.; Kripke, M.L.: Ultraviolet-induced carcinogenesis in mice treated with silica, trypan blue or pyran copolymer. J Reticuloendothel Soc *26:* 827–837 (1979).

128 Greene, M.I.; Sy, M.-S.; Kripke, M.L.: Impairment of antigen-presenting cell function by ultraviolet radiation. Proc Natl Acad Sci USA *76:* 6591–6595 (1979).

129 Magee, M.J.; Kripke, M.L.; Ullrich, S.E.: Suppression of the elicitation of the immune response to alloantigen by ultraviolet radiation. Transplantation *47:* 1008–1013 (1989).

130 Alcalay, J.; Goldberg, L.H.; Wolf, J.J.; Kripke, M.L.: Ultraviolet radiation-induced damage to human Langerhans cells in vivo is not reversed by ultraviolet A or visible light. J Invest Dermatol *95:* 144–146 (1990).

131 Welsh, E.A.; Kripke, M.L.: Murine Thy-1$^+$ dendritic epidermal cells induce immunologic tolerance in vivo. J Immunol *144:* 883–891 (1990).

132 Polla, L.; Margolis, R.; Goulston, C.; Parrish, J.A.; Granstein, R.D.: Enhancement of the elicitation phase of the murine contact hypersensitivity response by local ultraviolet radiation. J Invest Dermatol *86:* 13–17 (1986).

133 Simon, J.C.; Cruz, P.D.; Bergstresser, P.R.; Tigelaar, R.E.: Low-dose UVB-irradiated Langerhans cells preferentially activate CD4$^+$ cells of the TH1 subset. J Immunol *145:* 2087 (1990).

134 Simon, J.C.; Tigelaar, R.E.; Bergstresser, P.R.; Edelbaum, D.P.D.; Cruz, P.D.: Ultraviolet B radiation converts Langerhans cells from innunogenic to tolerogenic antigen-presenting cells. J Immunol *146:* 485–491 (1991).

135 Simon, J.C.; Edelbaum, D.; Bergstresser, P.R.; Cruz, P.D.: Distorted antigen-presenting function of Langerhans cells induced by tumor necrosis factor-alpha via a mechanism that appears different from that induced by ultraviolet B radiation. Photodermatol Photoimmunol Photomed *8:* 190–194 (1991).

136 Krutmann, J.K.; Kammer, G.M.; Toossi, Z.; Waller, R.L.; Ellner, J.J.; Elmets, C.A.: UVB radiation and human monocyte accessory function: Differential effects on premitotic events in T-cell activation. J Invest Dermatol *94:* 204–209 (1990).

137 Tang, A.: Udey, M.C.: Inhibition of epidermal Langerhans cell function by low dose ultraviolet B radiation. Ultraviolet B radiation selectively modulates ICAM-1 (CD54) expression by murine LC. J Immunol *146:* 3347–3355 (1991).

138 Vermeer, M.; Streilein, J.W.: Ultraviolet B light-induced alterations in epidermal Langerhans cells are mediated in part by tumor necrosis factor-alpha. Photodermatol Photoimmunol Photomed *7:* 258–265 (1990).

139 Euk, A.H.; Angeloni, V.L.; Udey, M.C.; Katz, S.I.: An essential role for Langerhans cell-derived IL-Iβ in the initiation of primary immune responses in skin. J Immunol *150:* 3698–3704 (1993).

140 Ding, L.; Shevach, E.M.: IL-10 inhibits mitogen-induced T cell proliferation by selectively inhibiting macrophage costimulatory function. J Immunol *148:* 3133–3139 (1992).

141 Fiorentino, D.F.; Zlotnik, A.; Vieira, P.; Mosmann, T.R.; Howard, M.; Moore, K.W.; O'Garra, A.: IL-10 acts on the antigen-presenting cell to inhibit cytokine production by Th1 cells. J Immunol *147:* 3815–3822 (1991).

142 Enk, A.H.; Katz, S.I.: Identification and induction of keratinocyte-derived IL-10. J Immunol *149:* 92–95 (1992).

143 Shimizu, J.; Zou, J.P.; Ikegame, K.; Katagiri, T.; Fujiwara, H.; Hamaoka, T.: Evidence for the functional binding in vivo of tumor rejection antigens to antigen-presenting cells in tumor-bearing hosts. J Immunol *146:* 1708–1714 (1991).

144 Shimizu, J.: Zou, J.P.; Ikegame, K.; Fujiwara, H.; Hamaoka, T.: Antigen-presenting cells constitutively bind tumor antigens in the tumor-bearing state in vivo to construct an effective immunogenic unit. Jpn J Cancer Res 82: 262–265 (1991).

145 Halliday, G. M.; Lucas, A.D.; Barnetson, R.S.C.: Control of Langerhans cell density by a skin tumour-derived cytokine. Immunology 77: 13–18 (1992).

146 Halliday, G.M.; Odling, K.A.; Ruby, J.C.; Muller, H.K.: Suppressor cell activation and enhanced skin allograft survival after tumor promotor but not initiator induced depletion of cutaneous Langerhans cells. J Invest Dermatol 90: 293–297 (1988).

147 Becker, Y.: Anticancer role of dendritic cells in human and experimental cancers – A review. Anticancer Res 12: 511–520 (1992).

148 Zou, J.; Shimizu, J.; Ikegame, K.; Yamamoto, N.; Ono, S.; Fujiwara, H.; Hamaoka, T.: Tumor-bearing mice exhibit a progressive increase in tumor antigen-presenting cell function and a reciprocal decrease in tumor antigen-responsive CD4$^+$ T cell activity. J Immunol 148: 648–655 (1992).

149 Restifo, N.; Esquivel, F.; Asher, A.; Stotter, H.; Barth, R.; Bennink, J.; Mule, J.; Yewdell, J.; Rosenberg, S.: Defective presentation of endogeneous antigens by a murine sarcoma. Implications for the failure of an antitumor immune response. J Immunol 147: 1453–1459 (1991).

150 Shimizu, J.; Suda, T.; Katagiri, T.; Fujiwara, H,; Hamaoka, T.: Tumor-specific T cell lines: Capacity to proliferate and produce interleukin-2 in response to various forms of tumor antigens. Jpn J Cancer Res 83: 184–193 (1992).

151 Shimizu, J.; Suda, T.; Yoshioka, T.; Kosugi, A.; Fujiwara, H.; Hamaoka, T.: Induction of tumor-specific in vivo protective immunity by immunization with tumor antigen-pulsed antigen-presenting cells. J Immunol 142: 1053–1059 (1989).

152 Bosco, M.; Giovarelli, M.; Forni, M.; Modesti, A.; Scarpa, S.; Masuelli, L.; Forni, G.: Low doses of IL-4 injected perilymphatically in tumor-bearing mice inhibit the growth of poorly and apparently nonimmunogenic tumors and induce a tumor-specific immune memory. J Immunol 145: 3136–3143 (1990).

153 Sakamoto, K.; Nakajima, H.; Shimizu, J.; Katagiri, T.; Kiyotaki, C.; Fujiwara, H.; Hamaoka, T.: The mode of recognition of tumor antigens by noncytolytic-type antitumor T cells: Role of antigen-presenting cells and their surface class I and class II H-2 molecules. Cancer Immunol Immunother 27: 261–266 (1988).

154 Alcalay, J.; Goldberg, L.H.; Kripke, M.L.; Wolf, J.E.J.: The sensitivity of Langerhans cells to simulated solar radiation in basal cell carcinoma patients. J Invest Dermatol 93: 746–750 (1989).

155 Cruz, P.D.; Bergstresser, P.R.: Ultraviolet radiation, Langerhans cells and skin cancer. Conspiracy and failure. Arch Dermatol 125: 975 (1989).

156 Schatten, S.; Granstein, R.D.; Drebin, J.A.; Greene, M.I.: Suppressor T cells and the immune response to tumors. CRC Crit Rev Immunol 4: 335–379 (1984).

157 Kern, D.; Klarnet, J.; Jensen, M.; Greenberg, P.: Requirement for recognition of class II molecules and processed tumor antigen for optimal generation of syngeneic tumor-specific class I-restricted CTL. J Immunol 136: 4303–4310 (1986).

158 Perry, L.; Dorf, M.; Bach, B.; Benaceraf, B.; Greene, M.: Mechanisms of regulation of cell-mediated immunity: Anti-I-A alloantisera interfere with induction and expression of T cell mediated immunity to cell bound antigen in vivo. Clin Immunol Immunopathol 15: 270–292 (1980).

159 Schatten, S.; Drebin, I.A.; Granstein, R.D.; Greene, M.I.: Differential antigen presentation in tumor immunity. FASEB J 43: 2460–2464 (1984).

160 Shimizu, J.; Zou, J.-P.; Ikegame, K.; Katagiri, T.; Fujiwara, H.; Hamaoka, T.: Evidence for the functional binding in vivo of tumor rejection antigens to antigen-presenting cells in tumor-bearing hosts. J Immunol 146: 1708–1714 (1991).

161 Yamashita, U.: Dysfunction of Ia-positive antigen-presenting cells in tumor-bearing mice. Jpn J Cancer Res *78:* 261–269 (1987).
162 Zou, J.P.; Shimizu, J.; Ikegame, K.; Takiuchi, H.; Fujiwara, H.; Hamaoka T.: Tumor immunotherapy with the use of tumor-antigen-pulsed antigen-presenting cells. Cancer Immunol Immunother *35:* 1–6 (1992).
163 Grabbe, S.; Bruvers, S.; Gallo, R.L.; Knisely, T.L.; Nazareno, R.; Granstein, R.D.: Tumor antigen presentation by murine epidermal cells. J Immunol *146:* 3656–3661 (1991).
164 Grabbe, S.; Bruvers, S.; Granstein, R.D.: Effects of immunomodulatory cytokines on the presentation of tumor-associated antigens by epidermal Langerhans' cells. J Invest Dermatol *99:* 66s–68s (1992).
165 Grabbe, S.; Bruvers, S.; Lindgren, A.M.; Hosoi, J.; Tan, K.C.; Granstein, R.D.: Tumor antigen presentation by epidermal antigen-presenting cells in the mouse: Modulation by granulocyte-macrophage colony stimulating factor, tumor necrosis factor-α, and ultraviolet radiation. J Leukocyte Biol *52:* 209–217 (1992).

Stephan Grabbe, MD, Department of Dermatology, University of Münster, D–48149 Münster (FRG)

Granstein RD (ed): Mechanisms of Immune Regulation.
Chem Immunol. Basel, Karger, 1994, vol 58, pp 314–330

Regulation of Immunity by Ultraviolet Radiation and Photosensitized Reactions

Maritza I. Perez, Richard L. Edelson

Yale University, Department of Dermatology, New Haven, Conn., USA

Photoimmunology is the discipline that evaluates the interaction of immune responses with ultraviolet light. This discussion of ultraviolet radiation influences on immune responses will focus on the effects of medium-wave ultraviolet radiation (UVB, 290–320 nm) and long-wave ultraviolet radiation (UVA, 320–400 nm) in the presence or absence of photoactivatable furocoumarin compounds such as 8-methoxypsoralen (8-MOP). In this chapter, we will discuss the effects of UVB irradiation of the skin in the development and propagation of such immune responses as contact hypersensitivity (CHS), delayed-type hypersensitivity (DTH, including alloreactive responses), and tumor rejection. Similarly, we will discuss the effects of photoactivation of psoralens by exposure to UVA on the skin (PUVA) or in blood components (extracorporeal photochemotherapy or photopheresis) (ECP); exposure of the skin to UVA by itself, and the influence of UVA on immune responses such as CHS, DTH, alloreactivity, and tumor rejection. Finally, we will summarize the immunologic effects observed in some other photosensitizing reactions as photodynamic therapy (PDT).

Medium-Wave Ultraviolet Light

The concept that photons or photoproducts might interfere with immune reactions occurring in the skin was first tested by Haniszko and Suskin [1] in 1963. They demonstrated that the elicitation of CHS reactions in guinea pigs was impaired by prior exposure of the test site to UVB radiation. Further research in this field led to the important observation by Toews et al. [2] in 1980 who demonstrated that epicutaneous application of a sensitizing hapten onto a skin area pretreated with suberythematous doses of UVB induced immunological tolerance that correlated with morphological alteration of Langerhans cells (LC) at the site of irradiation.

These morphological changes of LC have been reported in both murine and human skin [3]. They include depletion of cell surface markers such as the disappearance of surface expression of the MHC class II molecules and membrane ATPase activity [3–5]. New data demonstrates that MHC class II molecules internalize from the surface of LC in the formation of Birbeck granules [6, 7]. Another group of investigators has demonstrated that low-dose UVB radiation of LC abrogate their capacity to up-regulate intercellular adhesion molecule-1 (ICAM-1) expression [8] in short-term culture and decreased the survival of LC in short-term culture [9]. However, these alterations are transitory in vivo since LC recuperate from the UVB-induced damage in an early rapid recovery phase of proliferation (7–14 days after irradiation) and in a late recovery phase (42–56 days after irradiation) [10].

Subsequently, it was demonstrated that the disappearance of surface markers from LC was accompanied by an inhibition of their immunologic function during the induction phase of CHS, leading to the induction of hapten-specific suppressor T lymphocytes [11]. This finding provided a critical link between the effect of photons on the skin and the alteration of systemic immunity and focused attention on the antigen-presenting cells (APC) that determine the outcome of the encounter between T lymphocytes and haptens on the skin.

To evaluate the in vitro and in vivo effect of UVB radiation of LC and their ability to present antigens, epidermal cell suspensions were treated in vitro with UVB radiation, coupled with hapten, tested for activation upon exposure to the same hapten [12], and injected into mice that were then epicutaneously exposed to the hapten [13]. These UVB-treated epidermal cell suspensions demonstrated impaired ability to present antigen following UVB radiation [12], and when injected subcutaneously into mice, these cells induced suppressor T lymphocytes rather than CHS [13]. In addition, UVB radiation causes a selective, systemic suppression of immune responses [14–19]. This systemic impairment of immune responses such as CHS [14], DTH response to hapten-modified cells [15] or foreign erythrocytes and protein antigens [16] or allogeneic spleen cells [17, 18], and local graft-versus-host reaction (GVHR) [19] induced by UVB radiation requires higher doses of UVB radiation and abrogates the induction of the immune reaction occurring at the sites directly exposed to and distant from UVB exposure [14–18] without interfering with the elicitation of the immune response [20]. Indeed, in the mouse, exposure of the site of elicitation to UVB radiation enhances the CHS response [20]. Induction of these suppressor responses in vivo has been demonstrated for CHS in mice and in humans [14, 21], for DTH responses in mice [15–18], and for rejection of UVB-induced tumors in mice [22–28]. However, many other immune responses such as antibody production, allograft rejection, lymphocyte blastogenesis in response to antigen, and generation of cytotoxic T cells are unaffected by UVB radiation of the skin [22, 26].

Although the damaging effect of UVB radiation on epidermal LC appears to be universal in all strains of mice tested, and the generation of hapten-specific afferent T suppressor cells is a universal sequela to treatment of mice with UVB radiation and hapten, these suppressor T cells display differential ability to interrupt induction of effector mechanism that will manifest as suppression of the immune response [29]. For example, this alteration in response can be overcome by immunizing UVB-irradiated mice with hapten-coupled APC from normal syngeneic donors [30, 31] only for a DTH response but not for a CHS response [32] indicating that systemic suppression of CHS and DTH responses induced by UVB radiation in mice are mediated by different mechanisms. Also, the injection of supernatants from UVB-irradiated epidermal cells or a murine keratinocyte cell line had no effect on the induction of CHS to hapten, but significantly suppressed the DTH response to alloantigen [33]. These data indicate that soluble keratinocyte-derived suppressive factors are selectively involved in the induction of systemic immunosuppression of either CHS or DTH by UV radiation [33].

There are various possibilities (some of which are not mutually exclusive) to explain this selective, systemic suppression of immune responses induced by UVB irradiation of cutaneous LC. These possibilities include: (1) disturbance of antigen presentation and/or processing by UVB radiation exposure of LC; (2) UVB radiation-induced alteration of LC from immunogenic to tolerogenic; (3) the presence of distinct populations of APC (one that activates the effector pathway and one that activates the suppressor pathway) in the skin with different susceptibilities to UVB radiation; (4) exposure of LC to UVB radiation impeding the expression of a second signal for T cell activation and induction of clonal anergy; (5) redistribution of APC after UVB irradiation of the skin, or (6) generation of carrier-specific suppressor, inducer T cells. There is some experimental evidence supporting each of these possible explanations.

The concept that UVB radiation inhibits antigen presentation by LC comes from studies in which in vitro UVB irradiation of human and mice epidermal cell suspensions lose their ability to stimulate T cells against various antigens [13, 34]. A later study demonstrated that LC function was changed from immunogenic to tolerogenic when mice infused with UVB-irradiated and hapten-derivatized purified LC became incapable of mounting a CHS response to the same hapten even when the hapten was epicutaneously presented on unirradiated normal skin later [35].

Recent work in mice has suggested that there are two types of APC in the skin: I-A$^+$ LC that activate the effector pathway of CHS and are UVB-sensitive, and I-J restricted cells which activate the suppressor pathway [36, 37] and are more UVB-resistant [36]. This concept has been confirmed and expanded, with the demonstration that UVB-irradiated I-A$^+$ LC and irradiated or nonirradiated Thy-1$^+$ dendritic epidermal cells (EC) have the potential to deliver down-regulatory signals of CHS in mice [35]. In humans, UVB and UVC but not UVA radiation of the skin is associated with the appearance of OKM5$^+$T6$^-$DR$^+$ APC 72 h after irradiation, which have potent allostimulatory activity in the EC-lymphocyte reaction and also are capable of stimulation in the autologous mixed leukocyte reaction [38, 39].

Systemic suppression of CHS appears to be mediated by CD4+ T cells [40, 41] and/or CD8+ T cells [18]. Recent reports have indicated that LC, following low-dose UVB radiation, lose the capacity to stimulate the proliferation of CD4+ Th1 clones, but not of Th2 clones [42]. Th1 T cell subsets are responsible for mediating DTH responses, tumor lysis, and they secrete γ-interferon (IFN-γ), interleukin-2 (IL-2), IL-3, granulocyte-macrophage colony-stimulating factor (GM-CSF), and lymphotoxin [43]. By contrast, Th2 T cell subsets produce IL-4 and IL-5 that enable B cells to efficiently secrete antigen-specific antibodies [43]. Th1 and Th2 cells appear to counterregulate each other [44]: IFN-γ production by Th1 cells inhibits the proliferation of Th2 cells and the newly identified IL-10 produced by Th2 cells [45] and keratinocytes [46] can block the activation of Th1. Additional work has shown that this acquired unresponsiveness of Th1 cells induced by UVB-irradiated LC used as APC represents a long-lasting state of clonal anergy that resuls from a block in their ability to produce IL-2 [47]. Clonal anergy is induced when T cells receive the antigen-specific signal or first signal without the costimulatory signal or second signal leading to a blockade in the production of IL-2 [48]. The first signal is transduced when the MHC molecule, occupied by the specific antigenic peptide on the APC, becomes engaged with the T-cell receptor (TCR) complex on the selected T cell. A second signal is transduced when a costimulatory molecule from the APC engages with its receptor on the T cell. Simultaneous transduction of both signals enables T cells to secrete IL-2 and up-regulates the high affinity IL-2 receptors which, upon association with IL-2, induces T cells to proliferate [49, 50].

Moreover, mitogen-stimulated T cells cocultured with UVB-irradiated monocytes as accessory cells displayed increased intracellular free calcium, expressed low levels of IL-2 receptors, but failed to produce IL-2, suggesting induction of clonal anergy of those lymphocytes [50]. Furthermore, it has been proposed by three different groups of investigators that ICAM-1 is the putative costimulatory signal that is not expressed in UVB-irradiated LC, thus inducing clonal anergy of Th1 T cells [42, 47, 50, 51]. Th2 cells whose proliferative capacity is not affected by contact with UVB-irradiated LC [51] are the cells responsible for mediating suppression of Th1-mediated responses in vivo [47]. The fact that Th2 cells produce IL-10 [45] and UVB-irradiated keratinocytes also [46], and that IL-10 inhibits the activation of Th1 cells, makes this speculation attractive. However, it has been demonstrated that the UVB-induced inhibition of cytokine-induced ICAM-1 mRNA expression by keratinocytes is transient, and restored by IFN-γ but not by TNF-α [52], and is followed by an up-regulation of ICAM-1 mRNA expression after 48 h. Furthermore, this UVB effect on ICAM-1 expression on keratinocytes is biphasic, producing inhibition at 24 h and induction at 48, 72 and 96 h [53]. If the inhibition of ICAM-1 expression in short-term cultures of LC [9] is found to be biphasic and transitory also, and since clonal anergy might be overcome by exogenous sources of IL-2 [44], further experiments will be necessary to elucidate this possibility.

Another aspect of photoimmunology first evaluated by Kripke et al. [22] was that chronically UVB-irradiated mice are unable to reject highly antigenic skin cancers induced by UVB radiation and this impairment is associated with the development of suppressor T lymphocytes that prevent the rejection of transplanted and primary [22–28] UVB-induced skin cancers. Susceptibility to the growth of transplanted UVB-induced tumors persists indefinitely in the mice in which they were induced [22–30] as the chronically lasting suppression of CHS responses [14, 21]. In contrast, in mice chronically exposed to UVB, DTH responses [22, 30] are decreased initially and then eventually return to normal levels. Therefore, some investigators suggest that the systemic impairment of DTH responses appears to result from a transient redistribution of splenic APC in response to UVB-induced inflammation [30, 31, 54, 55]. However, others have demonstrated that suppression of DTH responses to alloantigens is adoptively transferred by CD8+ T cells [18].

Another group of investigators has proposed that since cutaneous APC (LC) and not splenic APC may be necessary for the induction of this long-lasting suppression of CHS and tumor rejection responses, UVB-radiation induces immunologic tolerance to modified 'skin antigens' serving as hapten carrier inducing the generation of carrier-specific, suppressor, inducer T cells [32]. More recently it has been proposed that both systemic and local suppression of CHS induced by UVB radiation might be mediated by TNF-α released in response to DNA damage in the form of pyrimidine dimers [56]. Others have shown that molecular alteration in DNA may be either an initiating or essential step in the development of the systemic suppression of CHS that is mediated by suppressor lymphocytes

[57]. The clinical relevance of the suppression of CHS interfering with the response to infectious agents is still under investigation [58]. The role of LC in the induction of immunity to tumors is explored in detail in Grabbe et al., pp

Other possible explanations for this systemic immunosuppression are: (1) generation of immunosuppressive prostaglandins during the irradiation of the skin [59] that interfere with the recovery of the APC function; (2) generation of immunomodulatory acute phase proteins [60] associated with pyrogen activity and increased IL-1 serum levels; (3) release of cytokines by UVB-treated epidermal cells which include an inhibitor of IL-1 [61], IL-1 [62] production which is followed by an increase in acute-phase reactant serum amyloid P component, α-melanocyte-stimulating hormone [63, 64] possibly inducing the release of glucocorticoids from the adrenal glands, a low-molecular-weight protein distinct from prostaglandins and leukotrienes that might participate in the regulation of UVB-mediated local and systemic suppression of CHS [65], IL-6 [66] which correlates with the fever course and the increase in acute-phase proteins, and IL-10 [46] which can counterregulate the activation of Th1 cells; (4) UVB-induced release by epidermal cells of propiomelanocorticotropin, the precursor molecule for α-melanocyte-stimulating factor, and of adrenocorticotropic hormone, which induces the release of glucocorticosteroids from the adrenal glands [63], and (5) isomerization of transurocanic acid in the skin to the cis-isomer, as a mediator of UVB- and PUVA-induced immunosuppression including suppression of organ allograft rejection and prevention of acute lethal GVDH in mice [67, 68].

Moreover, UVB radiation also suppresses mast cell degranulation [69], suppresses natural killer activity in humans [70], causes increases in dermal collagen type III in mice [71] and causes abnormal would healing in Sencar mice [72].

Long-Wave Ultraviolet Radiation and Psoralens

Cutaneous exposure to UVA radiation increased the mean epidermal and stratum corneum thickness, associated with increased activity of glucose-6-phosphate dehydrogenase [73]. UVA irradiation stimulates collagenase production in cultured human fibroblasts [74], and irradiation of murine keratinocytes with UVA resulted in the release of a factor that suppresses CHS but not DTH responses [32]. UVA irradiation also induces loss of ATPase activity followed by loss of class II expression of LC [75]. However, the major impact of UVA light on immunological responses is caused by the activation of psoralens with UVA.

Psoralens are a class of photosensitizing agents that intercalate between DNA base-pairs in the dark. Upon absorption of photons of the proper wavelength (in the UVA region) they covalently bind to one (monofunctional adduct) or both DNA chains (bifunctional adducts). Some psoralens can only produce monofunctional adducts while others form bifunctional adducts [76]. More recently, it has been demonstrated that 8-MOP also binds to cellular proteins and cell surface receptors [Gesparro, pers. comm.].

The use of psoralens plus UVA irradiation of the skin (known by the acronym PUVA) is useful in the treatment of a number of benign and malignant skin disorders, possibly due to cytotoxic, cytostatic or immuno-suppressive effects [76]. Exposure of murine skin to monofunctional and/or bifunctional psoralens followed by UVA exposure induces nearly total depletion of ATPase-, Ia-, and Thy-1-positive dendritic epidermal cells [75, 77]. These findings are associated with the induction of local [77, 78] and systemic [79, 80] suppression of CHS responses similar to that observed after exposure to UVB radiation. However, PUVA treatment of keratinocytes, unlike UVB, induces the production of factors that suppress both the DTH response to alloantigens and CHS response [80]. These findings imply that PUVA, unlike UVB treatment, can cause the release of multiple factors from keratinocytes and these factors may play a role in the induction of systemic immunosuppression following PUVA treatment. Also, murine splenocytes treated with PUVA demonstrated an impairment in the IL-2 production [81]. Moreover, it has been demonstrated recently that treatment with nonphysiologic doses of PUVA and UVB radiation, but not UVA radiation alone, was capable of activating HIV genes in a transgenic murine model [82]. However, the clinical relevance of this finding is still being researched and contradictory clinical observations have been reported [83].

PUVA has been used successfully also for prolongation of murine skin graft survival [84], rat renal allograft survival [85], rat cardiac allograft survival [86], and in the treatment of the cutaneous manifestations of GVDH in bone marrow transplant patients [87–89]. Also, PUVA has been shown to augment cyclosporine-mediated allograft survival in the rat heart transplant model [90]. Furthermore, PUVA inhibits degranulation of rat peritoneal mast cells [91] and induces suppression of human natural killer activity [70].

ECP involves the intravenous reinfusion of peripheral blood mononuclear cells that have been exposed to UVA light ex vivo after treatment with 8-methoxypsoralen (8-MOP) in vivo. In patients, ECP has been used successfully in the treatment of cutaneous T-cell lymphoma (CTCL) [92–94], progressive systemic sclerosis (PSS) [95], drug-resistant pemphigus vulgaris [96], rheumatoid arthritis [97], systemic lupus erythematosus (SLE) [98] and for inducing the reversal of acute [99] and chronic [100] heart transplant rejection. ECP has also been proposed as a potentially useful treatment for human immunodeficiency virus (HIV) infection previously referred to as AIDS-related complex [101]. It has been speculated that ECP may be useful in HIV infection by inducing increased immunity against HIV and/or by directly inactivating virus [101].

The mechanism by which ECP is efficacious in the treatment of certain malignant and autoimmune diseases has not yet been fully elucidated. However, several animal models have been helpful in understanding its mechanism of action [102–112], most significantly in a murine skin allograft system [102–106]. In that system, which involves transplantation of skin across major plus minor [102, 103] and across only major [104] histocompatibility barriers, it was found that the responses in vivo and in vitro to alloantigen can be attenuated in a donor-specific fashion. In that system, splenocytes containing expanded populations of effector T cells that mediated the relevant allograft rejection were first treated ex vivo with 8-MOP/UVA and were then infused into native syngeneic recipients which now, as a result, showed markedly enhanced skin allograft survival which was correlated with inhibition in vitro of the mixed leukocyte culture (MLC) response, the cytolytic T cell (CTL) response and the DTH response.

This suppressive response to alloantigen was augmented by cyclophosphamide pretreatment of the animals and inhibited by prednisolone therapy [105], suggesting that the immune response induced to the reinfused damage cells can be modified by immunosuppressive therapy.

Finally, it appears that specific suppression of the anti-skin allograft response, induced by infusions of 8-MOP/UVA-damaged spleen cells that include the effector cells of allograft recognition, is optimally transferred by radiosensitive Thy-1$^+$, Lyt-2$^+$ and L3T4$^-$ T lymphocytes [106]. This work was extended to primate cardiac allo- and xenotransplantation, demonstrating similar results [107]. Further studies demonstrated that prophylactic treatment of mice, suffering from an autoimmune disease similar to SLE, with syngeneic photoinactivated autoimmune splenocytes, improves the survival and inhibits the fulminant hyperoproliferation of abnormal T cells and the production of high titer anti-DNA antibodies invariably found in untreated mice [108]. Additional studies have been extended to demonstrate suppression of the induction of experimental murine GVDH [109], prolongation of skin allograft survival by infusions of 8-MOP/UVA-treated donor leukocytes [110] and suppression of CHS response [111].

More recently a murine model for T-cell lymphoma was developed using a highly malignant T-cell hybridoma [112]. Preliminary studies have demonstrated that the most efficacious method to induce protection against tumor development involves predamaging the tumor cells with UVA-activated 8-MOP to immunize mice prior to challenge with viable tumor cells that invariably killed all nontreated mice [112]. This induces cross-protective immunity against another T-cell hybridoma differing in the TCR specificity and also the parental thymoma. This protection is immune mediated, since pretreatment of nude mice in a similar fashion does not induce protection against tumor development. Furthermore, there is preliminary data that demonstrates the effectiveness of this form of therapy in treating established disease and adoptively transferring immunity against tumor.

There is evidence indicating that ECP induces a suppressive response involving mainly CD8+ T cells in the animal models, since the specific suppression of anti-skin allograft response was adoptively transferred by radiosensitive CD8+ T cells [106]. There is some circumstantial evidence of the suppressive response mediated by CD8+ T cells in the CTCL patients, since most of the long-term survivors to ECP were the patients who maintained normal levels of CD8+ T cells [94].

Recent technological advances in the field of immunology have made possible the isolation of peptides from the groove of class I molecules [113]. Future experiments using our animal model for T-cell lymphomas will characterize the antigenic peptides bound to the class I molecule of the tumor cells and then will be used to isolate cytotoxic/suppressor T cells from the immunoprotected animals. These experiments will further help to elucidate whether ECP induces the generation of CD8+ T cells by increasing, in some way, the availability of antigenic peptides in the class I molecule.

Other possibilities, as well, are under investigation. These include the possibility that 8-MOP/UVA induced mutagenesis causes the expression of new antigens that now are immunogenic to CD8+ T cells [114]. Another possibility suggests that photoadduct formation in the cell membrane DNA [115] by exposure to UVA-activated 8-MOP in some way is involved in the induction of suppression to alloantigen response [116]. However, the mechanism by which ECP is efficacious in the treatment of CTCL and some autoimmune diseases is yet to be determined.

Other Photosensitizing Reactions

Photodynamic therapy (PDT) involves the selective accumulation of a photosensitizer drug by certain tumors, followed by the activation of the drug with exposure to the appropriate wavelengths of nonionizing radiation [117–119]. One of these photosensitizing drugs is hematoporphyrin derivative (HPD), a porphyrin that has been shown to localize preferentially within malignant tumors [120]. Photodynamic excitation of HPD that accumulates within the tumor permits the detection of malignant cells because of the ability of porphyrins to fluoresce and therefore, permits the selective destruction of those tumors [117–119]. HPD is activated by visible light delivered by an argon pump dye laser [117, 118] or by high intensity mercury arc filtered through 9 mm of window glass [117].

Cutaneous [117] and peritoneal [118] PDT of mice have been shown to induce systemic suppression of CHS that is transferred by macrophages, is nonspecific in function [119], reversible 3 weeks after PDT and associated with leukocytosis and elevated serum levels of amyloid P protein [118].

In other studies, mice treated with rose bengal activated by exposure to visible radiation exhibited limited suppression of CHS that is not transferable by lymphocytes [57]. The mechanism of this immunosuppression is yet to be determined.

Conclusion

In this chapter we have discussed the past and current data evaluating the mechanism by which ultraviolet and visible radiation influences immune responses. Further, elucidation of the mechanism of UV-induced immunosuppression will help promote the interaction of immunology, dermatology, and oncology and expands the understanding of the molecular basis of tumor immunology. Further expansion of knowledge in these fields will then result in a better understanding of disease processes and help indicate steps where prevention may be possible.

Note Added in Proof

Since the submission of this manuscript it has been demonstrated that systemic suppression of CHS is mediated by TNF-α via a pathway in which urocaric acid acts as the photoreceptor but does not appear to be the primary mediator of tolerance [56a]. It has also been demonstrated that keratinocyte-derived IL-10 may play an important role in UV-induced suppression of DTH responses but not of CHS responses [46a].

References

1   Haniszko, J.; Suskin, R.R.: The effect of ultraviolet radiation on experimental cutaneous sensitization in guinea pigs. J. Invest. Dermatol. *40:* 183–190 (1963).

2   Towes, G.B.: Bergstresser, P.R.; Streilein, J.W.: Epidermal Langerhans cell density determines whether contact hypersensitivity or unresponsiveness follows skin painting with DNFB. J. Immunol. *124:* 445–453 (1980).

3   Aberer, W.; Schuler, G.; Stingl, G., et al.: Ultraviolet light depletes surface markers of Langerhans' cells. J. Invest. Dermatol. *76:* 202–210 (1981).

4   Aberer, W.; Romani, N.; Elbe, A., et al.: Effects of physiochemical agents on murine epidermal Langerhans' cells and Thy-1-positive dendritic epidermal cells. J. Immunol. *316:* 1210–1216 (1986).

5   Iacobelli, D.; Hashimoto, K.; Takahashi, S.: Effect of ultraviolet radiation on guinea pig epidermal Langerhans cell cytomembrane: Light and electron microscopic studies. Photodermatology *2:* 132–143 (1985).

6   Takahashi, S.; Hashimoto, K.: Deviation of Langerhans cell granules from cytomembrane. J. Invest. Dermatol. *84:* 469–471 (1985).

7   Bucana, C.D.; Munn, C.G.; Song, M.J., et al.: Internalization of Ia molecules into Birbeck granule-like structures in murine dendritic cells. J. Invest. Dermatol. *99:* 365–373 (1992).

8   Tang, A.; Udey, M.C.: Inhibition of epidermal Langerhans cell function by low dose ultraviolet B radiation: Ultraviolet B radiation selectively modulates ICAM-1 (CD54) expression by murine Langerhans' cells. J. Immunol. *146:* 3347–3355 (1992).

9   Tang, A.; Udey, M.C.: Effects of ultraviolet radiation on murine epidermal Langerhans' cells: Doses of ultraviolet radiation that modulate ICAM-1 (CD54) expression and inhibit Langerhans' cell function cause delayed cytotoxicity in vitro. J. Invest. Dermatol. *99:* 83–89 (1992).

10 Miyauchi, S.; Hashimoto, K.: Epidermal Langerhans' cells undergo mitosis during the early recovery phase under ultraviolet-B irradiation. J. Invest. Dermatol. *88:* 703–709 (1987).

11 Elmets, C.A.; Bergstresser, P.R.; Tigelaar, R. E., et al.: Analysis of the mechanism of unresponsiveness produced by haptens painted on skin exposed to low dose ultraviolet radiation. J. Exp. Med. *158:* 781–794 (1983).

12 Stingl, G.; Gazze-Stingl, L.A.; Aberer, W., et al.: Antigen presentation by murine epidermal Langerhans' cells and its alteration by ultraviolet B light. J. Immunol. *127:* 1707–1713 (1981).

13 Sauder, D.N.; Tamaki, K.; Moshell, A.N., et al.: Induction of tolerance to topically applied TCNB using TNP-conjugated ultraviolet light irradiated epidermal cells. J. Immunol. *127:* 261–265 (1981).

14 Noonan, F.P.; DeFabo, E.C.; Kripke, M.L.: Suppression of contact hypersensitivity by UV radiation: An experimental model. Springer Semin. Immunopathol. *4:* 293–304 (1981).

15 Greene, M.I.; Sy, M.S.; Kripke, M.L., et al.: Impairment of antigen presenting cell function by UV radiation. Proc. Natl Acad. Sci. USA *76:* 6591–6595 (1979).

16 Ullrich, S.E.; Azizi, E.; Kripke, M.L.: Suppression of the induction of DTH reactions in mice by a single exposure to UV radiation. Photochem. Photobiol. *43:* 633–638 (1986).

17 Ullrich, S.E.: Suppression of the immune response to allogeneic histocompatibility antigens by a single exposure to UV radiation. Transplantation *42:* 287–291 (1986).

18 Molendijk, A.; van Gurp, R.J.H.M.; Donselaar, I.G., et al.: Suppression of delayed-type hypersensitivity to histocompatibility antigens by ultraviolet radiation. Immunology *62:* 299–305 (1987).

19 Morison, W.L.; Pike, R.A.: Suppression of graft-versus-host reactivity in the mouse popliteal node by UVB radiation. J. Invest. Dermatol. *84:* 483–486 (1985).

20 Polla, L.; Margolis, R.: Goulston, C., et al.: Enhancement of the elicitation phase of the murine contact hypersensitivity response by prior exposure to local ultraviolet radiation. J. Invest. Dermatol. *86:* 13–17 (1986).

21 Tseng, C.; Hoffman, B.; Kurimoto, I., et al.: Analysis of effects of ultraviolet B radiation on induction of primary allergic reactions. J. Invest. Dermatol. *98:* 871–875 (1992).

22 Kripke, M.L.; Lofgreen, J.S.; Beard, J., et al.: In vivo immune responses of mice during carcinogenesis by ultraviolet radiation. J. Natl Cancer Inst. *59:* 1227–1230 (1977).

23 Kripke, M.L.; Fisher, M.S.: Immunologic parameters of ultraviolet carcinogenesis. J. Natl Cancer Inst. *57:* 211–215 (1976).

24 Fisher, M.S.; Kripke, M.L.: Systemic alteration induced in mice by ultraviolet light irradiation and its relationship to ultraviolet carcinogenesis. Proc. Natl Acad. Sci. USA *74:* 1688–1692 (1977).

25 Fisher, M.S.; Kripke, M.L.: Suppressor T lymphocytes control the development of primary skin cancers in ultraviolet-irradiated mice. Science *216:* 1133–1134 (1982).

26 Spellman, C.W.; Woodward, J.G.; Daynes, R.A.: Modification of immunologic potential by ultraviolet radiation. Immune status of short-term UV-irradiated mice. Transplantation *24:* 112–119 (1977).

27 Spellman, C.W.; Daynes, R.A.: Modification of immunologic potential by ultraviolet radiation: Generation of suppressor cells in short-term UV-irradiated mice. Transplantation *42:* 120–126 (1977).

28 Fisher, M.A.; Kripke, M.L.: Further studies on the tumor-specific suppressor cells induced by ultraviolet radiation. J. Immunol. *121:* 1139–1144 (1978).

29 Glass, M.J.; Bergstresser, P.R.; Tigelaar, R.E., et al.: UVB radiation and DNFB skin painting induce suppressor cells universally in mice. J. Invest. Dermatol. *94:* 273–278 (1990).

30 Jessup, J.M.; Hanna, N.; Palaszynski, E., et al.: Mechanisms of depressed reactivity to dinitrochlorobenzene and ultraviolet-induced tumors during ultraviolet carcinogenesis in BALB/c mice. Cell. Immunol. *38:* 105–115 (1978).

31 Greene, M.I.; Sy, M.S.; Kripke, M.L.: Impairment of antigen-presenting cell function by ultraviolet radiation. Proc. Natl Acad. Sci. USA *76:* 6591–6595 (1979).

32 Kripke, M.L.; Morison, W.L.: Studies on the mechanism of systemic suppression of contact hypersensitivity by UVB radiation. II. Differences in the suppression of delayed and contact hypersensitivity in mice. J. Invest. Dermatol. *86:* 543–549 (1986).

33 Kim, T.Y.; Kripke, M.L.; Ulrich, S.E.: Immunosuppression by factors released from UV-irradiated epidermal cells: Selective effects on the generation of contact and delayed hypersensitivity after exposure to UVA or UVB radiation. J. Invest. Dermatol. *94:* 26–32 (1990).

34 Austaad, J.; Braathen, L.R.: Effect of UVB on alloactivating and antigen presenting capacity of human epidermal Langerhans' cells. Scand. J. Immunol. *21:* 417–423 (1985).

35 Cruz, P.D.; Nixon-Fulton, J.L.; Tigelaar, R.E., et al.: Disparate effects of in vitro low-dose UVB irradiation on intravenous immunization with purified epidermal cell subpopulations for the induction of contact hypersensitivity. J. Invest. Dermatol. *92:* 160–165 (1989).

36 Granstein, R.D.; Askari, M.; Whitaker, D.; Murphy, G.F.: Epidermal cells in activation of suppression: Further characterization. J. Immunol. *138:* 4055–4062 (1987).

37 Halliday, G.M.; Wood, R.C.; Muller, H.K.: Presentation of antigen to suppression cells by a dimethylbenz(a)anthracene-resistant, Ia-positive, Thy-1⁻, I-J-restricted epidermal cell. Immunology *69:* 97–103 (1990).

38 Cooper, K.D.; Reises, G.R.; Katz, S.I.: Antigen-presenting OKM5+ melanophages appear in human epidermis after ultraviolet radiation. J. Invest. Dermatol. *86:* 363–370 (1986).

39 Baadsgaard, O.; Wulf, H.C.; Wantzin, G.L., et al.: UVB and UVC, but not UVA, potently induce the appearance of T6-DR+ antigen-presenting cells in human epidermis. J. Invest. Dermatol. *89:* 113–118 (1987).

40 Elmets, C.A.; Bergstresser, P.R.; Tigelaar, R.E., et al.: Analysis of the mechanism of unresponsiveness produced by haptens painted on skin exposed to low dose ultraviolet radiation. J. Exp. Med. *158:* 781–794 (1983).

41 Cruz, P.D., Jr.; Tigelaar, R.E.; Bergstresser, P.R.: Immunodominance, strain characterization, and MHC restriction of the unresponsiveness induced by UVB-irradiated Langerhans' cells (abstract). J. Invest. Dermatol. *94:* 516 (1990).

42 Simon, J.C.; Tigelaar, R.E.; Bergstresser, P.R., et al.: UVB radiation converts Langerhans' cells from immunogenic to tolerogenic antigen presenting cells. Induction of specific clonal anergy in CD4+ T helper 1 cells. J. Immunol. *146:* 485–491 (1991).

43 Mosmann, T.R.; Moore, K.W.: Th1 and Th2 cells: Different patterns of lymphokine secretion lead to different functional properties. Annu. Rev. Immunol. *7:* 145–173 (1989).

44 Mosmann, T.R.; Coffman, R.L.: Heterogeneity of cytokine secretion patterns and functions of helper T cells. Adv. Immunol. *46:* 111–147 (1989).

45 Mosmann, T.R.; Moore, K.W.: The role IL-10 in cross-regulation of Th1 and Th2 responses. Immunol. Today A49–A53 (1991).

46 Rivas, J.M.; Ullrich, S.E.: Keratinocyte-derived IL-10 (abstract). J. Invest. Dermatol. *98:* 578 (1992).

46a Rivas, J.M.; Ullrich, S.E.: Essential role of keratinocyte-derived interleukin-10 in the UV-induced suppression of delayed type hypersensitivity but not contact hypersensitivity. J. Invest. Dermatol. *100:* 522a (1993).

47 Simon, J.C.; Krutmann, J.; Elmets, C.A., et al.: Ultraviolet B-irradiated antigen-presenting cells display altered accessory signaling for T-cell activation: Relevance to immune responses initiated in skin. Invest. Dermatol. *98:* 66S–69S (1992).

48 Schwartz, R.H.: A cell culture model for T lymphocyte clonal anergy. Science *15:* 1349–1356 (1990).

49 Mueller, D.L.; Jenkins, M.K.; Schwartz, R.H.: An accessory cell-derived costimulatory signal acts independently of protein kinase c activation to allow T cell proliferation and prevent the induction of unresponsiveness. J. Immunol. *142:* 19203–19210 (1989).

50 Krutmann, J.; Kammer, G.M.; Toossi, Z., et al.: UVB radiation and human monocyte accessory function: Differential effects on pre-mitotic events in T-cell activation. J. Invest. Dermatol. *94:* 204–209 (1990).

51 Simon, J.C.; Cruz, P.D.; Bergstresser, P.R., et al.: Low-dose UBV-irradiated Langerhans' cells preferentially activate CD4 cells of the Th1 subset. J. Immunol. *145:* 2087–2091 (1990).

52 Krutmann, J.; Czech, W.; Parlow, F., et al.: Ultraviolet radiation effects on human keratinocyte ICAM-1 expression: UV-induced inhibition of cytokine-induced ICAM-1 mRNA expression is transient, differentially restored for IFN-γ versus TNF-α, and followed by ICAM-1 induction via a TNF-α-like pathway. J. Invest. Dermatol. *98:* 923–928 (1992).

53 Norris, D.A.; Lyons, B.; Middleton, M.J., et al.: Ultraviolet radiation can either suppress or induce expression of intracellular adhesion molecule 1 on the surface of cultured human keratinocytes. J. Invest. Dermatol. *95:* 132–138 (1990).

54 Sprangrude, G.J.; Bernhard, E.J.; Azioka, R.S., et al.: Alterations in lymphocyte homing patterns within mice exposed to ultraviolet radiation. J. Immunol. *130:* 2974–2981 (1983).

55 Chung, H.T.; Samlowski, W.E.; Kelsey, D.K., et al.: Alterations in lymphocyte recircu-
    lation within ultraviolet light-irradiated mice: Efferent blockade of lymphocyte egress
    from peripheral lymph nodes. Cell. Immunol. *101:* 571–585 (1986).
56 Kripke, M.L.; Cox, P.; Yarosh, D.: Local suppression of contact hypersensitivity by
    low dose UVB radiation involves a systemic component (abstract). J. Invest. Derma-
    tol. *98:* 594 (1992).
56a Tadamachi, S.; Arawam, XXX.; Ullrich, S.E.: UVB, TNF-α and urocanic acid induce
    tolerance to hapten via common, non-TNF-α dependent mechanism. J. Invest. Derma-
    tol. *100:* 565a (1993).
57 Morison, W.L.: Systemic suppression of contact hypersensitivity associated with sup-
    pressor lymphocytes: Is a lesion in DNA an essential step in pathway? Photodermatol.
    Photoimmunol. Photomed. *7:* 202–206 (1990).
58 Jeevan, A.; Evans, R.; Brown, E., et al.: Effect of local ultraviolet irradiation on
    infections of mice with *Candida albicans, Mycobacterium bovis* BCG, and *Schistosoma
    mansoni.* J. Invest. Dermatol. *99:* 59–64 (1992).
59 Jun, B.D.; Roberts, L.K.; Cho, B.H., et al.: Parallel recovery of epidermal antigen-pre-
    senting cell activity and contact hypersensitivity responses in mice exposed to ultravio-
    let irradiation: The role of a prostaglandin-dependent mechanism. J. Invest. Dermatol.
    90: 311–316 (1988).
60 Ansel, J.C.; Luger, T.A.; Green, I.: Fever and increased serum IL-1 activity as a
    systemic manifestation of acute phototoxicity in New Zealand white rabbits. J. Invest.
    Dermatol. *89:* 32–37 (1987).
61 Schwartz, T.; Urbanski, A.; Kirnbauer, R., et al.: Detection of a specific inhibitor of
    interleukin-1 in sera of UVB-treated mice. J. Invest. Dermatol. *91:* 536–540 (1988).
62 Griswold, D.E.; Cooper, J.R.; Dalton, B.J., et al.: Activation of the IL-1 gene in
    UV-irradiated mouse skin: Association with inflammatory sequelae and pharmacologic
    intervention. J. Invest. Dermatol. *97:* 1019–1023 (1991).
63 Kock, A.; Schauer, E.; Urbanski, A., et al.: Regulation of epidermal cell-derived
    alpha-melanocyte-stimulating hormone production (abstract). J. Invest. Dermatol. *96:*
    1029 (1991).
64 Schwarz, T.; Luger, T.A.: Effect of UV irradiation of epidermal cell cytokine produc-
    tion. J. Photochem. Photobiol. *4:* 1–13 (1989).
65 Schwartz, T.; Urbanski, A.; Gschnait, F., et al.: Inhibition of the induction of contact
    hypersensitivity by UV-mediated epidermal cytokine. J. Invest. Dermatol. *87:* 289–291
    (1986).
66 Urbanski, A.; Schwartz, T.; Neuner, P., et al.: Ultraviolet light induces increased
    circulating interleukin-6 in humans. J. Invest. Dermatol. *94:* 808–811 (1990).
67 Gruner, S.; Diezel, W.; Stoope, H., et al.: Inhibition of skin allograft rejection and
    acute graft-versus-host disease by cis-urocanic acid. J. Invest. Dermatol. *98:* 459–462
    (1992).
68 Gruner, S.; Oesterwitz, H.; Stoope, H., et al.: Cis-urocanic acid as a mediator of
    ultraviolet-light-induced immunosuppression. Semin. Hematol. *29:* 102–107 (1992).
69 Danno, K.; Fujii, K.; Tachibana, T., et al.: Suppressed histamine release from rat
    peritoneal mast cells by ultraviolet B irradiation: Decreased formation as a possible
    mechanism. J. Invest. Dermatol. *90:* 806–809 (1988).
70 Toda, K.; Miyachi, Y.; Nesumi, N., et al.: UVB/PUVA-induced suppression of human
    natural killer activity is reduced by superoxide dismutase and/or interleukin-2 in vitro.
    J. Invest. Dermatol. *86:* 519–522 (1986).
71 Plastow, S.R.; Harrison, J.A.; Young, A.R.: Early changes in dermal collagen of mice
    exposed to chronic UVB irradiation and the effects of a UVB sunscreen. J. Invest.
    Dermatol. *91:* 590–592 (1988).

72 Strickland, P.T.: Abnormal wound healing in UV-irradiated skin of Sencar mice. J. Invest. Dermatol. *86:* 37–41 (1986).

73 Pearse, A.N.; Gaskell, S.A.; Marks, D.: Epidermal changes in human skin following irradiation with either UVB or UVA. J. Invest. Dermatol. *88:* 83–87 (1987).

74 Petersen, M.J.; Hansen, C.; Craig, S.: Ultraviolet A irradiation stimulates collagenase production in cultured human fibroblasts. J. Invest. Dermatol. *99:* 440–444 (1992).

75 Alcalay, J.; Bucana, C.; Kripke, M.L.: Effect of psoralens and ultraviolet radiation on murine dendritic epidermal cells. J. Invest. Dermatol. *92:* 657–662 (1989).

76 Pathak, M.A.; Worden, L.R.; Kaufman, K.D.: Effect of structural alterations on the photosensitizing potency of furocoumarins (psoralens) and related compounds. J. Invest. Dermatol. *48:* 103–118 (1967).

77 Aubin, F.; Dall'Acqua, F.; Kripke, M.L.: Local suppression of contact hypersensitivity in mice by new bifunctional psoralens, 4,4′,5′-trimethylazapsoralen, and UVA radiation. J. Invest. Dermatol. *97:*50–54 (1991).

78 Alcalay, J.; Ullrich, S.E.; Kripke, M.L.: Local suppression of contact hypersensitivity in mice by a monofunctional psoralen plus UVA radiation. Photochem. Photobiol. *50:* 217–220 (1989).

79 Ullrich, S.E.: Systemic immunosuppression of cell-mediated immune reactions by a monofunctional psoralen plus ultraviolet A radiation. Photodermatol. Photoimmunol. Photomed. *8:* 116–122 (1991).

80 Activation of keratinocytes with psoralen plus UVA radiation induces the release of soluble factors that suppress delayed and contact hypersensitivity. J. Invest. Dermatol. *97:* 995–1000 (1991).

81 Okamoto, H.; Horio, T.; Maeda, M.: Alteration of lymphocyte functions by 8-methoxypsoralen and long-wave ultraviolet radiation. II. The effect of in vivo PUVA on IL-2 production. J. Invest. Dermatol. *89:* 24–26 (1987).

82 Wallace, B.M.; Lasker, J.S.: Awakenings . . . . UV light and HIV gene activation. Science *257:* 1211–1212 (1992).

83 Renki, A.N.; Puska, P.; Mattinen, S., et al.: Effect of PUVA on immunologic and virologic findings in HIV-infected patients. J. Am. Acad. Dermatol. *24:* 404–410 (1991).

84 Granstein, R.D.; Smith, L.; Parrish, J.A.: Prolongation of murine skin allograft survival by the systemic effects of 8-methoxypsoralen and long-wave ultraviolet radiation (PUVA). J. Invest. Dermatol. *88:* 424–429 (1987).

85 Osterwitz, H.; Kaden, J.; Scholz, D., et al.: Synergistic effect of donor pretreatment with 8-methoxypsoralen and ultraviolet irradiation of the graft plus azathioprine and prednisolone therapy in prolonging rat renal allograft survival. Urol. Res. *14:* 21–24 (1986).

86 Osterwitz, H.; Gruner, S.; Diezel, W.; Schneider, W.: Inhibition of rat heart allograft rejection by a PUVA treatment of the graft recipient. Transplant. Int. *3:* 8–11 (1990).

87 Eppinger, T.; Ehninger, G.; Steinert, M., et al.: 8-Methoxypsoralen and ultraviolet A therapy for cutaneous manifestations of graft-versus-host disease. Transplantation *50:* 807–811 (1990).

88 Volc-Platzer, B.; Honigsmann, H.; Hinterberger, W., et al.: Photochemotherapy improves chronic cutaneous graft-versus-host disease. J. Am. Acad. Dermatol. *23:* 220–228 (1990).

89 Jampel, R.M.; Farmer, E.R.; Vogelsang, G.B., et al.: PUVA therapy for chronic cutaneous graft-versus-host disease. Arch. Dermatol. *127:* 1673–1678 (1991).

90 Horvath, K.A.; Granstein, R.D.: PUVA augments cyclosporin A-mediated rat cardiac allograft survival. J. Surg. Res. *52:* 565–570 (1992).

91 Toda, K.; Danno, K.; Tachibana, T., et al.: Effect of 8-methoxypsoralen plus long-wave ultraviolet (PUVA) radiation on mast cells. II. In vitro PUVA inhibits degranulation of

rat peritoneal mast cells induced by compound 48/80. J. Invest. Dermatol. *87:* 113–116 (1986).

92 Edelson, R.; Berger, C.; Gasparro, F., et al.: Treatment of cutaneous T-cell lymphoma by extracorporeal photochemotherapy. N. Engl. J. Med. *316:* 297–303 (1987).

93 Armus, S.; Keyes, B.; Cahill, C., et al.: Photopheresis for the treatment of cutaneous T-cell lymphoma. J. Am. Acad. Dermatol. *23:* 898–902 (1990).

94 Heald, P.; Rook, A.; Perez, M., et al.: Treatment of erythrodermic cutaneous T-cell lymphoma with extracorporeal photochemotherapy. J. Am. Acad. Dermatol. *27:* 427–433 (1992).

95 Rook, A.H.; Freundlich, B.; Jegasothy, B.V., et al.: Treatment of systemic sclerosis with extracorporeal photochemotherapy: Results of a multicenter trial. Arch. Dermatol. *128:* 337–346 (1992).

96 Rook, A.; Jegasothy, B.; Heald, P., et al.: Extracorporeal photochemotherapy for drug-resistant pemphigus. Ann. Intern. Med. *112:* 303–305 (1990).

97 Malawista, S.E.; Trock, D.H.; Edelson, R.L.: A pilot study. Treatment of rheumatoid arthritis by extracorporeal photochemotherapy. Arthritis Rheum. *34:* 646–654 (1991).

98 Knobler, R.M.; Grainger, W.; Grainger, W., et al.: Extracorporeal photochemotherapy for the treatment of systemic lupus erythematosus. Arthritis Rheum. *35:* 319–324 (1992).

99 Costanzo-Nordon, M.R.; Hubbell, E.A.; O'Sullivan, E.J., et. al.: Successful treatment of heart transplant rejection with photopheresis. Transplantation *53:* 808–815 (1992).

100 Rose, E.A.; Barr, M.L.; Xu, H., et al.: Photochemotherapy in human heart transplant recipients at high risk for fatal rejection. J. Heart Lung Transplant. *11:* 746–750 (1992).

101 Bisaccia, E.; Berger, C.; Klainer, A.: Extracorporeal photopheresis in the acquired immune deficiency syndrome in related complex: A preliminary study. Ann. Intern. Med. *113:* 270–275 (1990).

102 Perez, M.; Edelson, R.; Laroche, L., et al.: Inhibition of antiskin allograft immunity by infusions with syngeneic photoinactivated effector lymphocytes. J. Invest. Dermatol. *92:* 669–676 (1989).

103 Perez, M.I.; Edelson, R.L.; John, L., et al.: Inhibition of antiskin allograft immunity induced by infusions with photoinactivated effector T lymphocytes (PET cells). Yale J. Biol. Med. *62:* 595–609 (1989).

104 Perez, M. I.; Berger, C.L.; Yamane, Y., et al.: Inhibition of antiskin allograft immunity induced by infusions with photoinactivated effector T lymphocytes (PET cells). The congenic model. Transplantation *51:* 1283–1289 (1991).

105 Yamane, Y.; Lobo, F.M.; John, L.A., et al.: Suppression of anti-skin-allograft response by photodamaged effector cells – The modulating effects of prednisolone and cyclophosphamide. Transplantation *54:* 119–124 (1992).

106 Perez, M.I.; Lobo, F.M.; John, L., et al.: Induction of a cell-transferable suppression of alloreactivity by photodamaged lymphocytes. Transplantation *54:* 896–903 (1992).

107 Pepino, P.; Berger, C.L.; Fuzesi, L., et al.: Primate cardiac allo- and xenotransplantation: Modulation of the immune response with photochemotherapy. Eur. Surg. Res. *21:* 105–113 (1989).

108 Berger, C.L.; Perez, M.; Laroche, L., et al.: Inhibition of autoimmune disease in a murine model of systemic lupus erythematosus induced by exposure to syngeneic photoinactivated lymphocytes. J. Invest. Dermatol. *94:* 52–57 (1990).

109 Ullrich, S.E.: Photoinactivation of T-cell function with psoralen and UVA radiation suppresses the induction of experimental murine graft-versus-host disease across major histocompatibility barriers. J. Invest. Dermatol. *96:* 303–308 (1991).

110   Gruner, S.; Noack, F.; Meffert, H.: Influence of a transfusion of donor leukocytes treated with 8-methoxypsoralen and long-wave ultraviolet light (PUVA) on skin allograft survival in mice. Biomed. Biochim. Acta 48: 477–485 (1989).

111   Van Iperen, H.P.; Beijersbergen van Henegouwen, G.M.J.: An animal mode for extracorporeal photochemotherapy based on contact hypersensitivity. J. Photochem. Photobiol. 15: 361–366 (1992).

112   Perez, M.I.; Lobo, F.M.; Yamane, Y., et al.: Induction of protection against tumor development (abstract). J. Invest. Dermatol. 98: 595 (1992).

113   Hunt, D.F.; Henderson, R.A.; Shabanowitz, J., et al.: Characterization of peptides bound to the class I MHC molecule HLA-A2.1 by mass spectrometry. Science 255: 1261–1263 (1992).

114   Malane, M.; Tigelaar, R.; Gasparro, F.: Treatment with 8-methoxypsoralen and ultraviolet A radiation enhances immune recognition in a murine mastocytoma model. J. Invest. Dermatol. 98: 554 (1992).

115   Gasparro, F.P.; Dall'Amico, R.; O'Malley, M., et al.: Cell membrane DNA: A new target for psoralen photoadduct formation. Photochem. Photobiol. 52: 315–321 (1992).

116   Perez, M. I.; Yamane, Y.; John, L., et al.: DNA associated with the cell membrane is involved in the inhibition of the skin rejection response induced by infusions of photodamaged alloreactive cells that mediate rejection of skin allograft. Photochem. Photobiol. 55: 839–849 (1992).

117   Elmets, C.A.; Bowen, K.D.: Immunological suppression in mice treated with hematoporphyrin derivative photoradiation. Cancer Res. 46: 1608–1611 (1986).

118   Jolles, C.J.; Ott, M.J.; Straight, R.C., et al.: Systemic immunosuppression induced by peritoneal photodynamic therapy. Am. J. Obstet. Gynecol. 158: 1446–1453 (1988).

119   Lynch, D.H.; Haddad, S.; King, V.J., et al.: Systemic immunosuppression induced by photodynamic therapy is adoptively transferred by macrophages. Photochem. Photobiol. 49: 453–458 (1989).

120   Lipson, R.L.; Baldes, E.G.; Olsen, A.M.: The use of a derivative of hematoporphyrin in tumor detection. J. Natl Cancer. Inst. 26: 1–8 (1961).

Maritza I. Perez, MD, Yale University School of Medicine,
Department of Dermatology, 333 Cedar Street, New Haven, CT 06510 (USA)

Subject Index